APPLIED CHEMISTRY

THEORY AND PRACTICE

SECOND EDITION

O P VERMANI

Retired Associate Professor
Department of Chemistry
Regional Engineering College
Kurukshetra

A K NARULA

Retired Professor
Department of Industrial Chemistry
Guru Jambheshwar University
Hisar

Publishing Globally

NEW AGE INTERNATIONAL (P) LIMITED, PUBLISHERS

LONDON • NEW DELHI • NAIROBI

Bangalore • Chennai • Cochin • Guwahati • Hyderabad • Kolkata • Lucknow • Mumbai

Visit us at **www.newagepublishers.com**

Published by New Age International (P) Ltd., Publishers
First Edition: 1995
Second Edition: 2006
Reprint: 2020

GLOBAL OFFICES

- **New Delhi** **NEW AGE INTERNATIONAL (P) LIMITED, PUBLISHERS**
  7/30 A, Daryaganj, New Delhi-110002, (INDIA)
  Tel.: (011) 23253771, 23253472, **Telefax:** 23267437, 43551305
  E-mail: contactus@newagepublishers.com • Visit us at www.newagepublishers.com
- **London** **NEW AGE INTERNATIONAL (UK) LTD.**
  27 Old Gloucester Street, London, WC1N 3AX, UK
  E-mail: info@newacademicscience.co.uk • Visit us at www.newacademicscience.co.uk
- **Nairobi** **NEW AGE GOLDEN (EAST AFRICA) LTD.**
  Ground Floor, Westlands Arcade, Chiromo Road (Next to Naivas Supermarket)
  Westlands, Nairobi, KENYA, **Tel.:** 00-254-713848772, 00-254-725700286
  E-mail: kenya@newagepublishers.com

BRANCHES

- **Bangalore** 37/10, 8th Cross (Near Hanuman Temple), Azad Nagar, Chamarajpet, Bangalore- 560 018
  Tel.: (080) 26756823, **Telefax:** 26756820, **E-mail: bangalore@newagepublishers.com**
- **Chennai** 26, Damodaran Street, T. Nagar, Chennai-600 017, **Tel.:** (044) 24353401
  Telefax: 24351463, **E-mail: chennai@newagepublishers.com**
- **Cochin** CC-39/1016, Carrier Station Road, Ernakulam South, Cochin-682 016
  Tel.: (0484) 2377303, **Telefax:** 4051304, **E-mail: cochin@newagepublishers.com**
- **Guwahati** Hemsen Complex, Mohd. Shah Road, Paltan Bazar, Near Starline Hotel
  Guwahati-781 008, **Tel.:** (0361) 2513881, **Telefax:** 2543669
  E-mail: guwahati@newagepublishers.com
- **Hyderabad** 105, 1st Floor, Madhiray Kaveri Tower, 3-2-19, Azam Jahi Road, Near Kumar Theater
  Nimboliadda Kachiguda, Hyderabad-500 027, **Tel.:** (040) 24652456, **Telefax:** 24652457
  E-mail: hyderabad@newagepublishers.com
- **Kolkata** RDB Chambers (Formerly Lotus Cinema) 106A, 1st Floor, S N Banerjee Road
  Kolkata-700 014, **Tel.:** (033) 22273773, **Telefax:** 22275247
  E-mail: kolkata@newagepublishers.com
- **Lucknow** 16-A, Jopling Road, Lucknow-226 001, **Tel.:** (0522) 2209578, 4045297, **Telefax:** 2204098
  E-mail: lucknow@newagepublishers.com
- **Mumbai** 142C, Victor House, Ground Floor, N.M. Joshi Marg, Lower Parel, Mumbai-400 013
  Tel.: (022) 24927869, **Telefax:** 24915415, **E-mail: mumbai@newagepublishers.com**
- **New Delhi** 22, Golden House, Daryaganj, New Delhi-110 002, **Tel.:** (011) 23262368, 23262370
  Telefax: 43551305, **E-mail: sales@newagepublishers.com**

ISBN: 978-81-224-0814-0
C-19-08-12214

Printed in India at A.P. Offset, Delhi.

NEW AGE INTERNATIONAL (P) LIMITED, PUBLISHERS
7/30 A, Daryaganj, New Delhi - 110002
Visit us at **www.newagepublishers.com**
(CIN: U74899DL1966PTC004618)

Contents

Preface to the First Edition

As not many books are available on Experiments in Applied Chemistry, the students of technical institutions have mainly to depend on class notes or spend much time and labour in collecting material from various sources. This book is an attempt to fill that void. We have endeavoured to cover a large number of experiments which are of interest to students of chemistry in technical institutions. The book will also be of assistance to teachers in planning laboratory work.

The book has been divided into eight chapters. The first chapter describes the preparation of solutions of reagents needed in the various tests included in the manual. The second chapter comprises a number of elementary experiments which are intended to give the students a practical knowledge of the concepts of chemical equilibrium and speeds of reactions and the factors that govern them. The chapters on Water, Lubricating Oils, Greases and Emulsions and Coal describe the methods of a large number of tests that are routinely carried out on these materials. The sixth chapter deals with the analysis of certain ores and alloys followed by a chapter that describes the analysis of a small number of industrially important materials. The last chapter provides solutions of exercises given at the end various experiments. It is hoped that this chapter will equip the students well for the viva-voce examination.

This book is the outcome of many years of experience in conducting laboratory classes in Applied Chemistry. A large number of books have also been consulted and are listed under Bibliography. The salient features of the book are:

(i) The theoretical principles on which the various tests are based have been elaboratively explained.

(ii) The significance and utility of the various tests have been discussed in detail.

(iii) For each method, important precautions and their 'why' have been given.

(iv) To help the students probe their grasp of the subject, a number of exercises have been included after every method.

This book could not have been possible without the able guidance and valuable suggestions of Dr. R.L. Kaushik, Professor and Chairman, Department of Chemistry, Regional Engineering College, Kurukshetra and Prof. R.N. Kapoor, Professor, Department of Chemistry, Delhi University, Delhi. We also acknowledge all the help and encouragement received from our friends and colleagues.

Despite our best efforts, errors may have crept in. We will be thankful to our readers for bringing them to our notice.

Suggestions for improvement will also be received with gratitude.

O.P. VERMANI
A.K. NARULA

Preface to the Second Edition

We express our profound gratitude to our readers whose overwhelming patronage to the first edition of this book has encouraged us to bring out a second edition.

New features of the current edition include a chapter on Polymers with a limited number of elementary experiments which has been added with a view to initiate the beginners into this vast field.

On the suggestion of friends and colleagues, some tables of physical constants have been incorporated as Appendix.

The topic of specific gravity has been added to the chapter on Lubricating Oils, Greases and Emulsions.

In order to make the contents of the book wholesome and to meet specific requirements of a segment of readers, a section on electroplating has been included in the chapter on Miscellaneous Topics.

We sincerely hope that the new edition will meet the aspirations of the readers. Suggestions for further improving the book would be gratefully received.

O.P. VERMANI
A.K. NARULA

1

PREPARATION OF SOLUTIONS OF THE REAGENTS

1.1 Alkali Solutions

Ammonia solution (1:1) : Mix equal volumes of concentrated ammonia solution (Sp. gr. 0.88) and boiled-out distilled water.

Ammonium hydroxide (5 N) : Dilute 333 ml of concentrated ammonia solution (15 N, 28.4%) to 1 litre with boiled-out distilled water.

Potassium hydroxide (5 N) : Dissolve 280 g of A.R. KOH (Mol. wt. 56) pellets in 1 litre of boiled-out distilled water.

Sodium hydroxide (5 N) : Dissolve 200 g of A.R. NaOH (Mol. wt. 40) pellets in 1 litre of boiled-out distilled water.

1.2 Buffer Solutions

Ammonium chloride – ammonium hydroxide buffer (pH ~ 10) : Dissolve 68 g of A.R. NH_4Cl in some boiled-out distilled water. Add 572 ml of concentrated ammonia solution and dilute to 1 litre.

Calcium precipitating buffer (pH ~ 8) : Dissolve 6.0 g of A.R. $(NH_4)_2 \cdot C_2O_4$ (ammonium oxalate) in about 100 ml of boiled-out distilled water. Add 144 g of A.R. NH_4Cl, 13 ml of concentrated ammonia solution and dilute to 1 litre.

Phosphate buffer (pH 6.2 – 6.5) : Dissolve 24 g of anhydrous disodium hydrogen phosphate (Na_2HPO_4), 46 g of anhydrous potassium dihydrogen phosphate (KH_2PO_4) and 0.8 g of EDTA (disodium salt) in 1 litre of distilled water.

1.3 Diluted Acids

Acetic acid (5 N) : Dilute 287 ml of A.R. glacial acetic acid (17.4 N, 99.5%) to 1 litre.

Hydrochloric acid (5 N) : Dilute 430 ml of concentrated A.R. hydrochloric acid (11.6 N, 36%) to 1 litre.

Nitric acid (5 N) : Dilute 309 ml of concentrated A.R. nitric acid (16.2 N, 72%) to 1 litre.

Sulphuric acid (5 N) : Add slowly 139 ml of concentrated A.R. sulphuric acid (36 N, 96%) to distilled water, cool and dilute to 1 litre.

1.4 Indicator Solutions

Calcon : Dissolve 0.2 g of the dyestuff [sodium-1-(2-hydroxy-1-naphthylazo)-2-naphthol-4-sulphonate] in 50 ml of methanol.

DPD (N,N-Diethyl-p-phenylenediamine): Dissolve 0.15 g of DPD oxalate or 0.15 g of DPD sulphate pentahydrate or 0.11 g of anhydrous DPD sulphate in boiled-out distilled water containing 1 ml of dilute sulphuric acid (1 part H_2SO_4 + 3 parts water) and 20 mg EDTA. Dilute to 100 ml.

Eriochrome Black T : Dissolve 0.2 g of the solid dyestuff in 15 ml of triethanolamine and 5 ml ethanol or dissolve 0.5 g of the dyestuff in 100 ml rectified spirit.

Ferric alum : Dissolve 25 g of $Fe_2(SO_4)_3{\cdot}(NH_4)_2SO_4{\cdot}24\,H_2O$ in about 100 ml of hot distilled water. Add 10 ml of A.R. HNO_3, boil (until the reddish brown colour changes to yellow) to expel nitrous acid. Cool and filter.

Ferroin : Dissolve 1.485 g of *o*-phenanthroline monohydrate and 0.695 g $FeSO_4{\cdot}7H_2O$ in boiled-out distilled water and dilute to 100 ml. (The indicator solution itself may be purchased from the market).

Methyl orange : Dissolve 50 mg of methyl orange (free acid) in distilled water and dilute to 100 ml. Filter, if a precipitate is formed.

Methyl red : Dissolve 0.1 g of methyl red (free acid) in 100 ml of hot distilled water and cool. Filter, if necessary.

or

Dissolve 0.1 g in 60 ml of distilled alchol and add 40 ml of distilled water.

p-Naphtholbenzoin : Dissolve 1 g of the indicator powder in 100 ml of isopropyl alcohol.

Phenolphthalein : Dissolve 0.5 g of phenolphthalein in 50 ml of ethyl alcohol and add 50 ml of boiled-out distilled water with constant stirring. Filter, if necessary.

Potassium chromate : Dissolve 5.0 g of A.R. potassium chromate in 100 ml of distilled water.

Sodium diphenylamine sulphonate : Dissolve 0.2 g of the substance in 100 ml of boiled-out distilled water.

Starch : Add a little distilled water to about 0.5 g of soluble starch (A.R.) taken in a beaker. Stir with a glass and heat to make a transparent paste. Pour into it about 100 ml of boiling distilled water with constant stirring and cool.

1.5 Standard Solutions

Alcoholic potassium hydroxide (N/100) : Dissolve about 6 g of A.R. KOH (Mol. wt. 56, Eq. wt. 56) pellets in 1 litre of isopropyl alcohol or ethyl alcohol (distilled from potassium hydroxide) by boiling. Allow to stand, filter through sintered glass crucible and standardise with N/10 oxalic acid solution.

Prepare N/100 solution by appropriate dilution with isopropyl alcohol or ethyl alcohol (the same solvent as used in the preparation of the solution).

Copper sulphate (N/10) : Dissolve 24.96 g of A.R. $CuSO_4 \cdot 5H_2O$ (Mol. wt. 249.6, Eq. wt. 249.6) in boiled-out distilled water and dilute to 1 litre.

EDTA (N/50, 1 ml EDTA ≅ 1 mg $CaCO_3$) : Dissolve 3.7225 g of A.R. disodium ethylenediamine tetra-acetate dihydrate ($Na_2H_2C_{10}H_{12}O_8N_2 \cdot 2H_2O$, Mol. wt. 372.25, Eq. wt. 186.125), dried at 80°C, in boiled-out distilled water and dilute to 1 litre.

Ethyl acetate (M/25) : Dilute 4.0 ml of pure ethyl acetate (Mol. wt. 88.1, sp. gr. 0.9005 g/ml at 20°C) to 100 ml with boiled-out distilled water. Measure 97.85 ml of the above solution and dilute to 1 litre with boiled-out distilled water.

Ferrous ammonium sulphate (N/4) : Dissolve 49 g of A.R. $Fe(NH_4)_2(SO_4)_2 \cdot 6H_2O$ in boiled-out distilled water containing 10 ml of concentrated sulphuric acid and dilute to 500 ml. Standardise with N/4 potassium dichromate solution.

Ferrous ammonium sulphate (N/100) : Prepare by appropriate dilution of the above solution.

Hard water (1 ml ≅ 1 mg $CaCO_3$) : Add slowly a small amount of dilute hydrochloric acid, through a funnel, to 1 g of anhydrous $CaCO_3$ (A.R. grade) taken in a conical flask. Boil gently to remove CO_2. Heat to dryness on a water bath. Dissolve in boiled-out distilled water and dilute to 1 litre.

Hydrochloric acid (N/2) : Dilute 45 ml of concentrated A.R. hydrochloric acid to 1 litre with distilled water and standardise with N/2 sodium carbonate solution using methyl orange indicator. Make exactly N/2 by appropriate dilution.

Hydrochloric acid (N/10) : Dilute 9 ml of concentrated A.R. hydrochloric acid to 1 litre with distilled water and standardise with N/10 Na_2CO_3. Make exactly N/10 by appropriate dilution.

Iodine solution (N/10) : Place about 10 g of A.R. KI (iodate-free) and 2–3 ml distilled water in a large weighing bottle fitted with a ground-glass stopper. Dissolve by gentle shaking and weigh accurately, after the bottle and the solution have reached room temperature. Introduce quickly 3.2–3.5 g of A.R. or resublimed iodine (weighed on a rough balance) into the weighing bottle, without splashing, and restopper it. Shake and let stand for sometime. After the temperature has reached equilibrium, reweigh the bottle accurately. The difference gives the weight of pure iodine. Dilute the solution in the weighing bottle and quickly transfer it quantitatively to a 250-ml measuring flask. Dilute upto the mark with distilled water. Stopper and keep in a dark place. The prepared solution is $\frac{W \times 4}{126.91}$ N, where W is the weight of iodine dissolved. Make exactly N/10 by appropriate dilution.

Oxalic acid (N/10) : Dissolve 6.303 g of A.R. oxalic acid dihydrate [$C_2O_4H_2 \cdot 2H_2O$ Mol. wt. 126.068, Eq. wt. 63.034] in boiled-out distilled water and dilute to 1 litre.

Potassium dichromate (N/4) : Dissolve 12.259 g of A.R. $K_2Cr_2O_7$ (Mol. wt. 294.22, Eq. wt. 49.035), dried at 100°C, in boiled-out distilled water and dilute to 1 litre.

Potassium dichromate (N/10) : Prepare by appropriate dilution of the above solution.

Potassium iodate (N/5) : Dissolve 10.7 g of A.R.KIO_3 (Mol. wt. 214.01, Eq. wt. 53.50 for Andrews titration) in boiled-out distilled water and dilute to 1 litre.

Potassium iodate (N/10) : Dissolve 3.567 g of A.R. KIO_3 (Eq. wt. 35.67 for thiosulpohate standardisation) in boiled-out distilled water and dilute to 1 litre.

Potassium permanganate (N/10) : Dissolve 3.2–3.25 g of A.R. $KMnO_4$ (Mol. wt. 158.037, Eq. wt. 31.607) in 1 litre of distilled water. Boil for 1 hour and cool, or keep overnight. Filter through a plug of purified glass wool placed in the neck of a funnel or through sintered glass crucible (G-4) into a dark brown-coloured glass bottle. [Standardisation : 20 ml N/10 sodium oxalate solution + 100 ml of about 2N sulphuric acid. Warm to 50–60°C and titrate with potassium permanganate solution. Make exactly N/10 by appropriate dilution with water that has been redistilled from alkaline permanganate].

Silver nitrate (N/20) : Dissolve 8.4945 g of A.R. $AgNO_3$ (Mol. wt. 169.89, Eq. wt. 169.89) in boiled-out distilled water and dilute to 1 litre.

Silver nitrate (N/50) : Prepare by appropriate dilution of the above soultion with boiled-out distilled water.

Sodium Carbonate (N/2) : Dissolve 13.25 g of A.R. Na_2CO_3 (Mol. wt. 106, Eq. wt. 53), dried at 250°C, in boiled-out distilled water and dilute to 500 ml.

Sodium Carbonate (N/10) : Dilute 50 ml of the above solution to 250 ml with boiled-out distilled.

Sodium Carbonate (N/50) : Prepare by appropriate dilution of N/2 Na_2CO_3 with boiled-out distilled water (10 ml to 250 ml)

Sodium hydroxide (N/10) : Dissolve 4–4.5 g of A.R. NaOH (Mol. wt. 40, Eq. wt. 40) pellets in 1 litre of boiled-out distilled water. Standardise with N/10 oxalic acid or against a known weight of potassium hydrogen phthalate ($KHC_8H_4O_4$, Mol. wt. 204.22 Eq. wt. 204.22) using phenolphthalein indicator. Make exactly N/10 by appropriate dilution with boiled-out distilled water.

Sodium hydroxide (M/25) : Prepare by appropriate dilution of the above solution with boiled-out distilled water.

Sodium hydroxide (N/50) : Prepare by appropriate dilution of N/10 sodium hydroxide solution with boiled-out distilled water.

Sodium thiosulphate (N/10) : Dissolve 25 g of A.R. $Na_2S_2O_3{\cdot}5H_2O$ (Mol. wt. 248.21, Eq. wt. 248.21) in boiled-out distilled water and dilute to 1 litre. [Standardisation: 20 ml N/10 KIO_3 solution + 2 g A.R. KI and shake. Add 10 ml of 1N HCl and titrate against thiosulphate solution using starch solution as indicator near the end point]. Make exactly N/10 by appropriate dilution.

Sodium thiosulphate (N/40) : Prepare by appropriate dilution of the above solution with boiled-out distilled water.

Sodium thiosulphate (N/100) : Prepare by appropriate dilution of N/10 sodium thiosulphate solution with boiled-out distilled water.

Sulphuric acid (N/2) : Add slowly with constant stirring, 14 ml of concentrated A.R. sulphuric acid to a large volume of distilled water. Cool and dilute to 1 litre. Standardise with N/2 Na_2CO_3 using methyl orange indicator. Make exactly N/2 by appropriate dilution.

Sulphuric acid (N/10) : Dilute 3 ml of A.R. concentrated sulphuric acid to 1 litre in the way as above and standardise with N/10 Na_2CO_3 solution. Make exactly N/10 by appropriate dilution.

Sulphuric acid (N/50) : Prepare by appropriate dilution of the above solution.

1.6 Solutions of Approximate Strength

Adipoyl chloride (~ 2% v/v) : Add 2 ml of adipoyl chloride (hexan-1, 6-dioyl dichloride) to 100 ml of CCl_4 and mix to dissolve.

Alcoholic potassium hydroxide (N/2) : Dissolve 28 g of A.R. KOH (Eq. wt. 56) pellets in 1 litre of 95% alcohol (freshly distilled from KOH). Mix thoroughly and let stand undisturbed, for any carbonate to settle down. Decant the clear supernatant solution and use.

Alkaline potassium iodide : Dissolve 700 g of KOH pellets and 150 g of A.R. KI in distilled water and dilute to 1 litre.

Aluminium sulphate (1.5% alum) : Dissolve 15 g of $Al_2(SO_4)_3 \cdot 18H_2O$ in 1 litre of distilled water.

Ammonium molybdate (5%) : Dissolve 50 g of ammonium heptamolybdate $[(NH_4)_6Mo_7O_{24} \cdot 4H_2O]$ in 1 litre of warm distilled water. Cool and filter, if necessary.

Ammonium oxalate (8%) : Dissolve 80 g of ammonium oxalate monohydrate $[(NH_4)_2 C_2O_4 \cdot H_2O]$ in 1 litre of distilled water.

Ammonium thiocyanate (10%) : Dissolve 100 g of A.R. NH_4SCN in 1 litre of distilled water.

Ammonium thiocyanate (N/20) : Dissolve 3.8 g A.R. NH_4SCN (Mol. wt. 76.11, Eq.wt. 76.11) in distilled water and dilute to 1 litre.

Barium chloride (5%) : Dissolve 50 g of $BaCl_2 \cdot 2H_2O$ in 1 litre of distilled water.

Bismuth chloride (O.1M) : Dissolve 31.5 g of $BiCl_3$ (Mol. wt. 315.5) in the minimum volume of 5N HC1 and dilute to 1 litre. Add dropwise, with stirring, sufficient hydrochloric acid to remove any precipitate.

Bromine solution (saturated) : Prepare a saturated solution by shaking 11 ml of liquid bromine with 1 litre of distilled water.

EDTA (N/50) : Dissolve 3.723 g of disodium ethylene diaminetetra-acetate dihydrate (Mol. wt. 372.25, Eq. wt. 186.125) in one litre of distilled water.

Ferric chloride (O.1M) : Dissolve 27 g of $FeCl_3 \cdot 6H_2O$ (Mol. wt. 270.5) in 1 litre of distilled water containing 50 ml of dilute HCl.

Ferric nitrate (O.1M) : Dissolve 40 g of $Fe(NO_3)_3 \cdot 9H_2O$ (Mol. wt. 404) in 1 litre of distilled water containing 50 ml of dilute nitric acid.

Ferric sulphate (0.5%) : Dissolve 5 g of $Fe_2(SO_4)_3$ (Mol. wt. 400) in 1 litre of distilled water.

Ferrous sulphate (N/2) : Dissolve 14 g of $FeSO_4 \cdot 7H_2O$ (Mol. wt. 278, Eq. wt. 278) in 100 ml of distilled water containing 10 ml of dilute sulphuric acid.

Hexamethylene diamine (~1% w/v) : Add 9–10 g of commercially available hexamethylene diamine to 100 ml of distilled water and shake to dissolve.

Hydrochloric acid (O.4N) : Dilute 36 ml of concentrated A.R. hydrochloric acid to 1 litre of distilled water.

Iodic acid (N/10) : Dissolve 3 g of iodic acid (HIO_3, Mol. wt. 176, Eq. wt. 29.33) in 1 litre of distilled water.

Manganous sulphate (O.1M) : Dissolve 18.7 g of $MnSO_4 \cdot 2H_2O$ (Mol. wt. 187) in distilled water. Filter and dilute to 1 litre.

Manganous sulphate (48%) : Dissolve 480 g of $MnSO_4 \cdot 2H_2O$ in distilled water. Filter and dilute to 1 litre.

Mercuric chloride (saturated) : Dissolve 80 g of $HgCl_2$ in 1 litre of hot distilled water. Cool to room temperature and filter.

Millon's Reagent : Dissolve 5 g of Hg in 5 g of fuming nitric acid by shaking at room temperature. Dilute with 20 ml distilled water. Filter to remove any precipitate formed.

Oxalic acid (N/10) : Dissolve 6.3 g of oxalic acid dihydrate in 1 litre of distilled water.

Potassium iodide (O.1M) : Dissolve 16.6 g of iodate-free KI (Mol. wt. 166) in 1 litre of boiled-out distilled water.

Potassium iodide (10%) : Dissolve 100 g of iodate-free KI in 1 litre of boiled-out distilled water.

Potassium oxalate (2%) : Dissolve 20 g of potassium oxalate [$K_2C_2O_4 \cdot H_2O$] in 1 litre of distilled water.

Potassium permanganate (N/10) : Dissolve 3.2 g of A.R. $KMnO_4$ in 1 litre of distilled water. Boil for 1 hour, cool and filter through a plug of glass wool placed in the neck of a funnel into a dark-brown-coloured glass bottle.

Potassium nitrate (O.1M) : Dissolve 9.1 g of pure KNO_3 (Mol. wt. 91) in 1 litre of distilled water.

Potassium thiocyanate (O.1M) : Dissolve 9.7 g of A.R. KSCN (Mol. wt. 97) in 1 litre of distilled water.

Saturated magnesium chloride (~ 475 g/l), L_2 : Add chemically pure anhydrous $MgCl_2$, in small amounts at a time and with constant stirring, to distilled water until a residue remains at the bottom. Store in a closed flask as the solution is hygroscopic.

Saturated zinc chloride (~1575 g/l), L_3 : Add chemically pure anhydrous $ZnCl_2$, in small amounts at a time and with constant stirring, to distilled water until a residue remains at the bottom. Store in a closed flask as the solution is hygroscopic.

Sebacoyl chloride (~3% v/v) : Add 3 ml of sebacoyl chloride to 100 ml of CCl_4 and mix to dissolve.

Silver sulphate–sulpluric acid reagent : Add 5.5 g of Ag_2SO_4 crystals to 1 kg of H_2SO_4 and let stand for 1–2 days for complete dissolution.

Sodium bisulphite (N/20) : Dissolve 2.6 g of $NaHSO_3$ (Mol. wt. 104, Eq. wt. 52) in 1 litre of distilled water containing 50 ml of 2N H_2SO_4.

Sodium thiosulphate (N/10) : Dissolve 24.82 g of $Na_2S_2O_3 \cdot 5H_2O$ ((Mol. wt. 248.21, Eq. wt. 248.21) in 1 litre of boiled-out distilled water.

Stannous chloride (O.1M) : Dissolve 2.26 g of $SnCl_2$. $2H_2O$ (Mol. wt. 225.7) in 5-6 ml of concentrated hydrochloric acid and dilute to 100 ml.

Stannous chloride (5%) : Dissolve 50 g of $SnCl_2 \cdot 2H_2O$ in 100 ml of concentrated hydrochloric acid and dilute to 1 litre.

Sucrose (10%) : Dissolve 100 g of granulated sugar in 1 litre of distilled water.

1.7 Titration Solvents for Acid Value of Oils

1. Mix together 500 ml toluene, 5 ml water and 495 ml isopropyl alcohol.
2. Mix together 500 ml chloroform, 5 ml water and 495 ml isopropyl alcohol.
3. Mix together equal volumes of ethyl alcohol and benzene.
4. Mix together equal volumes of alcohol and ether.
5. 95% alcohol–Add about 5 g of KOH to 1 litre of rectified spirit, boil for about half an hour and distil.
6. Pure distilled methyl alcohol.

1.8 Wij's Solution (0.2N Solution of ICl in Glacial Acetic Acid)

1. Commercially available Wij's solution, which gives satisfactory results, may be used.
2. Dissolve 16 g of commercially available iodine monochloride (ICl) in 1 litre of A.R. glacial acetic acid. Filter rapidly through a filter paper and store in amber-coloured glass-stoppered bottle in a dark place at a temperature below 30°C.
3. Dissolve 8 g of pure ICl_3 (iodine trichloride) and 9 g of resublimed iodine separately in 500 ml of A.R. glacial acetic acid each. Warm on a water bath, if dissolution is not complete at room temperature. Add iodine solution gradually to ICl_3 solution with constant shaking until a red-brown colour appears. Add 20–30 ml more. Heat on a water bath for about 15 minutes, cool and store.

4. Shake to dissolve 13 g of resublimed iodine in 1 litre of A.R. glacial acetic acid. Titrate 10 ml of the solution with N/10 $Na_2S_2O_3$ and set aside another portion (20–30 ml). Pass chlorine gas (dried by bubbling through concentrated H_2SO_4) into the remaining iodine solution until the volume of $Na_2S_2O_3$ used against 10 ml of the solution (after addition of 20 ml of 10% KI solution) is double the volume used before passing chlorine (Colour of the solution changes to orange at this stage). Neutralise any free chlorine with the original iodine solution. Heat on a water bath, cool and store.

Exercises

1. What is a primary standard? Give some examples.
2. What is the general procedure for preparing a solution of definite normality (say N/10) of a substance which is not a primary standard?
3. 3.35 g of A.R. sodium oxalate (Eq. wt. 67) are dissolved in distilled water and diluted to 1 litre. 20 ml of this solution requires 19 ml of an unknown potassium permanganate solution for complete titration. Calculate the volume (in ml) of the unknown solution that should be diluted to 500 ml so as to get exactly N/20 $KMnO_4$ solution.
4. What are the advantages of using boiled-out distilled water, instead of ordinary distilled water, for the preparation of standard solutions?
5. How is sodium hydroxide solution freed of its carbonate content?
6. Explain why the permanganate solution should not be (a) filtered through a filter paper, (b) used in a clip-burette (with rubber tubing and pinch-cock)?

2

CHEMICAL EQUILIBRIUM AND CHEMICAL KINETICS

When a mixture of CO(g) and NO_2 (g) is heated in a closed vessel, the reddish-brown colour of the mixture starts fading, indicating the progress of the following reaction:

$$\underset{\text{Colourless}}{CO(g)} + \underset{\text{Reddish brown}}{NO_2(g)} \longrightarrow \underset{\text{Colourless}}{CO_2(g)} + \underset{\text{Colourless}}{NO(g)} \qquad (2.1)$$

However, if we take the colourless mixture of CO_2(g) and NO(g) [the products of reaction (2.1)] and heat, it is seen that the mixture slowly develops a reddish-brown colour due to the formation of NO_2(g) as per the following reaction:

$$CO_2(g) + NO(g) \longrightarrow CO(g) + NO_2(g) \qquad (2.2)$$

which proceeds in a direction opposite to that of reaction (2.1). This is an example of a reversible reaction and is represented as:

$$CO(g) + NO_2(g) \rightleftharpoons CO_2(g) + NO(g) \qquad (2.3)$$

A chemical reaction is said to be Reversible under defined experimental conditions if, under these conditions, the products of the reaction can react with one another to reform the original reactant or reactants.

2.1 Chemical Equilibrium

If the reaction mixture in the above example is heated to a constant temperature in a thermostatic bath, it is seen that the colour of the reaction mixture, which changes in the beginning, becomes constant after some time, indicating that the composition of the reaction mixture is no longer changing. The reaction is said to have reached equilibrium.

Chemical equilibrium, at a given temperature, is characterised by constancy of macroscopic properties — observable properties such as colour, pressure, concentration, density, etc. — in a closed system (a system containing a constant amount of matter) at a uniform temperature. Microscopic processes (changes at the molecular level), however, continue but in a balance that yields no macroscopic changes. Thus, both forward and backward reactions are proceeding simultaneously but at equal rates so that no change is observed.

Any factor that affects the rate of any one of the reactions involved in equilibrium will affect the conditions at equilibrium. Concentration, temperature and pressure are such factors. The effect of changes in any one of these factors on the equilibrium concentrations can be qualitatively predicted by applying the generalisation known as Le Chatelier's Principle–If a system at equilibrium is subjected to a change, the equilibrium shifts in a direction that tends to counteract partially the imposed change. The exact quantitative relationship between the concentrations of the reactants and the products at equilibrium is given by the Law of Chemical Equilibrium according to which the ratio of the product of the equilibrium concentrations of the substances formed to the product of the equilibrium concentrations of the reacting substances is constant at constant temperature. Thus, if [A] and [B] represent the molar concentrations of the reactants A and B, while [C] and [D] be the molar concentrations of the products C and D at equilibrium for the hypothetical reaction:

$$A + B \rightleftharpoons C + D, \text{ then}$$

r_1, the rate of forward reaction = k_1 [A] [B], and

r_2, the rate of backward reaction = k_2 [C] [D]

where k_1 and k_2 are constants.

At equilibrium $$r_1 = r_2$$

or $$k_1 [A][B] = k_2 [C][D]$$

or $$\frac{[C]\ [D]}{[A]\ [B]} = \frac{k_1}{k_2} = K_C \quad (2.4)$$

K_C is known as the concentration equilibrium constant and is constant at constant temperature. If its numerical value is determined by measuring the concentration of all the species in a particular equilibrium solution, it can be used in the calculations for any other equilibrium state between the same species at that same temperature.

2.1.1 *To study the effect of addition of solutions of Fe $(NO_3)_3$, KSCN, $SnCl_2$ or KNO_3 to the equilibrium system obtained by mixing aqueous solutions of $Fe(NO_3)_3$ and KSCN*

Reagents Required

(1) Ferric nitrate solution (0.1M)

(2) Potassium thiocyanate solution (0.1M)

(3) Stannous chloride solution (0.1M)

(4) Potassium nitrate solution (0.1M)

Theory

When aqueous solutions of $Fe(NO_3)_3$ and KSCN are mixed, the following three equilibria exist simultaneously :

$$Fe(NO_3)_3 \rightleftharpoons Fe^{3+} + 3NO_3^- \quad (2.5)$$

$$KSCN \rightleftharpoons K^+ + SCN^- \quad (2.6)$$

$$Fe^{3+} + SCN^- \rightleftharpoons FeSCN^{2+} \quad (2.7)$$

(Pale yellow) (Colourless) (Reddish brown colour)

The equilibrium constant K_C for reaction (2.7) is given by

$$K_C = \frac{[FeSCN^{2+}]}{[Fe^{3+}]\,[SCN^-]} \quad (2.8)$$

and remains constant at constant temperature, irrespective of the source of Fe^{3+}, SCN^- and $FeSCN^{2+}$ ions. Addition of any substance that will tend to change the concentration of any one of these ions will disturb (shift) the equilibrium in a particular way so that K_C may remain constant.

(a) Addition of $Fe(NO_3)_3$ will increase the concentration of Fe^{3+} ions and in accordance with equation (2.8), in order to keep K_C constant, the concentration of $FeSCN^{2+}$ should automatically increase which will be evidenced by a deepening of the red colour.

(b) Addition of KSCN will increase the concentration of SCN^- and, in order that K_C may remain constant, more of $FeSCN^{2+}$ will be formed which means that red colour will again deepen.

(c) Addition of $SnCl_2$ will reduce Fe^{3+} ion to Fe^{2+} :

$$2Fe^{3+} + Sn^{2+} \longrightarrow 2Fe^{2+} + Sn^{4+} \quad (2.9)$$

The consequent decrease in the concentration of Fe^{3+} will cause $FeSCN^{2+}$ to dissociate so that K_C remains constant. Dissociation of $FeSCN^{2+}$ will result in fading away of the reddish-brown colour.

(d) Addition of KNO_3 will considerably increase the concentrations of K^+ and NO_3^- ions, thus simultaneously depressing the dissociation of $Fe(NO_3)_3$ and KSCN [reactions (2.5) and (2.6)] due to common ion effect. The resultant decrease in the concentrations of Fe^{3+} and SCN^- will make $FeSCN^{2+}$ dissociate and the intensity of the colour will therefore decrease.

Procedure

To about 40 ml of distilled water taken in a beaker, add 4 drops of $Fe(NO_3)_3$ solution. Add KSCN solution dropwise with shaking until the solution acquires a light reddish-brown colour. Mix well and pour equal volumes of the solution into 5 clean test tubes. Mark the tubes as A, B, C, D and E and set aside tube A for colour comparison. Add 10 drops of the solution of $Fe(NO_3)_3$ to tube B, KSCN to tube C, $SnCl_2$ to tube D and KNO_3 to tube E. After 2 – 3 minutes, compare the colours of solutions in tubes B, C, D and E with that of the solution in tube A, and record the observed changes in the intensity of the colour.

Observations

Tube	Initial colour	Solution added (10 drops)	Colour intensity
A	Reddish brown	—	No change
B	-do-	$Fe(NO_3)_3$	Increases
C	-do-	KSCN	-do-
D	-do-	$SnCl_2$	Decreases
E	-do-	KNO_3	-do-

Precautions

(i) Before pouring the reddish-brown solution into the test tubes, it should be well mixed so as to get a uniform concentration throughout.

(ii) As far as possible, the temperature should be kept constant (a thermostat may be used).

2.1.2 *To study the effect of addition of H_2O and HCl on the equilibrium represented by*

$$BiCl_3 + H_2O \rightleftharpoons BiOCl + 2HCl$$

and to show that the observations are in accord with Le Chatelier's Principle

Reagents Required

(1) Bismuth chloride solution (0.1M)

(2) Concentrated HCl

Theory

The equilibrium constant K_C for the reaction

$$BiCl_3 + H_2O \rightleftharpoons \underset{\text{white precipitate}}{BiOCl} + 2HCl \qquad (2.10)$$

is given by

$$K_C = \frac{[BiOCl]\,[HCl]^2}{[BiCl_3]\,[H_2O]} \qquad (2.11)$$

Addition of water will increase the denominator in the above expression. Therefore, in order to keep K_C constant, the numerator should also increase, i.e., the reaction (2.10) will proceed in the forward direction with the formation of more of BiOCl which will be indicated by an increase in the amount of white precipitate. Addition of a large excess of water will, in fact lead to complete hydrolysis of $BiCl_3$. However, addition of HCl will increase the numerator in the expression (2.11). In order that K_C may remain constant, BiOCl will dissolve in HCl to increase concentration of $BiCl_3$ and water. In presence of a large excess of HCl, the reaction will go to completion in the backward direction, i.e., the whole of the precipitate of BiOCl will disappear.

Procedure

To about 5 ml of $BiCl_3$ solution taken in a 100-ml beaker, add distilled water drop by drop, stirring throughout with a glass rod, until a white precipitate is observed. Continue the addition of water until an increase in the amount of white precipitate is no longer observed. Now add dropwise concentrated HCl, again stirring with a glass rod. Record the changes observed and continue the addition of HCl until the precipitate is completely dissolved. Again add water to observe the reversal of the changes.

Observations

(1) With the addition of more and more water, the amount of white precipitate goes on increasing.

(2) With the addition of more and more of concentrated HCl, the amount of white precipitate goes on decreasing.

Accord with Le Chatelier's Principle

By adding water, we tend to increase the concentration of one of the reactants. To reduce the effect of this change, the equilibrium shifts in the forward direction as evidenced by increase in the amount of precipitate. Similarly, addition of concentrated HCl tends to increase the concentration of one of the products. A portion of this HCl is utilised in dissolving BiOCl, thus partially counteracting the imposed change.

Exercises

7. (a) Write down the expressions for K_C and K_P for general reaction:

$$aA + bB + \cdots \rightleftharpoons lL + mM + \cdots \qquad (2.12)$$

(b) How are K_C and K_P related?

8. What specifically is 'equal' in a chemical reaction that has attained a state of equilibrium?

9. What is a steady state? With the help of an example, explain how it is different from chemical equilibrium.

10. (a) Using Le Chatelier's Principle, predict the effect of changing temperature on the equilibrium represented by the reaction:

$$N_2O_4(g) \rightleftharpoons 2NO_2(g), \quad \Delta H = +14.1 \text{ K cal} \quad (2.13)$$

(b) What experiment would you perform to prove the above predictions?

11. What is the technological importance of Le Chatelier's Principle and Law of Chemical Equilibrium? Illustrate your answer by applying Le Chatelier's Principle to the following equilibrium:

$$N_2(g) + 3H_2(g) \rightleftharpoons 2NH_3(g), \quad \Delta H = -22 \text{ K cal} \quad (2.14)$$

12. Giving examples, describe some ways in which equilibria can be shifted to bring about essentially complete reactions.
13. Explain why Cu^{2+} and Zn^{2+} cannot be completely removed from industrial wastes by precipitation with NH_4OH?
14. What is meant by an irreversible reaction? Give some examples.

2.2 Rate of a Reaction

Reactions proceed at different rates. The rate of a reaction refers to the amount of a reactant consumed or a product formed in a reaction in a definite unit of time. The quantity consumed or produced is expressed in concentration units (gram-moles per litre) if the substance is in solution, or in partial pressure units if the substance is a gas. The time may be expressed in microseconds for very rapid reactions such as explosion of household gas and oxygen; in seconds or minutes for reactions proceeding at moderate rates at room temperature such as decomposition of H_2O_2 or oxidation of oxalic acid by permanganate; in days or months for slow reactions and in years for very slow reactions such as half-life period of $_{46}Ra^{226}$ (1590 years).

2.2.1 *Factors affecting the rate of a reaction*

The rate of a reaction depends on a large number of factors such as

(a) Concentration of the reactants.

(b) Temperature of the system.

(c) Nature of the reactants and the products.

(d) Presence of a catalyst.

(e) Surface area of the reactants, and

(f) Exposure to radiation.

By a knowledge of how various factors influence the rate of a reaction, it becomes possible to bring the reaction under control, i.e., the speed of a reaction can be regulated to gain the desired effect. The economic viability of the process can therefore be ascertained.

(a) Effect of Concentration

This effect is best summarised in terms of the Law of Mass Action which states that "the rate at which a substance reacts is proportional to its 'active mass' and the rate of a chemical reaction is proportional to the product of the 'active masses' of

the reacting substances". The active mass of solids is taken as unity and that of gases and solutions as the number of gram-moles present per litre.

A reaction between reacting molecules or ions can take place only when they come sufficiently close together. The rate of reaction will thus depend on the frequency with which the reacting particles collide (Collision Theory). The increase in concentration of one or more reactants by admitting their additional amounts (or by increasing the pressure in case of gaseous reactants) will increase the frequency of collision and is therefore expected to increase the rate of the reaction.

(b) Effect of Temperature

Rise in temperature causes only a slight increase in the frequency of molecular collisions (which are proportional to $\sqrt{T}$) but experimental measurements indicate a manifold rise in the rate of reaction — A 10°C rise at room temperature causes about 1.5% increase in total number of binary collisions of HI molecules but the same change in temperature almost doubles the rate of dissociation.

Explanation: In a methane–oxygen mixture under ordinary conditions, although a methane molecule collides with an oxygen molecule a billion times in one second, there is no noticeable reaction. It has been postulated that a collision will result into a reaction only if the colliding particles possess an energy equal to or higher than a certain minimum (different for different reactions) known as the Threshold Energy. At room temperature, the number of such particles may constitute only a negligible fraction of the total number and thus the reaction may not be noticeable at all. However, even with a small rise in temperature, the number of particles having energy equal to or in excess of Threshold Energy increases appreciably, resulting in a large increase in the number of successful collisions.

(c) *Nature of Reactants* has a considerable influence on the rate of reaction. Under identical experimental conditions, reactions that involve considerable bond rearrangement (breaking of old bonds in the reactant molecules and formation of new ones in the product molecules) are expected to be slow while those that do not involve much bond rearrangement are expected to be fast.

Thus oxidation of NO by O_2 which involves breaking of two bonds in nitric oxide and formation of four bonds in nitrogen peroxide is much faster at room temperature than oxidation of methane with O_2, which involves breaking of four bonds in methane and formation of six new bonds (2 in CO_2 and 4 in $2H_2O$).

$$2\,NO + O_2 \longrightarrow 2\,NO_2 \qquad (2.15)$$

$$CH_4 + 2O_2 \longrightarrow CO_2 + 2H_2O \qquad (2.16)$$

(d) Effect of Presence of a Catalyst

The additional energy that the reactant molecules having energy less than Threshold Energy must acquire so that their collision results in the formation of the products is known as Activation Energy.

Activation Energy = Threshold Energy – Average Energy possessed by the molecules

Addition of a catalyst provides a new reaction path with a lower energy barrier (Activation Energy). Since the energy barrier is reduced, a larger number of the molecules of the reactants can get over it (i.e., have energy equal to or higher than Threshold Energy) and consequently the rate of the reaction increases.

(e) Surface Area of Reactants has a bearing on the rate of a reaction taking place in a heterogeneous system (reaction between two solids, a solid and a liquid, two immiscible liquids, etc.) since reaction can take place only at the surface at which the reactants are in contact. Surface area for a given mass of a substance increases with decrease in particle size. Smaller particles, therefore, react more rapidly than larger particles. Thus, in air, coal dust burns much more rapidly than a large lump of coal.

2.2.2 *To study the effect of concentration and temperature on the rate of reaction between iodic acid and sodium bisulphite*

Reagents Required

1. Iodic acid solution (N/10).
2. Sodium bisulphite solution (N/20).
3. Freshly prepared starch solution.

Theory

The reaction between iodic acid and bisulphite, which is carried out in presence of a small amount of freshly prepared starch solution to act as indicator, essentially consists of the following consecutive steps:

$$3HSO_3^- + HIO_3 \xrightarrow{\text{slow}} 3HSO_4^- + HI \qquad (2.17)$$

$$5HI + HIO_3 \xrightarrow{\text{fast}} 3I_2 + 3H_2O \qquad (2.18)$$

The iodine liberated in step (2.18) readily reacts with any unconsumed bisulphite present

$$HSO_3^- + I_2 + H_2O \xrightarrow{\text{fast}} HSO_4^- + 2HI \qquad (2.19)$$

and so no blue colour is observed. However, when step (2.17) has gone to completion, step (2.18) takes place almost instantaneously producing iodine which is evidenced by the appearance of characteristic blue-black colour. The sudden colour change thus indicates complete oxidation of bisulphite. The time elapsed, which is measured using a stop watch, between the mixing of the solutions of bisulphite and iodic acid and the sudden appearance of the blue-black colour is a measure of the rate of reaction.

Procedure

1. Take three 250-ml conical flasks and mark them as A, B and C.
2. With a measuring cylinder, pour 170 ml distilled water into flasks A and C and 160 ml distilled water into flask B.
3. Add with a pipet 20 ml of bisulphite solution to all the flasks.
4. Add 1 ml of freshly prepared starch solution to all the flasks.
5. With a pipet, add 10 ml of iodic acid solution to flask A, starting a stop-watch simultaneously. Mix and allow to stand at room temperature. Record the time when a blue-black colour appears.
6. Repeat the process with flask B, but adding 20 ml of iodic acid.
7. Add 10 ml of iodic acid to flask C, mix and keep the flask in a hot water bath and note the time for the development of blue colour.

Observations

Flask	Vol. of H_2O(ml)	Vol. of Bisulphite(ml)	Vol. of HIO_3(ml)	Temperature	Time in Seconds
A	170	20	10	room temp.	t_1 =
B	160	20	20	room temp.	t_2 =
C	170	20	10	hot water bath	t_3 =

Discussion of Results

1. Blue colour appears in flask B quicker than in flask A ($t_2 < t_1$) because the reaction proceeds at a faster rate due to higher concentration of iodic acid in flask B.
2. Appearance of blue colour in flask C is quicker than in flask A ($t_3 < t_1$) because the reaction is carried out at higher temperature which increases the rate of reaction.

2.2.3 *To study the effect of nature of reactants on the rate of reaction*

Reagents Required

1. Oxalic acid solution (N/10)
2. Ferrous sulphate solution (N/10)
3. Potassium permanganate solution (N/10)
4. Dilute H_2SO_4

Theory

$KMnO_4$, in acidic medium, oxidises both ferrous ion and oxalate ion as per the following reactions:

$$Fe^{2+} \longrightarrow Fe^{3+} + e\,] \times 5$$

$$MnO_4^- + 8H^+ + 5e \longrightarrow Mn^{2+} + 4H_2O$$

$$5Fe^{2+} + MnO_4^- + 8H^+ \longrightarrow 5Fe^{3+} + Mn^{2+} + 4H_2O \qquad (2.20)$$

$$C_2O_4^{2-} \longrightarrow 2CO_2 + 2e\,] \times 5$$

$$MnO_4^- + 8H^+ + 5e \longrightarrow Mn^{2+} + 4H_2O\,] \times 2$$

$$5C_2O_4^{2-} + 2MnO_4^- + 16H^+ \longrightarrow 10CO_2 + 2Mn^{2+} + 8H_2O \qquad (2.21)$$

As the permanganate ion is common to the two reactions, any difference in their rates will evidently be due to the difference in the specific characteristics of ferrous ion and oxalate ion. The rates of the two reactions are compared by noting the times required for decolourisation of equal amounts of acified permanganate solution by equivalent amounts of ferrous ion and oxalate ion solutions at room temperature.

Procedure

(1) Take two test tubes and mark them A and B.

(2) Transfer with a graduated pipet 1 ml of oxalic acid solution to tube A and 1 ml of ferrous sulphate solution to tube B.

(3) To 20 ml of distilled water taken in a beaker, add 5 ml of dilute H_2SO_4 and 1 drop of $KMnO_4$ solution and mix well.

(4) Add 1 ml of the above diluted solution to each of the two test tubes A and B.

(5) Allow to stand at room temperature and note the time needed for the purple colour of permanganate to disappear in the tubes A and B and record in the table.

Observations

Tube	Oxalic acid solution taken	Ferrous sulphate solution taken	$KMnO_4$ solution added	Decolourisation time in seconds
A	1 ml	–	1 ml	t_1 =
B	–	1 ml	1 ml	t_2 =

Results and Discussion

The observations show that the colour of $KMnO_4$ disappears much faster in the tube containing ferrous sulphate than in the tube containing oxalic acid ($t_2 < t_1$). This is as expected since ferrous is a simple ion whereas oxalate is a polyatomic ion containing many covalent bonds that have to be broken.

2.2.4 *To study the catalysing effect of Mn^{2+} ions on the rate of reaction between oxalate and permanganate ions in acid medium*

Reagents Required

1. Oxalic acid solution (N/10)
2. Manganous sulphate solution (0.1M)
3. Potassium permanganate solution (N/10)
4. Dilute H_2SO_4

Theory

Oxalate decolourises permanganate in acid medium (reaction 2.21). The time required is a measure of the rate of reaction. The effect of the presence of Mn^{2+} ions on the rate of reaction is seen by comparing the times needed for decolourisation in the absence and in the presence of Mn^{2+} ions. Other factors like the concentration of oxalate, permanganate and acid, and the temperature are kept constant.

Procedure

(1) Take two clean test tubes and mark them A and B.

(2) Pour with a graduated pipet 1 ml of oxalic acid solution into each of the two tubes.

(3) Add 1 ml of manganous sulphate solution to tube B.

(4) Pour 1 ml of very dilute solution of acidified $KMnO_4$ (as prepared in the previous experiment) into each of the tubes.

(5) Shake the contents of the two tubes and allow to stand at room temperature.

(6) Record the time needed for purple colour of $KMnO_4$ to disappear in the two tubes.

Observations

Tube	Oxalic acid	$KMnO_4$	$MnSO_4$	Time for decolourisation (seconds)
A	1 ml	1 ml	–	t_1 =
B	1 ml	1 ml	1 ml	t_2 =

Results and Discussion

Observations indicate that t_2 is less than t_1. Manganese(II) ions present in tube B catalyze the reaction between oxalate and permanganate ions (reaction 2.21) and so decolourisation of $KMnO_4$ takes place at a faster rate in tube B.

2.2.5 *To study the effect of surface area on the rate of a reaction*

Reagents Required

1. Potassium thiocyanate solution (0.1M)
2. Ferric chloride solution (0.1M)

3. Dilute HCl
4. Steel wool
5. Iron nail

Theory

Addition of $FeCl_3$ solution to the acidified solution of KCNS produces a reddish-brown colour:

$$Fe^{3+} + SCN^- \rightleftharpoons Fe(SCN)^{2+} \quad (2.22)$$

The iron added reacts with HCl to produce nascent hydrogen which

$$Fe + 2H^+ \longrightarrow Fe^{2+} + 2H \quad (2.23)$$

reduces ferric to ferrous:

$$Fe^{3+} + H \longrightarrow Fe^{2+} + H^+ \quad (2.24)$$

This decrease in the concentration of Fe^{3+} ions shifts the equilibrium (reaction 2.22) to left thus reducing the intensity of the colour.

Procedure

Take 10–15 ml of dilute HCl in a beaker. Add 1 ml of KSCN solution and a drop of $FeCl_3$. A red colour appears. Add equal volumes of this solution to three clean test tubes. Mark them A, B and C and set aside tube A for colour comparison. Add a small wad of steel wool to the tube B and an iron nail to tube C. Set aside the tubes and observe their colour during the next 10 minutes.

Observations

Tube	Initial colour	Physical state of iron added	Colour after 10 minutes
A	Reddish-brown	–	No change in colour intensity
B	-do-	Steel wool	Least intense colour
C	-do-	Iron nail	Less intense than A but more intense than B

Results and Discussion

The reddish-brown colour fades faster in tube B (to which steel wool has been added) than in tube C (to which iron nail is added). Because of the much larger surface area of steel wool than that of iron nail, rate of liberation of nascent hydrogen in tube B is higher which leads to faster rate of reduction of ferric ions and hence faster fading of the colour.

Precautions

(1) The weights of steel wool and iron nail should, as far as possible, be the same.

Exercises

15. In experiment 2.2.2, why is 10 ml less water taken in flask B than in flasks A & C?
16. In the above experiment, what effect would you expect on the reaction time if more water is added to the reaction mixture at room temperature?
17. What is meant by the rate-determining step? Point out the rate-determining step in the reaction between HIO_3 and $NaHSO_3$ (Experiment 2.2.2).
18. What is meant by a clock reaction?
19. Explain why there is a danger of explosion in a saw mill.
20. Predict the observations when a few (8–10) crystals of KI and an equal amount of $HgCl_2$ crystals are placed in a mortar and
    (i) Mixed with a glass rod.
    (ii) Ground with a pestle.
    (iii) About 10 ml distilled water is added to the above mixture, and
    (iv) A concentrated solution of KI is then added while stirring with a glass rod.
21. Give some examples of reactions that are catalysed by radiation.
22. Why does the rate of a reaction fall with time?
23. Why is the speed of a reaction important?

2.3 Velocity Constant and Order of a Reaction

Consider the hypothetical reaction

$$A \longrightarrow \text{Products}$$

The rate of this reaction at a particular temperature is given by the expression $r = k\,[A]$, called the rate equation, where [A] represents the molar concentration of the only reactant A and k is a constant characteristic of the reaction, and is known as the Rate Constant or the Velocity Constant of the reaction at the given temperature. If [A] = 1, $r = k$. Thus, at a given temperature, the rate constant or velocity constant of a reaction involving a single reactant is equal to the rate of the reaction when the molar concentration of the reactant is unity. It is also called the Specific Reaction Rate and when multiplied by the concentration of the reactant at any instant gives the rate of the reaction at that instant. For a reaction involving more than one reactant, the Rate Constant is equal to the rate of the reaction when the concentration of each of the reactants is unity.

Order of a Reaction is equal to the sum of the powers of the concentration terms of the experimentally determined rate equation of a reaction. Thus a reaction is said to be of First Order if its experimentally determined rate can be represented by an expression of the type $r = k\,C_A$ and of Second Order if the rate can be expressed as

$$r = k \cdot C_A^2$$

or

$$r = k\,C_A\,C_B$$

where C_A and C_B represent the molar concentrations of the reactants A and B, and so on.

2.3.1 *Kinetics of a First Order reaction*

Let the initial concentration of the reactant A for the reaction

$$A \longrightarrow \text{Products}$$

be '*a*' gram-moles/litre, out of which '*x*' gram-moles/litre is converted into products in '*t*' seconds. Thus after '*t*' seconds from the start, the concentration of the reactant will be $(a - x)$ gram-moles/litre and the rate of the reaction at this instant will be given by $\frac{dx}{dt} = k(a - x)$. Integrating this equation between the time limits $t = 0$ and $t = t$, and using the fact that when $t = 0$, $x = 0$, we get

$$k = \frac{l}{t} \ln \frac{a}{a - x} = \frac{2.303}{t} \log \frac{a}{a - x} \qquad (2.25)$$

If x_1 and x_2 be respectively the number of gram-moles/litre of the reactant changed into the products up to times t_1 and t_2 from the start, then the above equation takes the form

$$k = \frac{2.303}{t_2 - t_1} \log \frac{a - x_1}{a - x_2} \qquad (2.26)$$

2.3.2 *To determine the rate constant of hydrolysis of methyl acetate by dilute HCl and to show that it is a First Order reaction*

Reagents Required

1. Hydrochloric acid (0.4N)
2. Sodium hydroxide solution (N/10)
3. Pure methyl acetate
4. Phenolphthalein indicator

Theory

The hydrolysis of methyl acetate in presence of an acid may be represented as

$$CH_3COOCH_3 + H_2O \xrightarrow[(HCl)]{H^+} CH_3COOH + CH_3OH \qquad (2.27)$$

HCl merely acts as a catalyst and so its concentration, which remains constant throughout, will not affect the order of the reaction. Also, since water is present in large excess, its active mass (molar concentration) remains practically unchanged during the reaction. The rate of hydrolysis is therefore determined only by the concentration of methyl acetate:

$$r = \frac{dx}{dt} = k\,[CH_3COOCH_3] = k(a - x)$$

or

$$k = \frac{2.303}{t} \log \frac{a}{a - x} \qquad (2.25)$$

$$= \frac{2.303}{t_2 - t_1} \log \frac{a - x_1}{a - x_2} \tag{2.26}$$

where x_1 and x_2 are the gram-moles/litre of methyl acetate hydrolysed at times t_1 and t_2 respectively from the beginning.

As acetic acid is produced on hydrolysis, the progress of the reaction is followed by withdrawing definite volumes of the reaction mixture at suitable time intervals and titrating against a standard alkali solution. The increase in the volume of alkali used is a measure of methyl acetate hydrolysed. If V_{t_1}, V_{t_2} and V_∞ respectively be the volumes of standard alkali used against the same definite volume of the reaction mixture at times t_1, t_2 and t_∞ (the time required for the reaction to be completed), then

$V_{t_1} \propto$ amount of HCl present originally + amount of acetic acid formed or methylacetate hydrolysed at time t_1 (x_1),

and $V_\infty \propto$ amount of HCl present originally + amount of CH_3COOH formed at the end of reaction or initial concentration of methylacetate(a).

Therefore, $V_\infty - V_{t_1} \propto (a - x_1)$, i.e., the concentration of unhydrolysed CH_3COOCH_3 at time t_1.

Similarly, $V_\infty - V_{t_2} \propto (a - x_2)$, i.e. the concentration of unhydrolysed CH_3COOCH_3 at time t_2.

Therefore, equation (2.26) can be written as

$$k = \frac{2.303}{t_2 - t_1} \log \frac{V_\infty - V_{t_1}}{V_\infty - V_{t_2}} \tag{2.28}$$

Procedure

(1) Take about 100 ml of the HCl solution in a clean conical flask and suspend it in a thermostat maintained at 25° C. Take about 15 ml of pure methylacetate in a clean dry tube, cork it and suspend in the same thermostat.

(2) Take 8 conical flasks, each of 100-ml capacity; pour about 25 ml of distilled water into each and keep them immersed in an ice-water bath.

(3) Wash a burette and rinse and fill it with NaOH solution.

(4) When the two liquids (methyl acetate and HCl) have attained the temperature of the bath, pipet out 10 ml of methyl acetate and pour it into the acid contained in the flask.

(5) Mix by shaking and pipet out 5 ml of the reaction mixture into one of the flasks containing 25 ml of ice-cold water, starting a stop watch when the pipet has been half discharged. Add 1 drop of phenolphthalein indicator to this solution and titrate quickly with NaOH solution until a pink colour appears. Record the titre value as V_t. Time for this value is zero.

(6) Similarly, withdraw 5-ml portions of the reaction mixture at times 5, 10, 20, 30, 50 and 80 minutes and pour into flasks containing 25 ml of ice-cold water and titrate quickly against the same alkali solution.

(7) For finding V_∞, transfer about 20 ml of the reaction mixture to clean dry flask, stopper it tightly and keep immersed in a water bath maintained at 50° C for 80–90 minutes so as to complete the hydrolysis. Cool to room temperature. Pipet out 5 ml of the solution to a flask containing 25 ml of ice-cold water and titrate as before.

Precautions

(1) As the rate of reaction is very sensitive to temperature changes, the temperature of the bath should be maintained constant within ± 0.5° C throughout the course of the experiment.

(2) For every titration, the time should be recorded when the pipet has been half discharged into the flask.

(3) The titration should be made as quickly as possible—for this the bulk of the expected volume of alkali should be run in quickly and then there should be dropwise addition to the first appearance of pink colour that persists for at least 10 seconds.

Observations and Calculations

Temperature of the bath = t °C

Time in min t	Titre Value V_t ml	Concentration of unhydrolysed CH_3COOCH_3 at time t $(V_\infty - V_t)$	$\log (V_\infty - V_t)$	$k = \frac{2.303}{t_2 - t_1} \log \frac{V_\infty - V_{t_1}}{V_\infty - V_{t_2}}$
t_0 =	V_{t_0} =	$V_\infty - V_{t_0}$		
t_1 =	V_{t_1} =	$V_\infty - V_{t_1}$		
t_2 =	V_{t_2} =	$V_\infty - V_{t_2}$		
t_3 =	V_{t_3} =	$V_\infty - V_{t_3}$		
t_4 =	V_{t_4} =	$V_\infty - V_{t_4}$		
t_5 =	V_{t_5} =	$V_\infty - V_{t_5}$		
t_6 =	V_{t_6} =	$V_\infty - V_{t_6}$		
t_∞ =	V_∞ =			

Results and Discussion

(1) Since the values of k, as calculated by substituting the various values of $V_\infty - V_t$ for the corresponding t values in the kinetic equation for the first order, i.e.,

$$k = \frac{2.303}{t_2 - t_1} \log \frac{V_\infty - V_{t_1}}{V_\infty - V_{t_2}}$$

remain almost constant, the hydrolysis of methyl acetate catalysed by acid is a first order reaction.

(2) The average of the various values of k is the velocity constant of the reaction at the temperature of the experiment.

Graphical Method

The rate equation for the first order reaction

$$t = \frac{2.303}{k} \log \frac{a}{a - x}$$

$$= \frac{2.303}{k} \log a - \frac{2.303}{k} \log (a - x)$$

$$= \text{constant} - \frac{2.303}{k} \log (a - x) \qquad (2.29)$$

indicates that a plot of t against $\log (a - x)$ or $\log (V_\infty - V_t)$ will be a straight line whose slope is given by $-\frac{2.303}{k}$.

Thus $$-\frac{2.303}{k} = \tan \theta = -\frac{OA}{OB}$$

or $$k = 2.303 \times \frac{OB}{OA}$$

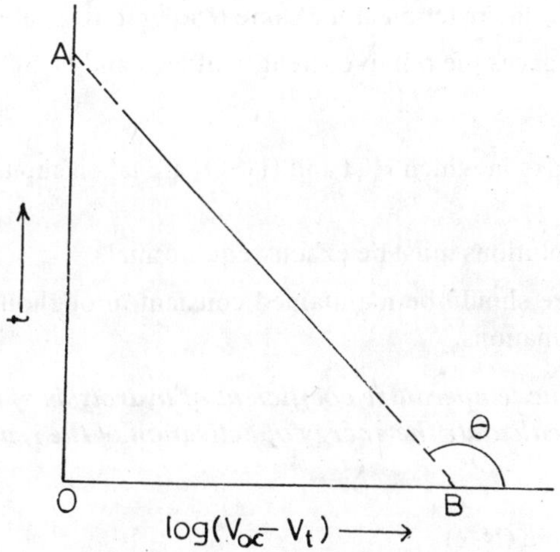

Fig. 2.1 Plot of t versus $\log (V_\infty - V_t)$

2.3.3 *To compare the strengths of HCl and H_2SO_4 by studying their catalysing action on hydrolysis of methyl acetate*

Reagents Required

1. Standard hydrochloric acid (N/2)
2. Standard sulphuric acid (N/2)
3. Sodium hydroxide solution (N/10)
4. Pure methyl acetate
5. Phenolphthalein indicator

Theory

The hydrolysis of methyl acetate is catalysed by an acid (reaction 2.27) and the rate of hydrolysis is found to be approximately proportional to the concentration or more correctly activity of H^+ ions. Therefore, the velocity constants of the reaction in presence of equal concentrations (equinormal solutions) of two acids will be proportional to their degrees of dissociation or their strengths:

$$\text{Relative strengths} = \frac{\text{Strength of HCl}}{\text{Strength of } H_2SO_4} = \frac{\alpha_1}{\alpha_2} = \frac{k_1}{k_2} \qquad (2.30)$$

where α_1 and α_2 are the degrees of dissociation of HCl and H_2SO_4 and k_1 and k_2 are the velocity constants of the hydrolysis of methyl acetate in presence of their equinormal solutions.

Procedure

(1) Determine the velocity constant k_1 of the hydrolysis of methyl acetate by taking exactly 100 ml of N/2 HCl (standard) at a definite temperature maintained by means of a thermostat, as described in the earlier experiment.

(2) In the same way, determine the velocity constant k_2 by taking exactly 100 ml of N/2 H_2SO_4 (standard) and the same exact amount of methyl acetate, and carrying out the reaction at the same temperature as above.

(3) The ratio k_1/k_2 gives the relative strength of HCl and H_2SO_4.

Precautions

1. The conical flasks in which HCl and H_2SO_4 are taken should be clean and dry.
2. The two acid solutions must be exactly equinormal.
3. The temperature should be maintained constant throughout the course of the two determinations.

2.3.4 *To determine the temperature coefficient of hydrolysis of methyl acetate by acid and to calculate the energy of activation of the reaction*

Reagents Required

1. Hydrochloric acid (N/2)
2. Sodium hydroxide solution (N/10)

3. Pure methyl acetate
4. Phenolphthalein indicator solution

Theory

As discussed earlier (p.15), a rise in temperature markedly increases the rate of a reaction. The ratio of the velocity constants of a reaction at two temperatures differing by 10° C is known as temperature coefficient of the reaction. The temperatures usually selected are 25° C and 35° C and the value of the temperature coefficient for a majority of the reactions varies from 2 to 3.

$$\text{Temperature coefficient} = \frac{\text{Velocity constant at } 35^\circ\text{ C}}{\text{Velocity constant at } 25^\circ\text{ C}}$$

$$= \frac{k_{35}}{k_{25}} = 2 \text{ to } 3 \qquad (2.31)$$

The temperature coefficient usually decreases with rise in temperature. The variation of velocity constant with temperature is best expressed by the following empirical equation proposed by Arrhenius :

$$k = A \exp(-E_a/RT) \qquad (2.32)$$

where A = a constant; E_a = Energy of activation; R = Gas constant and T = Absolute temperature

Taking logarithms

$$\ln k = -\frac{E_a}{RT} + \underset{\text{(constant)}}{\ln A}$$

If k_1 and k_2 be the velocity constants of the reaction at temperatures $T_1°K$ and $T_2°K$, respectively, then assuming E_a to be independent of temperature,

$$\ln k_2 - \ln k_1 = -\frac{E_a}{R}\left[\frac{1}{T_2} - \frac{1}{T_1}\right]$$

or

$$\ln\frac{k_2}{k_1} = \frac{E_a}{R}\left[\frac{1}{T_1} - \frac{1}{T_2}\right]$$

or

$$\log\frac{k_2}{k_1} = \frac{E_a}{2.303\,R}\left[\frac{1}{T_1} - \frac{1}{T_2}\right] \qquad (2.33)$$

Thus, by measuring velocity constants k_1 and k_2 of the hydrolysis of methyl acetate at two different temperatures T_1 and T_2, and substituting the values in the above equation, E_a, the Energy of Activation for the reaction can be calculated.

Procedure

(1) Determine the velocity constant k_{25} of the hydrolysis of methyl acetate catalysed by HCl, at a temperature of 25°C (298°K) maintained by a thermostat, as described in experiment 2.3.2.

(2) Determine the velocity constant k_{35} for the hydrolysis of methyl acetate, by taking exactly the same amounts of HCl and methyl acetate as in (1) above, at a temperature of 35°C (308°K) maintained with the help of a thermostat.

(3) Calculate the ratio $\frac{k_{35}}{k_{25}}$ and report as temperature coefficient of the hydrolysis of methyl acetate catalysed by acid.

(4) Substitute the values of k_{35}, k_{25}, T_1 (298° K), T_2 (308° K) and R = 2 cal/degree/mole or 8.314 joules/K/mole in the expression

$$\log \frac{k_{35}}{k_{25}} = \frac{E_a}{2.303R}\left[\frac{1}{T_1} - \frac{1}{T_2}\right] \quad (2.34)$$

and calculate the value of E_a.

Exercises

24. Why is the reaction mixture dropped into ice-cold water before titration?
25. Give some examples of First Order reactions.
26. What is meant by molecularity of a reaction? How is it different from order of the reaction?

2.3.5 *Kinetics of a Second Order Reaction*

A second order reaction in which there is only one reactant is represented by $2A \longrightarrow$ Products. Let 'a' gram-moles/litre be the initial concentration of A, out of which 'x' gram-moles/litre is converted into products in 't' minutes. Thus after t minutes from the start, the concentration of the reactant will be $(a - x)$ gram-moles/litre, and the rate of reaction at this instant will be given by

$$\frac{dx}{dt} = k(a-x)^2$$

Integrating this equation between the limits $t = 0$ and $t = t$, and using the fact that when $t = 0$, $x = 0$, we get

$$k = \frac{1}{at} \cdot \frac{x}{(a-x)} \quad (2.35)$$

The above equation can be rearranged to give

$$\frac{1}{(a-x)} = kt + \frac{1}{a} \quad (2.36)$$

If the reaction involves two different reactants, i.e., $A + B \longrightarrow$ Products, but both the reactants have the same initial concentration — 'a' gram-moles/litre — the rate constant k is given by equation (2.35). If, however, the initial concentration of A and B is 'a' gram-moles litre and 'b' gram-moles/litre, respectively, the rate equation will be $\frac{dx}{dt} = k(a-x)(b-x)$ and the rate constant will be given by

$$k = \frac{2.303}{t(a-b)} \log \frac{b(a-x)}{a(b-x)} \quad (2.37)$$

In the expressions for k (2.35 or 2.37), there is one additional concentration term in the denominator. Hence, the value of the second order rate constant depends on the units in which the concentration of the reactants is expressed. If the concentration is expressed in moles/litre and the time in minutes, the units of k will be $(\text{moles/litre})^{-1}$ minute^{-1}.

2.3.6 *To determine the rate constant of hydrolysis of ethyl acetate by NaOH (saponification) and to show that it is a Second Order reaction*

Reagents Required

1. Standard ethyl acetate solution (M/25)
2. Standard sodium hydroxide solution (M/25)
3. Standard sodium hydroxide solution (M/50)
4. Standard hydrochloric acid (M/50)
5. Phenolphthalein indicator solution

Theory

The hydrolysis of ethyl acetate by sodium hydroxide, also known as saponification of ethyl acetate, may be represented as

$$CH_3COOC_2H_5 + NaOH \longrightarrow CH_3COONa + C_2H_5OH$$

or

$$CH_3COOC_2H_5 + OH^- \longrightarrow CH_3COO^- + C_2H_5OH \qquad (2.38)$$

The rate of saponification may be represented by

$$\gamma = \frac{dx}{dt} = k\,[CH_3COOC_2H_5]\,[OH^-]$$

$$= k(a - x)^2,$$

when the initial concentration of both ethyl acetate and sodium hydroxide is 'a' gram-moles/litre and 'x' gram-moles/litre of each reacts in time t. The rate constant k is then given by equation (2.35):

$$k = \frac{1}{at} \cdot \frac{x}{(a - x)}$$

As one mole of sodium hydroxide is consumed for every mole of ethyl acetate hydrolysed, the progress of the reaction is followed by withdrawing definite fixed volumes of the reaction mixture at suitable time intervals and determining the amount of residual alkali.

Procedure

(1) Take 50 ml (with a pipet) of M/25 ethyl acetate solution and 70–80 ml of M/25 NaOH in separate clean and dry 250-ml conical flasks. Stopper the flasks and suspend in a thermostat maintained at 25° C.

(2) Take 6 conical flasks each of 100 ml capacity. Pour with a pipet, 20 ml of M/50 HCl into each flask and immerse in ice-water bath.

(3) Wash a burette and rinse and fill it with M/50 NaOH.

(4) When the two liquids (M/25 ethyl acetate and M/25 NaOH) have attained the temperature of the bath, pipet out 50 ml of sodium hydroxide solution. Pour this solution into the flask containing ethyl acetate solution, starting a stop watch when the pipet has been half discharged. This should be taken as zero time.

(5) Immediately mix by shaking and after 5 minutes from the start, withdraw 10 ml of the reaction mixture and pour it into one of the flasks containing 20 ml of ice-cold M/50 HCl. Record the time when the pipet has been half discharged into the flask. Take this as the time t_1 of arresting the reaction.

(6) Immediately add 2 drops of phenolphthalein indicator and titrate quickly against M/50 NaOH solution until a pink colour appears. Record the titre value as V_{t_1}.

(7) Similarly, withdraw 10-ml portions of the reaction mixture at times 10, 20, 30, 50, and 80 minutes and pour into flasks containing 20 ml of ice-cold HC1 and titrate quickly against the standard (M/50) alkali solution. Record the titre values as V_{t_2}, V_{t_3}, V_{t_4}, V_{t_5}, and V_{t_6} respectively.

(8) As the reaction advances, the amount of residual alkali decreases and so the volume of M/50 NaOH required to back titrate the excess of acid increases and will approach 20 ml when the saponification reaction is complete.

Precautions

1. As the rate of reaction is very sensitive to temperature changes, the temperature of the bath should be maintained constant within ± 0.5°C throughout the course of the experiment.
2. For every titration, the time should be recorded when the pipet has been half discharged into the flask.
3. The titration should be made as quickly as possible — for this the bulk of the expected volume of alkali should be run in quickly and then there should be dropwise addition to the first appearance of pink colour that persists for at least 10 seconds.

Observations and Calculations

Temperature of the bath = t °C

Volume of M/25 ethyl acetate taken = 50 ml

Volume of M/25 NaOH taken = 50 ml

Total volume of the reaction mixture = 100 ml

Therefore, 'a' the initial concentration of ethyl acetate

Or NaOH in the reaction mixture $= M/25 \times \frac{50}{100}$

$= M/50$

Volume of reaction mixture withdrawn = 10 ml

Volume of M/50 HCl added = 20 ml

Time in min. t	Titre value V_t (ml)	$k = \frac{1}{at} \cdot \frac{V_t - 10}{20 - V_{t_1}}$ (moles/litre)$^{-1}$ min^{-1}	$(a - x) = \frac{20 - V_t}{500}$	$\frac{1}{(a - x)}$	$k = \tan\theta$
t_1 =	V_{t_1} =				
t_2 =	V_{t_2} =				
t_3 =	V_{t_3} =				
t_4 =	V_{t_4} =				
t_5 =	V_{t_5} =				
t_6 =	V_{t_6} =				

Let the volume of M/50 NaOH needed, after t minutes from the start, to back titrate excess acid = V_t ml

V_t ml M/50 NaOH = V_t ml M/50 HCl

Therefore, excess acid = V_t ml of M/50 NaOH

Acid used for residual NaOH in 10 ml of the reaction mixture = $(20 - V_t)$ ml of M/50 NaOH

Residual concentration of NaOH in the reaction mixture at time t $= \frac{(20 - V_t)}{10} \times M/50$

Thus $(a - x)$ at time t $= \frac{20 - V_t}{500} M$

Therefore, x $= \frac{1}{50} - \frac{20 - V_t}{500} M = \frac{V_t - 10}{500} M$

and $= \frac{x}{a - x} = \frac{V_t - 10}{500} \div \frac{20 - V_t}{500}$

$$= \frac{V_t - 10}{20 - V_t} \tag{2.39}$$

Substituting $\frac{V_t - 10}{20 - V_t}$ for $\frac{x}{a - x}$ in equation (2.35), we get

$$k = \frac{1}{at} \cdot \frac{V_t - 10}{20 - V_t} \tag{2.40}$$

where $a = M/50$

Results and Discussion

1. Since the values of k, as calculated by substituting the various values of V_t for the corresponding t values in equation (2.40) (the kinetic equation for second order) remain almost constant, the saponification of ethyl acetate is a second order reaction.

2. The average of various values of k is the velocity constant or rate constant of the saponification at the temperature of the experiment.

Graphical Method

The rate equation (2.36) for the second order reaction indicates that a plot of $\frac{1}{a-x}$ against t will be a straight line whose slope (tan θ) gives the value k:

$$k = \tan\theta = \frac{BC}{AB} \tag{2.41}$$

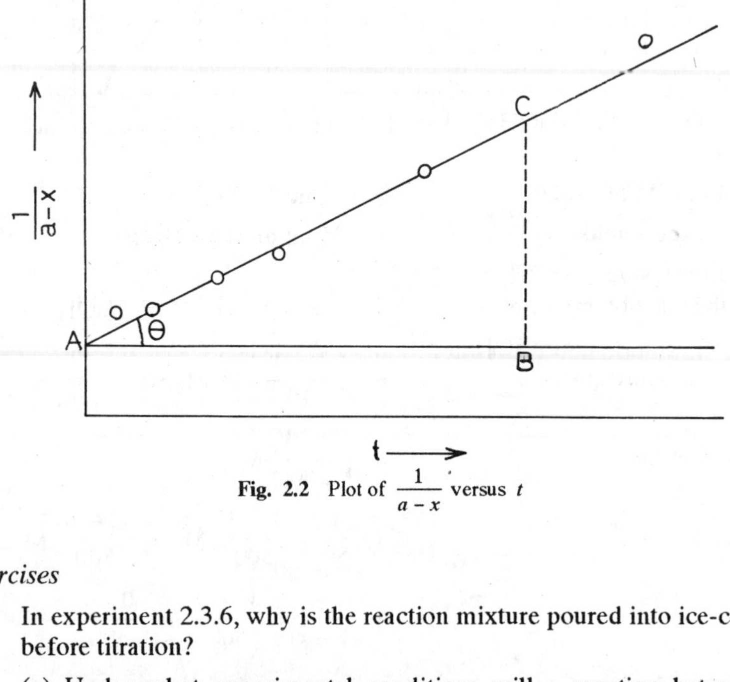

Fig. 2.2 Plot of $\frac{1}{a-x}$ versus t

Excercises

27. In experiment 2.3.6, why is the reaction mixture poured into ice-cold HCl before titration?

28. (a) Under what experimental conditions will a reaction between two reactants be of first order?

    (b) Can you arrive at the same result mathematically?

29. Give some examples of second order reactions.

30. Write the expression for the rate constant of a third order reaction starting with equal concentration of all the reactants.

31. Why are Third or Higher Order Reactions very rare?

32. Give some examples of Third Order Reactions and one example of a Fourth Order Reaction.

33. Can the order of a reaction be zero? If so, give some examples.
34. What is meant by half-life period of a reaction?
35. What is the relationship between the half-life period of a reaction and the initial concentration of the reactant(s)?
36. Outline the usefulness of the study of chemical kinetics.
37. Giving one example, explain how measuring the order of a reaction helps in understanding its mechanism.

3

WATER

Water is one of the most essential substances needed to sustain human life, animals and plants. It is needed for drinking, cooking, bathing, washing (cleaning), for sanitary disposal of domestic and human wastes, and for agriculture. It is also one of the most important engineering materials and is used for steam generation, as a coolant in power plants, for air-conditioning and fire-fighting, and in buildings and other concrete constructions. Water has a unique position in industry. It is needed for the production of such a wide variety of materials as steel and other metals, paper, textiles, beverages, dairy products, petroleum and coal, rubber and plastics, automobile industry, and as a solvent in chemical processing. In fact, production units not using water for some purpose or the other may be hard to find. It is for this reason that before setting up a production unit at a particular site, the quantity of water needed, the character and quality of water available and the effect of impurities in water on the process must be considered.

Although available in abundance, water from almost all the sources is associated with some kind of impurities; their nature and amount varying with the source of water. *Rain water*, the purest form of natural water, usually contains dust particles and gases dissolved from atmosphere, such as O_2, N_2, CO_2 (ammonium salts, H_2S, or H_2SO_4 may be present in rain waters in industrial areas). *Surface waters* (from rivers, ponds, etc.) are usually rich in turbidity, suspended impurities of decaying organic matter (vegetable and/or animal), sand and finely divided clay, micro-organisms and bacteria, and small amounts of mineral salts (mainly Ca^{2+}, Mg^{2+}, Na^+, SO_4^{2-}, Cl^-, etc.) dissolved from top soil. *Underground waters* (from springs, wells or bore-holes) usually contain negligible amounts of suspended and organic impurities (removed while passing through sand layers in earth) but may contain appreciable amounts of mineral impurities (Ca^{2+}, Mg^{2+},K^+, Na^+, Fe^{2+}, Al^{3+}, Mn^{2+}, HCO_3^-, CO_3^{2-}, SO_4^{2-},Cl^-, NO_3^-, finely divided clay, etc.) brought into solution due to disintegration of mineral deposits and insoluble carbonate or aluminosilicate rocks by the combined action of high underground temperature, hydration, dissolved O_2 and CO_2, and organic acids produced by aerobic and/or anaerobic decay of organic matter with which water has been in contact. *Sea water* is highly impure, containing around 3.5% dissolved minerals; NaCl alone is present to the extent of 2.6%.

Use of impure waters often leads to many problems like health complications, decrease in the efficiency of plants, increased cost, etc. However, of the wide range of impurities present, only a few may affect a particular process. Therefore, removal or reduction in the concentration of only these impurities is necessary; the rest of the impurities which do not adversely affect the process may be allowed to remain as such. For example, water for laundries should be soft and free from colour; water for steam generation should be free from all dissolved solids and dissolved corrosive gases but removal of bacteria is not required. Specifications for drinking water are much more stringent with respect to the presence of pathogenic bacteria and certain toxic substances (Phenols < 0.005 ppm; As, Cr, Pb, Ag, Mn < 0.05 ppm; Cd, Se, cyanide < 0.01 ppm; etc.) but may contain large amounts of substances like Ca^{2+}, Mg^{2+}, Na^+, K^+, Cl^{1-}, SO_4^{2-}, etc.

The treatment methods depend on the nature of impurities present, which can be determined by analysis. A complete analysis, however, is usually not done. The extent of analysis is governed by the purpose for which water is to be used and the specifications laid down for the purpose. Analysis of the treated water is also carried out to compare the efficacy of different treatment processes and to choose the most efficient and economical process.

With the rapid growth in population, improvement in living standards and proliferation of industry, the demand for adequate supplies of suitable waters is increasing. So the water used in most of the industries and in sanitary disposal of human and domestic wastes, the water which is not actually consumed but is polluted, has to be reused. The treatment methods for recycling and re-using water have to depend on the nature and extent of contamination and pollution which can be determined by analysing the used water. A few of the routine tests that are carried out on water and waste water are discussed in the following pages.

3.1 Acidity and Alkalinity of Water

The acidity of water sample is its capacity to neutralise a base whereas its alkalinity is a measure of its capacity to neutralise acids. Natural waters may be acidic or alkaline, depending upon the source of water and the extent and nature of pollutants from industry and municipal sanitary disposal.

Most natural waters contain dissolved CO_2. It may enter water by absorption from atmosphere or may be produced by biological oxidation of organic matter with which water has been in contact, e.g.,

$$\underset{\text{Hexose}}{C_6H_{12}O_6} + 6O_2 \xrightarrow[\text{bacteria}]{\text{aerobic}} 6CO_2 + 6H_2O \tag{3.1}$$

$$\underset{\text{Hexose}}{C_6H_{12}O_6} \xrightarrow[\text{bacteria}]{\text{anaerobic}} 2C_2H_5OH + 2CO_2 \tag{3.2}$$

CO_2 content of surface waters is normally much less than 10 mg/l but some ground waters may contain between 30 to 50 mg/l of CO_2. When waters containing high concentration of CO_2 are exposed to atmosphere, most of CO_2 escapes from the solutions as the partial pressure of CO_2 in atmosphere is normally much less.

Waters may be acidic due to the presence of CO_2 and/or organic acids formed by the decay of organic matter, e.g.,

$$\underset{\text{Hexose}}{C_6H_{12}O_6} \xrightarrow[\text{bacteria}]{\text{anaerobic}} \underset{\text{propanoic acid}}{CH_3CH_2COOH} + \underset{\text{acetic acid}}{CH_3COOH} + \underset{\text{formic acid}}{HCOOH} \quad (3.3)$$

Acidity may also arise due to the presence of mineral acids produced by the hydrolysis of salts of certain heavy metals such as $FeCl_3$ or $Al_2(SO_4)_3$:

$$FeCl_3 + 3H_2O \rightleftharpoons Fe(OH)_3 + 3H^+ + 3Cl^{1-} \quad (3.4)$$

$$Al_2(SO_4)_3 + 6H_2O \rightleftharpoons 2Al(OH)_3 + 6H^+ + 3SO_4^{2-} \quad (3.5)$$

Waters may be alkaline due to the presence of a wide variety of salts of weak acids such as carbonates, bicarbonates, borates, silicates, phosphates, etc., and also due to the presence of weak and strong bases (due to contamination with industrial wastes). The major portion of alkalinity in natural water is, however, caused by presence of bicarbonates that are formed in appreciable amounts when water containing free CO_2 percolates through soils containing $CaCO_3$ and/or $MgCO_3$:

$$CaCO_3(s) + CO_2 + H_2O \longrightarrow Ca^{2+} + 2HCO_3^- \quad (3.6)$$

$$MgCO_3(s) + CO_2 + H_2O \longrightarrow Mg^{2+} + 2HCO_3^- \quad (3.7)$$

Waters softened by lime-soda process always contain excess amounts of carbonate and hydroxide, added to ensure complete precipitation of calcium and magnesium. Waters in which algae are present contain appreciable amounts of carbonate and hydroxide as algae remove both free and combined CO_2 :

$$2HCO_3^- \xrightarrow[\text{(removed by algae)}]{-CO_2} CO_3^{2-} + H_2O \quad (3.8)$$

$$CO_3^{2-} + H_2O \xrightarrow[\text{(removed by algae)}]{-CO_2} 2OH^- \quad (3.9)$$

With borates, silicates and phosphates making only an insignificant contribution, the alkalinity of natural waters may be taken as an indication of the concentration of (i) Hydroxides, (ii) Carbonates, and (iii) Bicarbonates.

3.1.1 *Determination of free CO_2 in a sample of water*

Significance

The CO_2 content of a water sample is important as it contributes to corrosion. When present in boiler-feed water, it may cause corrosion of even those parts that lie at a distance from the boiler, such as superheater or the blades of steam turbines. In lime-soda process of water-softening, measurement of free CO_2 is necessary for calculating additional amounts of lime to be added. Because of its corrosive nature and the unacceptable taste it imparts to water, CO_2 is to be removed from municipal water supplies by aeration or by neutralisation with lime; the choice of the method depending upon the amounts of free CO_2 present in the sample.

Reagents Required

1. Standard Na_2CO_3 solution (N/50)
2. Phenolphthalein indicator solution.

Theory

Free CO_2 can be determined by titrating with a standard solution of NaOH or Na_2CO_3 with which it reacts to give $NaHCO_3$:

$$CO_2 + OH^- \longrightarrow HCO_3^- \quad (3.10)$$

$$CO_2 + CO_3^{2-} + H_2O \longrightarrow 2HCO_3^- \quad (3.11)$$

Phenolphthalein is used as indicator and titration is carried to pH 8.3 indicated by the appearance of pink colour.

Procedure

Collect a water sample in a 100-ml graduated cylinder by means of a submerged rubber tubing. To remove air bubbles, allow the sample to overflow for sometime. Remove the excess of the sample with the help of pipet and adjust the level to 100-ml mark. Add 5 drops of the phenolphthalein indicator (appearance of pink colour indicates absence of CO_2) and titrate against N/50 Na_2CO_3 solution until the pink colour persists for at least 30 seconds. Take concordant readings and record the volume used as A ml.

Precautions

1. To minimise the loss of CO_2

   (a) The titration should be carried out rapidly and the reaction mixture should be stirred only gently during the titration.

   (b) In subsequent readings, the full amount of alkali used in the first titration should be first added without stirring. The indicator should then be added and the titration completed to the desired endpoint by adding extra alkali.

2. The same amount of the indicator solution should be added while taking different readings.

Observations and Calculations

Volume of water sample taken for each titration = 100 ml

Concordant volume of N/50 Na_2CO_3 used = A ml

$$\underset{(CO_2)}{N_1V_1} = \underset{(Na_2CO_3)}{N_2V_2}$$

$$N_1 \times 100 = \frac{1}{50} \times A$$

Therefore, normality of the sample w.r.t. $CO_2 = \dfrac{A}{50 \times 100}$

$$\text{Strength} = \frac{A \times 22}{50 \times 100} \text{ g/l}$$

or

$$\text{Free } CO_2 = \frac{A \times 22 \times 1000}{50 \times 100} \text{ g/l}$$

$$= 4.4 \times A \text{ mg/l}$$

3.1.2 *Determination of acidity of a water sample*

Significance

Measurement of acidity is important as acidic waters are corrosive, and corrosion-producing substances have to be controlled or removed. In the lime-soda process of water softening, amounts of reagents to be added have to be adjusted to account for the acidity of the sample. In the phosphate conditioning of boiler-feed water, the pH has to be adjusted to about 9.5; the choice of a suitable phosphate therefore depends on the measurement of acidity of the sample. Measurement of methyl orange acidity is important in calculating the quantities of chemicals required to neutralise the mineral acidity of the industrial wastes before they can be discharged into receiving waters (rivers, streams or lakes).

Reagents Required

1. Standard sodium hydroxide solution (N/50)
2. Methyl orange indicator solution
3. Phenolphthalein indicator solution
4. Sodium thiosulphate solution (N/10)

Theory

The acidity of a sample may be determined by titrating it against a standard alkali solution upto the stoichiometric equivalence point. The accurate identification of the stoichiometric equivalence point being very difficult, the titration is carried to an arbitrary end-point pH. The acidity, therefore, becomes a measure of the amount of a base required to neutralise a given sample to a specific pH :

$$OH^- + H^- \longrightarrow H_2O \tag{3.12}$$

For mineral acids, the titration is carried to a pH of about 4.5 by using methyl orange indicator, giving a colour change from red to yellow. The acidity thus determined is called Methyl Orange Acidity. Total Acidity or Phenolphthalein Acidity is determined by carrying the titration to phenolphthalein end-point of pH 8.3 (thus measuring mineral acids, organic acids and free CO_2) indicated by appearance of pink colour. The results are expressed as parts of equivalent $CaCO_3$ per million parts of water.

Procedure

(A) Methyl Orange Acidity: Pipet out 100 ml of the sample and discharge it near the bottom of a titration flask. Add 1 drop of N/10 $Na_2S_2O_3$ solution to destroy any residual chlorine. Add 2 drops of methyl orange indicator and titrate with N/50

NaOH solution until the red colour changes to yellow. Take a number of readings and record the concordant volume as A ml.

(B) Phenolphthalein Acidity: Proceed as above using 2 drops of phenolphthalein (in place of methyl orange) indicator and titrate the sample until the pink colour persists for at least 30 seconds. Take a number of readings and record the concordant volume as B ml.

Precaution

(1) To avoid loss of CO_2, the titration should be carried out quickly and vigorous shaking should be avoided.

Observations and Calculations

Volume of water sample taken for each titration = 100 ml

Volume of N/50 NaOH used in presence of methyl orange indicator = A ml

Volume of N/50 NaOH used in presence of phenolphthalein indicator = B ml

(A) Methyl Orange Acidity

$$\underset{\text{(Sample)}}{N_1V_1} = \underset{\text{(N/50NaOH)}}{N_2V_2}$$

$$N_1 \times 100 = \frac{1}{50} \times A$$

or

$$N_1 = \frac{A}{50 \times 100}$$

$$\text{Methyl Orange Acidity (as } CaCO_3) = \frac{A \times 50}{50 \times 100}\ \text{g/l}$$

$$= \frac{A \times 50 \times 1000}{50 \times 100}\ \text{ppm}$$

$$= A \times 10\ \text{ppm}$$

$$\text{(B) Phenolphthalein Acidity} = \frac{B \times 50 \times 1000}{50 \times 100}\ \text{ppm}$$

$$= B \times 10\ \text{ppm}$$

3.1.3 *Determination of alkalinity of a water sample*

Significance

Highly alkaline waters being usually unpalatable, upper limits with respect to phenolphthalein alkalinity and total alkalinity have been specified for municipal water supplies. Alkaline waters when used in boilers for steam generation may lead to precipitation of sludges, deposition of scales and cause caustic embrittlement. A knowledge of the kinds of alkalinity present in water and their magnitudes is important

1. In calculating the amounts of lime and soda needed for water-softening.
2. In maintaining a pH range where acids produced by hydrolysis of salts (like $MgCl_2$, $FeCl_2$) and coagulants $[Al_2(SO_4)_3$ and $Fe_2(SO_4)_3]$ may be effectively neutralised.

3. In corrosion control, and
4. In internal conditioning of boiler-feed water.

Caustic alkalinity of industrial wastes has to be neutralised before discharging them into rivers or other receiving waters. Waters containing alkalinity in excess of alkaline earth concentrations are not suitable for irrigation purposes.

Reagents Required

1. Standard sulphuric acid solution (N/50).
2. Phenolphthalein indicator solution.
3. Methyl orange indicator solution.

Theory

The alkalinity of a water sample may be determined volumetrically by titrating it with a standard acid to an arbitrary pH using an indicator. When the pH of the sample is above 8.3, titration is first carried out using phenolphthalein indicator. At the end point, when the indicator changes from pink to colourless, the pH is lowered to about 8.3. The volume of the acid (A ml) used upto this point corresponds to the complete neutralisation of hydroxide and conversion of all the corbonate into bicarbonate:

$$\left.\begin{array}{ll} OH^- + H^+ \longrightarrow H_2O & (3.13) \\ CO_3^{2-} + H^+ \longrightarrow HCO_3^- & (3.14) \end{array}\right] \text{A ml}$$

The alkalinity measured upto this point is called Phenolphthalein Alkalinity. Beyond the phenolphthalein end-point, titration is continued using methyl orange indicator. The colour change from yellow to red occurs at a pH of about 4.5 and the additional volume of acid (B ml) used corresponds to the complete neutralisation of all the bicarbonate whether present originally or obtained from CO_3^{2-} (Reaction 3.14):

$$HCO_3^- + H^+ \longrightarrow H_2O + CO_2 \qquad (3.15)\ ]\ \text{B ml}$$

The total volume of acid [(A + B) ml] used in the two titrations, therefore, corresponds to the neutralisation of hydroxide, carbonate and bicarbonate and is, thus, a measure of Total Alkalinity, also called Methyl Orange Alkalinity. Alkalinity is expressed as parts of equivalent $CaCO_3$ per million parts of water.

From the measurements for Phenolphthalein Alkalinity and Methyl Orange Alkalinity, it is possible to calculate the magnitudes of various forms of alkalinity, namely, Hydroxide Alkalinity, Carbonate Alkalinity and Bicarbonate Alkalinity. If it is assumed (with only negligible error) that hydroxide and bicarbonate cannot co-exist, as they react to give carbonate:

$$OH^- + HCO_3^- \longrightarrow CO_3^{2-} + H_2O \qquad (3.16)$$

We are left with the following five possible situations:

(1) Hydroxide only
(2) Carbonate only
(3) Hydroxide and carbonate

(4) Carbonate and bicarbonate
(5) Bicarbonate only.

(1) When the sample contains *only Hydroxide*, the neutralisation is complete at the phenolphthalein end-point (Reaction 3.13) and so B, the additional volume of acid needed for neutralisation to methyl orange end-point, will be zero. In this case, Hydroxide Alkalinity is equal to Phenolphthalein Alkalinity, which is also the Total Alkalinity.

(2) For samples containing *only Carbonate*, titration to phenolphthalein end-point corresponds to the complete conversion of carbonate to bicarbonate or what is also known as half-neutralisation of carbonate (Reaction 3.14). This means that exactly the same amount of additional acid will be required for further titration to methyl orange end-point (Reaction 3.15).

(3) For samples containing *Hydroxide and Carbonate*, titration to phenolphthalein end-point corresponds to complete neutralisation of hydroxide and half nuetralisation of carbonate (Reaction 3.13 and 3.14), whereas titration from phenolphthalein end-point to methyl orange end-point corresponds to neutralisation of bicarbonate (Reaction 3.15) obtained from carbonate in reaction 3.14.

Hence in this case, A > B. Also, the acid required to neutralise the hydroxide (Reaction 3.13) is (A – B) ml and that required for the complete neutralisation of carbonate (Reactions 3.14 and 3.15) is 2B ml.

(4) For samples containing *Carbonate and Bicarbonate*, titration to phenolphthalein end-point (A ml) corresponds to half-neutralisation of carbonate, i.e., conversion of carbonate to bicarbonate (Reaction 3.14) and the titration from phenolphthalein end-point to methyl orange end-point (B ml) represents half neutralisation of carbonate [i.e. the neutralisation of bicarbonate obtained from carbonate, which requires another A ml of the acid] and neutralisation of bicarbonate originally present in the sample.

$$\underset{\text{(obtained from } CO_3^{2-})}{HCO_3^-} + H^+ \longrightarrow CO_2 + H_2O \qquad \text{(3.15a) A ml]}$$

B ml

$$\underset{\text{(originally present)}}{HCO_3^-} + H^+ \longrightarrow CO_2 + H_2O \qquad \text{(3.15b)]}$$

For such a mixture, evidently, B > A. Carbonate Alkalinity is equivalent to 2A ml of the acid and Bicarbonate Alkalinity is equivalent to (B – A) ml of the acid.

(5) When the sample contains *only Bicarbonate*, it does not give a pink colour with phenolphthalein indicator. Therefore A, the volume of acid consumed with phenolphthalein indicator is zero. Titration to methyl orange end-point represents the complete neutralisation of bicarbonate (Reaction 3.15). Bicarbonate Alkalinity is therefore equal to Total or Methyl Orange Alkalinity.

Procedure

With a pipet, transfer 100 ml of the water sample into a titration flask. Add 2 drops of phenolphthalein indicator and titrate against N/50 H_2SO_4 until the pink colour just disappears. Record the volume of acid consumed as A ml. To the same solution add 2–3 drops of methyl orange indicator and titrate further until the colour changes from yellow to red. Record the additional volume of acid consumed as B ml. Repeat the whole process a number of times to get concordant readings.

Observations and Calculations

Volume of sample taken for each titration = 100 ml

Volume of N/50 H_2SO_4 used to phenolphthalein end-point = A ml

Additional volume of N/50 acid used to methyl orange end-point = B ml

$$\underset{\text{(sample)}}{N_1V_1} = \underset{\text{(N/50acid)}}{N_2V_2}$$

$$N_1 \times 100 = \frac{1}{50} \times A$$

or

$$N_1 = \frac{1}{50} \times A \times \frac{1}{100}$$

$$\text{Strength (in terms of } CaCO_3) = \frac{1}{50} \times A \times \frac{1}{100} \times 50 \text{ g/l}$$

or

$$\text{Phenolphthalein Alkalinity} = \frac{1}{50} \times A \times \frac{1}{100} \times 50 \times 1000 \text{ mg/l}$$

$$= 10 \times A \text{ ppm}$$

Similarly, Total Alkalinity

or

$$\text{Methyl Orange Alkalinity} = \frac{1}{50} \times (A + B) \times \frac{1}{100} \times 50 \times 1000 \text{ mg/l}$$

$$= 10 \times (A + B) \text{ ppm}$$

$$\text{Hydroxide Alkalinity} = \frac{1}{50} \times (A - B) \times \frac{1}{100} \times 50 \times 1000 \text{ mg/l}$$

$$= 10 \times (A - B) \text{ ppm}$$

$$\text{Hydroxide ion concentration} = \frac{1}{50} \times (A - B) \times \frac{1}{100} \times 17 \times 1000 \text{ mg/l}$$

$$= 10 \times (A - B) \times 0.34 \text{ ppm}$$

$$= \text{Hydroxide Alkalinity} \times 0.34 \text{ ppm}$$

$$\text{Carbonate Alkalinity} = \frac{1}{50} \times 2B \times \frac{1}{100} \times 50 \times 1000$$

$$= 10 \times 2B \text{ ppm when } A > B$$

and

$$= \frac{1}{50} \times 2A \times \frac{1}{100} \times 50 \times 1000$$

$$= 10 \times 2A \text{ ppm when } A \leq B$$

$$\text{Carbonate ion concentration} = 10 \times 2B \times \frac{30}{50}$$

$$= 6 \times 2B \text{ ppm when } A > B$$

and

$$= 6 \times 2A \text{ ppm when } A \leq B$$

Bicarbonate Alkalinity $= \frac{1}{50} \times (B - A) \times \frac{1}{100} \times 50 \times 1000$

$= 10 \times (B - A)$ ppm

Bicarbonate ion concentration $= 10 \times (B - A) \times \frac{61}{50}$ ppm

= Bicarbonate Alkalinity × 1.22 ppm

Exercises

38. Write the structural formula of methyl orange. In what forms does it exist in acidic and alkaline medium ?
39. Write the structural forms of methyl red indicator in alkaline and acidic medium.
40. Explain the action of phenolphthalein as an acid-base indicator.
41. In the determination of free CO_2, why is it preferable to use a standard solution of Na_2CO_3 instead of NaOH ?
42. In the presence of heavy metal salts, what modification would you suggest to complete the titration rapidly ?
43. What is the source of mineral acidity in water ?
44. In acidity and alkalinity measurements, the titrant used (Na_2CO_3 or H_2SO_4) is usually N/50. Why ?

3.2 Chloride Content

Chlorides usually occur as NaCl, $CaCl_2$ and $MgCl_2$, and in widely varying concentrations, in all natural waters. They enter waters

(a) by solvent action of water on salts present in soil,

(b). from polluting materials like sewage (containing the salt used in household) and trade wastes (containing chloride used in manufacturing), and

(c) in areas around sea, from (i) salt-water droplets carried inland, by wind, as spray from the sea, (ii) invasion (upstream flow) by sea waters into rivers that drain into sea.

Significance

When present at concentrations above 250 ppm, chlorides impart an unacceptable taste to waters although no adverse effects have been observed on human beings regularly consuming waters with much higher concentrations of chloride. Exceptionally high concentration of chloride in a water (as compared to that in other waters in the general vicinity and known to be unpolluted) may be considered as an indication of contamination by domestic waste water. Chloride ion concentration should be known

(i) for deciding the type of desalting apparatus when brackish waters have to be used.

(ii) for treating industrial wastes before discharging them into natural bodies of water.

(iii) for controlling pumping of ground water from sea coasts where invasion by sea water can take place, and

(iv) for applying correction factor in COD determination (where chloride interferes).

3.2.1 *Determination of chloride content of a water sample by Mohr's Method (Argentometric)*

Reagents Required

1. Standard silver nitrate solution (N/50)
2. Potassium chromate indicator solution (K_2CrO_4)
3. Solid $CaCO_3$.

Theory

Chloride ions in a neutral or faintly alkaline solution can be estimated by titration with a standard solution of $AgNO_3$ using K_2CrO_4 as indicator. The pH must be in the range of 7 to 8 because Ag^+ ion is precipitated as AgOH at higher pH

$$Ag^+ + OH^- \longrightarrow AgOH\ (K_{sp} = 2.3 \times 10^{-8}) \qquad (3.17)$$

and CrO_4^{2-} is converted to $Cr_2O_7^{2-}$ at lower pH

$$2\,CrO_4^{2-} + 2H^+ \rightleftharpoons 2HCrO_4^- \rightleftharpoons Cr_2O_7^{2-} + H_2O \qquad (3.18)$$

$HCrO_4^-$ being a weak acid, CrO_4^{2-} ion concentration is decreased necessitating higher concentration of Ag^+ for the solubility product of Ag_2CrO_4 to be exceeded, thus leading to higher results. The required pH range can be achieved easily by adding a pinch of pure $CaCO_3$ to the pink or red-coloured solution obtained at the end-point of the Methyl Orange Alkalinity determination (experiment 3.1.3):

$$2H^+ + CaCO_3 \longrightarrow Ca^{2+} + CO_2 + H_2O \qquad (3.19)$$

Excess $CaCO_3$,being insoluble, does not interfere.

As $AgNO_3$ solution is added from the burette to the chloride ion sample containing CrO_4^{2-}, Ag^+ reacts with both Cl^- and CrO_4^{2-} forming the respective precipitates:

$$Ag^+ + Cl^- \rightleftharpoons \underset{\text{white ppt.}}{AgCl}\ (K_{sp} = 3 \times 10^{-10}) \qquad (3.20)$$

$$2Ag^+ + CrO_4^{2-} \rightleftharpoons \underset{\text{reddish-brown ppt.}}{Ag_2CrO_4}\ (K_{sp} = 5 \times 10^{-12}) \qquad (3.21)$$

But the red colour formed by the addition of each drop of $AgNO_3$ disappears because of the large concentration of Cl^- ions in the solution:

$$Ag_2CrO_4 + 2Cl^- \longrightarrow 2AgCl + CrO_4^{2-} \qquad (3.22)$$

As the concentration of Cl^- ions decreases, the red colour disappears more slowly and when all the chloride has been precipitated, a faint reddish or pinkish tinge persists in the white precipitate even after brisk shaking.

Procedure

Add a pinch of pure $CaCO_3$ to the solution obtained at the end of Methyl Orange Alkalinity determination, so that the red or pink colour vanishes. Add 1–1.5 ml of K_2CrO_4 indicator solution and titrate against N/50 $AgNO_3$ taken in the burette until a permanent pink tinge persists in the white precipitate. Repeat the whole process a number of times to get concordant readings. Record the concordant volume as A ml.

Precautions

(1) The whole apparatus must be washed with distilled water.

(2) The same amount of the indicator must be added each time.

(3) The reaction mixture should be briskly shaken during the titration.

Observations and Calculations

Volume of the sample taken each time = 100 ml

Concordant volume of N/50 $AgNO_3$ used = A ml

$$\underset{(Cl^- \text{ ion sample})}{N_1V_1} = \underset{(AgNO_3)}{N_2V_2}$$

$$N_1 \times 100 = \frac{1}{50} \times A$$

Therefore,

$$N_1 = \frac{1}{50} \times A \times \frac{1}{100}$$

Strength of Cl^- ions

$$= \frac{1}{50} \times A \times \frac{1}{100} \times 35.5\text{g/l}$$

$$= \frac{1}{50} \times A \times \frac{1}{100} \times 35.5 \times 1000 \text{ mg/l}$$

$$= 7.1 \times A \text{ ppm}$$

Exercises

45. How is the pH adjusted when the alkalinity is not being measured?

46. How does the presence of NH_4^+ ions in the sample affect the determination?

47. What is meant by indicator-blank correction? How can it be avoided in the chloride ion estimation?

48. What is the effect of temperature on the determination?

49. How is the test performed on waters having (i) colour and turbidity, and (ii) H_2S odour?

3.3 Hardness

Water samples that do not readily produce lather with soap, or deposit scale on the walls of the container when there is appreciable change in temperature, are called Hard Waters. Principal hardness-causing cations are Ca^{2+} and Mg^{2+}. Other divalent metallic cations that cause hardness are Fe^{2+}, Mn^{2+} and Sr^{2+} but these are usually present in small amounts. The cations react with soaps (sodium or potassium salts of fatty acids) to form precipitates:

$$\underset{\substack{\text{Sodium stearate}\\\text{(soap)}}}{2C_{17}H_{15}COONa} + \underset{\substack{\text{Hardness causing}\\\text{cation}}}{M^{2+}} \longrightarrow \underset{\text{ppt.}}{(C_{17}H_{35}COO)_2M} + 2Na^+ \quad (3.23)$$

Lather is produced only after sufficient soap has been added to precipitate all the hardness causing ions.

At higher concentrations, these ions associate themselves with certain anions (CO_3^{2-}, SO_4^{2-}, Cl^-, NO_3^-, SiO_3^{2-}, etc.) and produce scale.

Al^{3+} and Fe^{3+} may also contribute to the hardness of water but their ionic concentrations in natural waters are generally negligible. Surface waters contain very small amounts of dissolved impurity. However, rain water percolates into soil, it dissolves CO_2 (released by bacterial action on organic matter) and attacks basic materials like limestone formations, dolomite, etc., and dissolves the insoluble carbonates present:

$$MCO_3 + CO_2 + H_2O \longrightarrow M(HCO_3)_2 [M = Ca, Mg, Sr] \quad (3.24)$$

Sulphates, chlorides, silicates, etc., of Ca, Mg, Na, K, Al, etc., present as impurities in the limestone become exposed to the solvent action of water and pass into solution, thereby adding to the hardness of water.

Hardness is generally reported in terms of calcium carbonate equivalent and is calculated by the general formula:

$$\text{Hardness (in mg/l)} = M^{2+} \text{ (in mg/l)} \times \frac{50}{\text{eq. wt. of } M^{2+}} \quad (3.25)$$

as $CaCO_3$

[50 being the eq. wt of $CaCO_3$]

where M^{2+} is any divalent metallic ion.

Units of Hardness

1. Parts per million (ppm) is the most commonly used unit of hardness and represents the number of parts of calcium carbonate, equivalent to hardness causing ions present, per million parts of water.

1 ppm = 1 part $CaCO_3$ equivalent hardness in 10^6 parts of water

= 1 g $CaCO_3$ equivalent hardness in 10^6g water

= 1 mg $CaCO_3$ equivalent hardness in 10^6 mg or 10^3g water

= 1 mg $CaCO_3$ equivalent hardness in 1 litre of water.

2. *Degree Clark* (°Cl) is the number of grains $\left(1 \text{ grain} = \frac{1}{7000} \text{ lb}\right)$ of $CaCO_3$ equivalent to hardness causing ions in one gallon (10 lbs) of water.

Thus $1°Cl = \frac{1}{7000}$ parts of $CaCO_3$ equivalent hardness in 10 part of water

$= \frac{1}{70000}$ parts of $CaCO_3$ equivalent hardness in 1 part of water

$$= \frac{10^6}{70000} \text{ parts of } CaCO_3 \text{ equivalent hardness in } 10^6 \text{ parts of water}$$

$$= \frac{100}{7} = 14.3 \text{ ppm}$$

3. *Degree French* (°F) is the number of parts of $CaCO_3$, equivalent to hardness causing ions in 10^5 parts of water. Thus 1°F = 1 part $CaCO_3$ equivalent hardness in 10^5 parts of water

$$= \frac{10^6}{10^5} \text{ parts } CaCO_3 \text{ equivalent hardness in } 10^6 \text{ parts of water}$$

$$= 10 \text{ ppm}$$

Relationship between various units: 1°F = 10 ppm = 0.7°Cl

Water is generally called:

Soft, when the degree of hardness is < 75 ppm

Moderately hard, when the degree of hardness is between 75 – 150 ppm

Hard, when the degree of hardness is between 150 – 300 ppm

Very hard, when the degree of hardness is > 300 ppm.

Types of Hardness

The portion of the total hardness that can be largely removed by boiling is known as Temporary Hardness. Boiling converts the bicarbonates present into insoluble carbnonates and hydroxides, which can be removed by filtration.

$$Ca(HCO_3)_2 \longrightarrow CaCO_3 + H_2O + CO_2 \quad (3.26)$$

$$Mg(HCO_3)_2 \longrightarrow Mg(OH)_2 + 2CO_2 \quad (3.27)$$

The portion of the hardness that cannot be removed by boiling is termed as Permanent Hardness.

In order to relate the hardness to the chemical species, it is customary to refer the part of the total hardness that is chemically equivalent to the bicarbonate plus carbonate alkalinities as Carbonate Hardness (CH) and the amount of hardness which is in excess of CH as Non-Carbonate Hardness (NCH).

With respect to the metallic ion present, hardness is classified into Calcium Hardness and Magnesium Hardness (other cations being neglected as their contribution to hardness is not appreciable).

3.3.1 *Outline of various methods*

(1) Hardness is best calculated from a complete mineral analysis of water.

Total hardness in ppm

$$= 2.5 \times \text{mg/l of } Ca^{2+} + 4.12 \times \text{mg/l of } Mg^{2+} + (1.79 \times \text{mg/l of } Fe^{2+} + 1.82 \times \text{mg/l of } Mn^{2+} + 1.14 \times \text{mg/l of } Sr^{2+}) \quad (3.28)$$

[Fe = 55.85, Ca = 40, Mg = 24.3, Mn = 54.9, Sr = 87.6]

The last three make only an insignificant contribution to total hardness.

The method is applicable to all waters and gives results of high accuracy. However, such a detailed analysis is not done in routine work and other methods have to be used.

As both alkalinity and hardness are expressed in terms of $CaCO_3$, CH and NCH may be calculated as follows:

When alkalinity < total hardness,

$$\text{CH (in ppm)} = \text{Alkalinity (in ppm).} \tag{3.29}$$

When alkalinity ≥ total hardness (3.30)

$$\text{CH (in ppm)} = \text{Total hardness (in ppm)}$$

and

$$\text{NCH} = \text{Total hardness} - \text{CH.}$$

(2) *Hehner's Method:* Boiling removes temporary hardness by converting bicarbonates into insoluble carbonates and hydroxides [reactions 3.26 and 3.27] which are removed by filtration, thereby causing an equivalent decrease in alkalinity. Measurement of alkalinity before and after boiling thus affords a method of determination of temporary hardness.

(3) *Determination of Permanent Hardness using a Mixture of Na_2CO_3 and NaOH:* An aliquot of the hard water sample is boiled with a known excess of alkali mixture (Na_2CO_3 + NaOH). This precipitates permanent hardness due to Ca^{2+} and Mg^{2+}:

$$Ca^{2+} + CO_3^{2-} \longrightarrow CaCO_3 \tag{3.31}$$

$$Mg^{2+} + 2OH^- \longrightarrow Mg(OH)_2 \tag{3.32}$$

The excess of the unused alkali is back-titrated with a standard acid using methyl orange indicator. The permanent hardness of the sample can be calculated from the amount of the alkali mixture consumed. $Ca(HCO_3)$ and $Mg(HCO_3)_2$ present in the sample do not consume any alkali as an equivalent amount of Na_2CO_3 is produced.

$$Ca(HCO_3)_{2+} + 2OH^- \longrightarrow CaCO_3 + CO_3^{2-} + 2H_2O \tag{3.33}$$

$$Mg(HCO_3)_2 + 4OH^- \longrightarrow Mg(OH)_2 + 2CO_3^{2-} + 2H_2O \tag{3.34}$$

It is customary, however, to remove temporary hardness by boiling before the addition of alkali mixture.

(4) *Determination of Magnesium Hardness using NaOH/lime water:* The carbonate hardness (CH) of the hard water sample is converted into non-carbonate hardness (NCH) by neutralising a definite volume of the sample with HCl to methyl orange end-point, or best the solution at the end-point of methyl orange alkalinity determination experiment is taken and heated to boil off CO_2. A known excess of NaOH (Lime water is preferred as NaOH may contain Na_2CO_3 as impurity is added and the reaction mixture is boiled gently for 10–15 minutes. Magnesium hardness is precipitated as $Mg(OH)_2$:

$$Mg^{2+} + 2OH^- \longrightarrow Mg(OH)_2 \tag{3.35}$$

The precipitate is filtered and the excess of NaOH (or lime) left unconsumed is determined by titrating the filtrate against standard acid. The amount of NaOH (or lime) consumed is a measure of the magnesium hardness of the simple.

(5) *Soap Titration*: The fact that sufficient soap solution has to be added to precipitate all the hardness causing ions from a hard water sample before lather is formed is utilized in the determination of hardness. A suitable aliquot of the hard water sample is taken in a stoppered flask and after successive additions of a soap solution (previously standardised by titration against standard hard water) from a burette, the mixture is vigorously shaken. The stage where the lather formed persists for 2–3 minutes is taken as the end-point of the titration. The hardness of the water sample is equivalent to the volume of the soap solution used minus lather factor (volume of soap solution used to produce lather with a volume of distilled water equal to the aliquot of the hard water sample).

3.3.2 *Determination of hardness by EDTA*

Significance of Hardness Determination

Though hard waters are as palatable as soft waters, a knowledge of the magnitude and type of hardness is important in determining the suitability of a water for domestic and industrial purposes. Use of hard waters for cleansing purposes is unsatisfactory as they increase the consumption of soaps. This drawback has been partly overcome by the use of synthetic detergents but for personal hygiene, soap is preferred and hard waters remain objectionable. Hard waters offer difficulties in dyeing of textiles, are uneconomical and even hazardous in steam generation and impart many of the undesirable characteristics to the finished products in paper industry, beverages, dairies, and allied industries. Soft waters are required for an innumerable number of other industries also.

In devising an efficient and economical softening process, determination of total hardness and the relative amounts of CH & NCH and calcium & magnesium hardness is important.

Reagents Required

1. Standard hard water (N/50, 1 ml ≅ 1 mg $CaCO_3$)
2. EDTA solution (N/50)
3. Ammonium chloride-ammonium hydroxide buffer (pH 10).
4. Eriochrome Black T indicator solution.

Theory

The disodium salt of ethylenediamine tetreaacetic acid (EDTA)

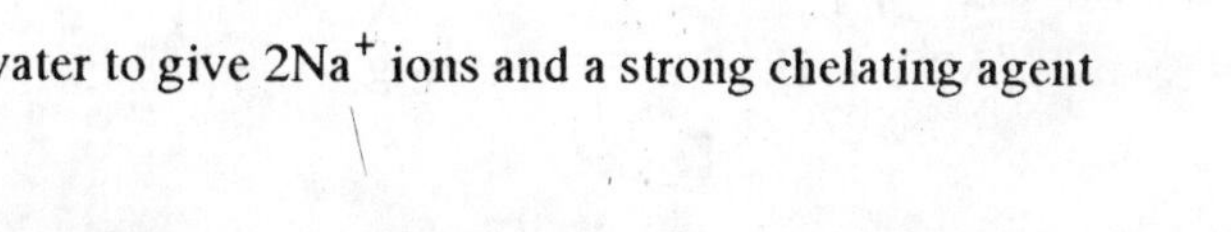

ionises in water to give $2Na^+$ ions and a strong chelating agent

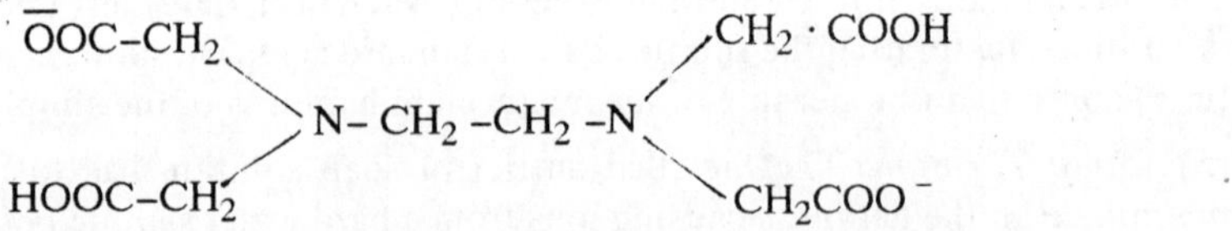

which for simplicity can be represented by H_2Y^{2-}. It forms complexes with Ca^{2+} and Mg^{2+} and other divalent or higher valent cations repersented by the reactions:

$$M^{2+} + H_2Y^{2-} \longrightarrow MY^{2-} + 2H^+ \quad (3.36)$$

$$M^{n+} + H_2Y^{2-} \longrightarrow MY^{(n-4)+} + 2H^+ \quad (3.37)$$

The dissociation of these complexes is governed by the pH of the solution and the complexes with hardness causing divalent ions are stable in alkaline medium (pH range 8–10).

The Indicator used is a complex organic compound [sodium-1-(1-hydroxy-2-napthylazo)-6-nitro-2-naphthol-4-sulphonate]:

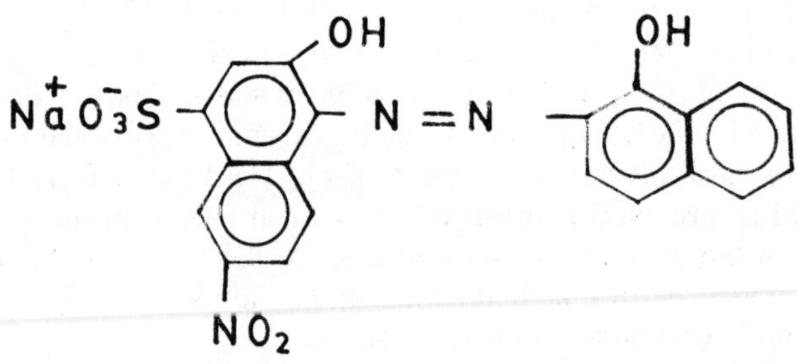

commonly known as Eriochrome Black T. It has two ionisable phenolic hydrogen atoms and for simplicity is represented as $Na^+H_2In^-$:

$$\underset{\text{Red}}{H_2In^-} \underset{5.5}{\overset{7.0}{\rightleftharpoons}} \underset{\text{Blue}}{HIn^{2-}} \underset{11.0}{\overset{11.5}{\rightleftharpoons}} \underset{\text{Yellowish Orange}}{In^{3-}} \quad (3.38)$$

In the pH range 7–11, the indicator reacts with the metal ion to form a weak complex with a wine red colour:

$$\underset{\text{Blue}}{HIn^{2-}} + M^{2+} \rightleftharpoons \underset{\text{Wine red}}{MIn^-} + H^+ \quad (3.39)$$

Buffer Solution: The optimum pH for the experiment is 10.0 ± 0.1 and is adjusted by $NH_4OH + NH_4Cl$ buffer.

When a small amount of the indicator solution is added to a hard water sample whose pH has been controlled by the addition of the buffer solution, the indicator reacts with Mg^{2+} to produce

$$M^{2+} + HI^{2-} \longrightarrow MgIn^- + H^+ \quad (3.40)$$

wine red colour. As EDTA (H_2Y^{2-}) is added, free Ca^{2+} ions are first complexed to CaY^{2-}, this being the most stable complex:

$$Ca^{2+} + H_2Y^{2-} \longrightarrow CaY^{2-} + 2H^+ \qquad (3.41)$$

Free Mg^{2+} ions then react to give Mg-EDTA complex which is less

$$Mg^{2+} + H_2Y^{2-} \longrightarrow MgY^{2-} + 2H^+ \qquad (3.42)$$

stable than Ca-EDTA complex but more stable than Mg-indicator complex. Therefore, if an extra drop of EDTA is added after all the free Ca^{2+} and Mg^{2+} ions have been complexed, EDTA takes up Mg^{2+} from the weak Mg-indicator complex to form stable Mg-EDTA complex simultaneously liberating the indicator in the free form:

$$\underset{\text{(wine red)}}{MgIn^-} + H_2Y^{2-} \longrightarrow MgY^{2-} + \underset{\text{(Pure blue)}}{HIn^{2-}} + H^+ \qquad (3.43)$$

Completion of the above reaction makes the end-point of the titration.

Procedure

(a) *Standardisation of EDTA Solution with Standard Hard Water*

(1 ml S.H.W. ≅ 1 mg $CaCO_3$)

Pipet out 10 ml of standard hard water into a conical flask. Add to it 40 ml distilled water with a measuring cylinder, 2 ml of buffer solution and 2 drops of Eriochrome black T indicator. A wine red colour appears. Titrate against EDTA solution, taken in the burette, to a colour change from wine red to pure blue. Record the volume of EDTA used as A ml. Take three concordant readings.

(b) *Determination of Total Hardness*: Pipet out 50 ml of the hard water sample into a conical flask. Add 2 ml buffer solution and 2 drops of the indicator. Titrate against EDTA till the wine red colour changes to pure blue. Record the volume of EDTA used as B ml. This corresponds to the total hardness of the water sample.

(c) *Determination of Temporary and Permanent Hardness*: Take 250 ml of the hard water sample in a 500-ml beaker and boil gently for about one hour. Cool, filter into a 250-ml measuring flask and make the volume up to the mark. Take 50 ml of this solution and proceed in the same way as in (b) above. The volume of EDTA used (C ml) corresponds to permanent hardness of the water sample. Temporary hardness is calculated by subtracting permanent hardness from total hardness.

(d) *Determination of Calcium and Magnesium Hardness*

Method I

Reagents Required

1. Standard EDTA solution (N/50, 1 ml ≅ 1 mg $CaCO_3$) or standardise as in (a) above
2. Diethylamine
3. Calcon indicator solution.

Theory

Mg^{2+} in the hard water sample is precipitated as $Mg(OH)_2$ by adding diethylamine which raises the pH of the solution to about 12.5.

$$Mg^{2+} + 2\,OH^- \longrightarrow Mg(OH)_2 \qquad (3.44)$$

Ca^{2+} can then titrated with EDTA using calcon indicator:

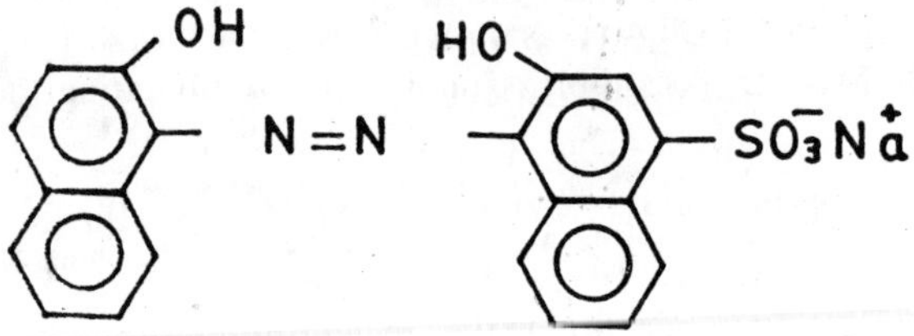

[Calcon, Solochrome Dark Blue or Eriochrome Blue Black R-Chemical name is Sodium 1-(2-hydroxy-1-naphthylazo)-2-naphthol-4-sulphonate]. The colour change at the end-point is from pink to pure blue.

Procedure

Pipet out 50 ml of the hard water sample into a conical flask. Add 3 ml of diethylamine and 4 drops of calcon indicator. Keep the flask on a magnetic stirrer and titrate against standard EDTA solution until the colour changes from pink to pure blue. The volume of EDTA used [D_1 ml] corresponds to calcium hardness. Magnesium hardness is obtained by subtracting calcium hardness from total hardness.

Method II

Reagents Required

1. Standardised EDTA solution
2. Calcium precipitating buffer solution
3. Ammonium chloride-ammonium hydroxide buffer (pH ~ 10)
4. Eriochrome Black T indicator solution.

Theory

Ca^{2+} in the hard water sample is precipitated as calcium oxalate by adding calcium precipitating buffer solution $\left[\begin{array}{l} COONH_4 \\ | \\ COONH_4 \end{array} + NH_4Cl + NH_4OH\right]$

$$C_2O_4^{2-} + Ca^{2+} \longrightarrow \begin{array}{l} COO \\ | \quad\;\; \rangle Ca \\ COO \end{array} \qquad (3.45)$$

White ppt.

The precipitate is filtered and Mg^{2+} in the filtrate is titrated with standard EDTA using Eriochrome Black T indicator.

Procedure

Measure out 200 ml of the hard water sample into a 500 ml dry beaker and add 40 ml of the calcium precipitating buffer solution while constantly stirring the mixture with a glass rod. Allow the precipitate to settle down for about 1 hour and filter through a dry funnel fitted with two pieces of Whatman filter paper No.42 into a dry flask. Do not wait for the filtration to be completed. Measure 60 ml of the filtrate into a conical flask, add 50 ml of distilled water, 5 ml NH_4OH/NH_4Cl buffer (pH ~ 10), 4 drops of Eriochrome black T indicator and titrate against standard EDTA solution. The volume of EDTA used (D_2 ml) corresponds to magnesium hardness. Calcium hardness is obtained by subtracting magnesium hardness from the total hardness.

Calculations

(a) Standardisation of EDTA solution:

1 ml S.H.W. $\cong$ 1 mg $CaCO_3$

A ml EDTA $\cong$ 10 ml S.H.W

$\cong$ 10 mg $CaCO_3$

1 ml EDTA $\cong \frac{10}{A}$ mg $CaCO_3$

(b) Total Hardness:

50 ml hard water $\cong$ B ml EDTA

$\cong B \times \frac{10}{A}$ mg $CaCO_3$

1 ml hard water $\cong B \times \frac{10}{A \times 50}$ mg $CaCO_3$

1000 ml hard water $\cong B \times \frac{10}{A} \times \frac{1000}{50}$ mg $CaCO_3$

$= \frac{B}{A} \times 200$ mg $CaCO_3$

or Total hardness $= \frac{B}{A} \times 200$ ppm

(c) Permanent Hardness:

50 ml hard water $\cong$ C ml EDTA

$\cong C \times \frac{10}{A}$ mg $CaCO_3$

1000 ml hard water $\cong C \times \frac{10}{A} \times \frac{1000}{50}$ mg $CaCO_3$

$\cong \frac{C}{A} \times 200$ mg $CaCO_3$

or Permanent hardness $= \frac{C}{A} \times 200$ ppm

Temporary hardness $=$ Total hardness $-$ Permanent hardness

$$= \left[\frac{B}{A} \times 200 - \frac{C}{A} \times 200\right] \text{ ppm}$$

(d) Method I-Calcium Hardness

50 ml hard water $\cong D_1$ ml EDTA

$$\cong D_1 \times \frac{10}{A} \text{ mg } CaCO_3$$

1000 ml hard water $\cong D_1 \times \frac{10}{A} \times \frac{1000}{50}$ mg $CaCO_3$

$$\cong \frac{D_1}{A} \times 200 \text{ mg } CaCO_3$$

Therefore, Calcium hardness $= \frac{D_1}{A} \times 200$ ppm

Magnesium hardness $= \left[\frac{B}{A} \times 200 - \frac{D_1}{A} \times 200\right]$ ppm

Method II-Magnesium hardness

240 ml filtrate $\cong$ 200 ml hard water

1 ml filtrate $\cong \frac{200}{240}$ ml hard water

60 ml filtrate $\cong \frac{200 \times 60}{240}$ ml hard water

$\cong$ 50 ml hard water

Therefore, hardness of 60 ml of filtrate is equal to the magnesium hardness of 50 ml of hard water sample

50 ml hard water $= D_2$ ml EDTA

$$= D_2 \times \frac{10}{A} \text{ mg } CaCO_3$$

1000 ml hard water $= D_2 \times \frac{10}{A} \times \frac{1000}{50}$ mg $CaCO_3$

or Magnesium hardness $= \frac{D_2}{A} \times 200$ ppm

Calcium hardness $= \left[\frac{B}{A} \times 200 - \frac{D_2}{A} \times 200\right]$ ppm

Exercises

50. Which is the best method of hardness determination and why?

51. What is the effect on the end-point with Eriochrome black-T indicator if the hard water sample does not contain Mg^{2+}?

52. To obtain a sharp end-point with Eriochrome black-T indicator, how can Mg^{2+} ions be added without introducing any error.

53. Why and how is the pH value adjusted to about 10?

54. Describe and calculate the hardness of water samples giving the following analysis:

	Impurity	Amount in mg/l or ppm	Impurity	Amount in mg/l or ppm
(a)	Ca^{2+}	20	SO_4^{2-}	96
	K^+	39	HCO_3^-	183
	Na^+	23	Cl^-	35.5
	Mg^{2+}	48.6		
(b)	Mg^{2+}	24.3	Cl^-	35.5
	Na^+	23	SO_4^{2-}	48
	Ca^+	40	HCO_3^-	305
	K^+	78		

55. 100 ml of a water sample required 10 ml of N/50 HCl for neutralisation to methyl orange end-point. Another 100 ml of the same sample, after being boiled and filtered, required 0.2 ml of N/50 HCl for complete neutralisation. Calculate temporary hardness of the water sample (Hehner's method).

56. 100 ml of a hard water sample were taken in a 250-ml beaker and boiled to precipitate temporary hardness. 20 ml of an alkali mixture solution (NaOH + Na_2CO_3), whose 10 ml required 8 ml of N/10 HCl for neutralisation to methyl orange end-point, were then added. The reaction mixture was heated to boiling, cooled and filtered. The filtrate required 13 ml of N/10 HCl for complete neutralisation. Calculate permanent hardness of the water sample.

57. 100 ml of a hard water sample were neutralised with HCl using methyl orange indicator. The solution was then boiled with 25 ml of N/50 NaOH and filtered. The filtrate required 21 ml of N/50 HCl for complete neutralisation. Calculate magnesium hardness of the water sample.

58. 70 ml of a standard hard water (containing 0.28 g $CaCO_3$/litre) required 20.1 ml of a given soap solution for producing lather that persisted for 5 minutes. 70 ml of tap water under the same conditions required 14.5 ml of

soap solution. 70 ml of the same water sample, after being boiled, filtered and diluted with distilled water to original volume, required 5.4 ml of the soap solution. If 70 ml of distilled water requires 0.5 ml of the same soap solution for producing lather, calculate total, permanent and temporary hardness of tap water in degrees Clark. Also report the result in degrees French and in ppm.

3.4 Dissolved Oxygen (D.O.) and Oxygen Demand

Oxygen is poorly soluble in water. Since it does not react with water chemically, its amount present in water at saturation at any given temperature and pressure can be calculated by Henry's law. Under a pressure of one atmosphere, the solubility of oxygen of air in distilled or fresh waters with low solid concentrations varies from 14.5 mg/l at 0°C to about 7.5 mg/l at 30°C. The solubility is less in saline waters and at a given temperature decreases with increase in the concentration of the impurities.

3.4.1 *Determination of oxygen dissolved in a water sample*

Significance of the Test

In steam generation, D.O. is one of the important factors causing corrosion of the boiler material. From boiler-feed water, D.O. is removed by physical and chemicals means. Measurement of D.O. is necessary to control these processes.

If sufficient D.O. is not present in polluted water, nuisance conditions are developed due to anaerobic degradation of the pollutants. D.O. test is thus vital as a means of controlling the rate of aeration to ensure aerobic conditions and also to prevent excessive use of air. The test also helps in maintaining dissolved oxygen levels that will support the desired aquatic life (fish, etc.).

Reagents Required

1. Standard sodium thiosulphate solution (N/40)
2. Potassium permanganate solution (N/10)
3. Potassium oxalate solution (2%)
4. Manganous sulphate solution (48%)
5. Alkaline potassium iodide solution
6. Freshly prepared starch solution
7. Concentrated sulphuric acid.

Theory

The iodometric (Winkler) method is based on the fact that in alkaline medium D.O. oxidises Mn^{2+} to Mn^{4+}, which in acidic medium oxidises I^- to free iodine. The amount of iodine released, which can be titrated with a standard solution of sodium thiouslphate, is thus equivalent to the D.O. originally present.

In the modified Winkler method, the interference due to certain oxidising agents such as NO_2^- or reducing agents such as Fe^{2+} or SO_3^{2-} is removed by treating the sample with an excess of $KMnO_4$ in acid medium. The following reactions take place:

$$NO_2^- + H_2O \longrightarrow NO_3^- + 2H^+ + 2e] \times 5$$

$$MnO_4^- + 8H^+ + 5e \longrightarrow Mn^{2+} + 4H_2O] \times 2$$

$$5NO_2^- + 2MnO_4^- + 6H^+ \longrightarrow 5NO_3^- + 2Mn^{2+} + 3H_2O \quad (3.46)$$

$$SO_3^{2-} + H_2O \longrightarrow SO_4^{2-} + 2H^+ + 2e] \times 5$$

$$MnO_4^- + 8H^+ + 5e \longrightarrow Mn^{2+} + 4H_2O] \times 2$$

$$5SO_3^{2-} + 2MnO_4^- + 6H^+ \longrightarrow 5SO_4^{2-} + 2Mn^{2+} + 3H_2O \quad (3.47)$$

$$Fe^{2+} \longrightarrow Fe^{3+} + e] \times 5$$

$$MnO_4^- + 8H^+ + 5e \longrightarrow Mn^{2+} + 4H_2O$$

$$5Fe^{2+} + MnO_4^- + 8H^+ \longrightarrow 5Fe^{3+} + Mn^{2+} + 4H_2O \quad (3.48)$$

NO_3^- and SO_4^{2-} formed in reactions (3.46) and (3.47) do not interfere. Fe^{3+} interferes only when present in amounts above 10 mg/l. This interference is avoided by converting Fe^{3+} into poorly dissociated complex FeF_6^{3-} by adding KF:

$$Fe^{3+} + 6F^- \longrightarrow FeF_6^{3-} \quad (3.49)$$

Excess of $KMnO_4$ is destroyed by adding potassium oxalate:

$$MnO_4^- + 8H^+ + 5e \longrightarrow Mn^{2+} + 4H_2O] \times 2$$

$$C_2O_4^{2-} \longrightarrow 2CO_2 + 2e] \times 5$$

$$2MnO_4^- + 16H^+ + 5C_2O_4^{2-} \longrightarrow 2Mn^{2+} + 10CO_2 + 8H_2O \quad (3.50)$$

$MnSO_4$ and alkaline KI are then added when a white precipitate of $Mn(OH)_2$ is formed.

$$Mn^{2+} + 2OH^- \longrightarrow Mn(OH)_2 \quad (3.51)$$

The oxygen dissolved in the sample oxidises some of the Mn^{2+} to Mn^{4+} which is precipitated as brown hydrated manganese dioxide:

$$Mn^{2+} + 4OH^- \longrightarrow MnO(OH)_2 + H_2O + 2e] \times 2$$

$$O_2 + 2H_2O + 4e \longrightarrow 4OH^-$$

$$2Mn^{2+} + O_2 + 4OH^- \longrightarrow 2MnO(OH)_2 \text{ or } (2MnO_2\,H_2O) \quad (3.52)$$

(This process is sometimes called fixation of oxygen).

On acidifying the solution, manganese in the higher oxidation state oxidises I^- to free iodine:

$$MnO(OH)_2 + 4H^+ + 2e \longrightarrow Mn^{2+} + 3H_2O$$

$$2I^- \longrightarrow I_2 + 2e$$

$$MnO(OH)_2 + 4H^+ + 2I^- \longrightarrow Mn^{2+} + I_2 + 3H_2O \quad (3.53)$$

The iodine liberated is titrated with standard $Na_2S_2O_3$ solution using freshly prepared starch solution as indicator:

$$I_2 + 2e \longrightarrow 2I^-$$

$$2S_2O_3^{2-} \longrightarrow S_4O_6^{2-} + 2e$$

$$I_2 + 2S_2O_3^{2-} \longrightarrow 2I^- + S_4O_6^{2-} \quad (3.54)$$

Procedure

Collect the water sample in a 300-ml glass-stoppered (water tight) bottle. With the help of a graduated pipet, add 0.7 ml conc. H_2SO_4 and 0.2 ml (4 drops) $KMnO_4$ solution. Stopper the bottle and mix the contents of the bottle by inverting it a few times. If the permanganate colour disappears within 5 minutes, add additional amount of $KMnO_4$. Add 0.5 ml of potassium oxalate solution, stopper and mix well. Add additional amounts of oxalate solution if the permanganate colour is not discharged within 10 minutes. Add 2 ml of $MnSO_4$ solution followed by 3 ml of alkaline KI solution. Stopper and mix by rotating and inverting the bottle 10 to 15 times and allow to stand. After about 2 minutes, shake and allow the precipitate to settle. When some portion of the liquid below the stopper is clear, add 1 ml conc. H_2SO_4. Stopper and mix until the precipitate is completely dissolved. Take 204.4 ml (with a measuring cylinder) of this solution in a titration flask and titrate slowly against (N/40) $Na_2S_2O_3$ solution. When the colour of the solution is very light yellow, add about 2 ml of freshly prepared starch solution and continue the titration to the first disappearance of the blue colour. Repeat the whole experiment two to three times and record the mean volume of $Na_2S_2O_3$ used as A ml.

Precautions

(1) As far as possible, the sample should not be allowed to come in contact with air.

In most cases of interest, the D.O. level of the sample is below saturation, and exposure to air will lead to higher results. For reducing the contact of the sample with air

(a) the sample should be collected from the tap using a rubber tubing whose other end should be lowered to the bottom of the bottle. Water should be allowed to overflow for sometime.

(b) the stopper of the bottle should be removed only when some reagent is to be added.

(c) while replacing the stopper, care should be taken to exclude air bubbles.

(2) While adding various reagents, the tip of the graduated pipet should be dipped below the surface of the liquid.

(3) The flocculated material formed on addition of alkaline KI should be moved throughout the solution to enable all the oxygen to react. The reaction [which may also be represented as stoichiometric absorption of dissolved molecular oxygen by $Mn(OH)_2$:

$$2Mn(OH)_2 + O_2 \longrightarrow 2MnO(OH)_2 \qquad (3.52a)]$$

being slow, sufficient time should also be allowed.

4. While adding the reagents and replacing the stopper, bottle should be placed in a trough or basin as some liquid will always overflow.

Observations

Total volume of the sample taken = 300 ml

Volume of reagents added during the preparation of iodine solution = (0.7ml H_2SO_4 + 0.2ml $KMnO_4$ + 0.5ml $K_2C_2O_4$ + 2 ml $MnSO_4$ + 3 ml alkaline KI) = 6.4 ml.

Volume of prepared solution (iodine) taken for titration = 204.4 ml

Mean volume of N/40 $Na_2S_2O_3$ used = A ml

Calculations

6.4 ml of the reagents have been added under such conditions that approximately equal volume of the sample is displaced. This dilutes the sample and so a correction is needed. Volume of the original sample that will be equivalent to 204.4 ml of the prepared (diluted) solution is given

$$\frac{\text{Total volume of the sample} - \text{volume displaced by reagents}}{\text{Total volume of the sample}} \times 204.4$$

$$= \frac{300 - 6.4}{300} \times 204.4 = 200 \text{ ml}$$

Thus the iodine content of 204.4 ml of the prepared solution (diluted sample) is equivalent to the D.O. of 200 ml of the original sample. Then using the normality equation

$$\underset{\text{(Oxygen solution)}}{N_1V_1} = \underset{\text{(Hypo solution)}}{N_2V_2}$$

$$N_1 \times 200 = \frac{1}{40} \times A$$

or N_1 (Normality of the sample w.r.t. D.O.) $= \frac{1}{200} \times \frac{1}{40} \times A$

$$\therefore \quad \text{Strength} = \frac{1}{200} \times \frac{1}{40} \times A \times 8 \text{ g/l}$$

$$= A \text{ mg/l} = A \text{ ppm}$$

Exercises

59. What is the effect of the presence of oxidising impurities like NO_2^- and Fe^{3+} (if not removed) on the D.O. results ?

60. How do reducing impurities such as Fe^{2+}, SO_3^{2-}, S^{2-}, etc., effect the D.O. determination ?

61. What do you understand by the phrase 'fixation of dissolved oxygen'?

62. Why is the $Na_2S_2O_3$ solution of N/40 strength selected for titration ?

63. State Henry's law for the solubility of a gas in a liquid. Outline the conditions under which the law is applicable.

3.4.2 *Biochemical Oxygen Demand (BOD)*

Micro-organisms can utilize carbohydrates, proteins and oils and fats as food and oxidize them to obtain energy for their life processes. Some bacteria can also utilize reduced inorganic materials like Fe^{2+}, S^{2-} and ammonia as sources of energy. In the biological degradation (of sewage and other waste material) which is brought about by a diverse group of living organisms or bacteria, organic matter is first converted to fatty acids which, by successive oxidation at β-carbon atom, are 'broken down' into fragments consisting of acetic acid. When sufficient oxygen is present, that is in aerobic systems, oxygen is the ultimate electron acceptor which is reduced while organic matter is being oxidised to CO_2 and H_2O:

$$CH_3COO^- + 2O_2 \xrightarrow{\text{bacteria}} CO_2 \;\; HCO_3^- + H_2O \tag{3.55}$$

Inorganic materials like S^{2-} and reduced nitrogen (NH_3 or NH_4^+), if present, are oxidised to sulphate and nitrate:

$$S^2 + 2O_2 \xrightarrow{\text{bacteria}} SO_4^{2-} \tag{3.56}$$

$$NH_3 + 2O_2 \xrightarrow[\text{bacteria}]{\text{Nitrifying}} NO_3^- + H^+ + H_2O \tag{3.57}$$

When the availability of oxygen is insufficient (anoxic conditions), organic matter is oxidised by using nitrate as electron acceptor:

$$5CH_3COO^- + 8NO_3^- + 8H^+ \xrightarrow[\text{bacteria}]{\text{facultive}} 5CO_2 + 5HCO_3^- + 4N_2 + 9H_2O \tag{3.58}$$

[Nitrate may also be reduced to the amine stage

$$CH_3COO^- + NO_3^- \xrightarrow{\text{bacteria}} CO_2 + HCO_3^- + NH_2^- \tag{3.59}$$

giving odour resembling that of decaying rotten fish.]

Under strictly anaerobic conditions, i.e., in the absence of oxygen, SO_4^{2-}, PO_4^{3-} and CO_2, whatever available, can act as electron acceptors and are reduced to H_2S,

HS^- (mercaptans—rotten egg small), phosphine (poisonous and wormy odour), methane (suffocating), etc.:

$$2CH_3COO^- + 2SO_4^{2-} + 3H^+ \longrightarrow 2CO_2 + 2HCO_3^- + H_2S + HS^- + 2H_2O \quad (3.60)$$

$$CH_3COO^- + PO_4^{3-} + 3H^+ \longrightarrow CO_2 + HCO_3^- + PH_3 + H_2O \quad (3.61)$$

$$CH_3COO^- + H_2O \longrightarrow CH_4 + HCO_3^- \quad (3.62)$$

Sewage and other oxygen-demanding wastes are termed as water pollutants because:

(a) their decay leads to depletion of dissolved oxygen, which affects (and even kills) fish and other equatic life,

(b) they produce annoying odours,

(c) they impair domestic and livestock water supplies by affecting taste, odours and colours and

(d) they promote conditions favourable for the growth of infectious agents (pathogenic bacteria).

To avoid development of nuisance or septic conditions, sufficient oxygen must be present or supplied externally to maintain aerobic conditions throughout.

The amount of oxygen required by a mixed population of micro-organisms in oxidising organic matter present in a sample, under strictly aerobic conditions, is generally known as B.O.D. and is directly related to the extent of pollution (by sewage or other oxygen-demanding wastes).

The rate of bacterial oxidation at any instant is proportional to the amount of the oxygen-demanding wastes left at that instant, i.e., the reaction follows First Order Kinetics and theoretically should be completed in infinite time. It has however been found that nearly 70 to 80% of the total BOD is exerted in the first five days. The sample is therefore incubated for 5 days at 20°C (roughly a mean value of temperatures of natural bodies of water) and the BOD values are reported as BOD_5. A polluted sample may consume more oxygen in 5 days than present in water (nearly 9 mg/l at 20°C) and so, before analysis, is diluted with a specially prepared 'Dilution Water'. The diluted sample is taken in two bottles. The dissolved oxygen (D.O.) in one bottle is determined immediately and in the other after 5 days of incubation. The BOD of the sample is then given by

$$BOD = \frac{D_1 - D_2}{A} \times B \text{ mg/l} \quad (3.63)$$

where D_1 = D.O. of the sample in mg/l at the start of the experiment

D_2 = D.O. of the sample in mg/l after 5 days

A = ml of the sample before dilution

B = ml of the sample after dilution.

Exercises

64. 10 ml of a polluted water sample was diluted to 600 ml and equal volumes were poured into two B.O.D. bottles. D.O. in one bottle was determined immediately (Blank) by Winkler's method and 204.4 ml solution required 3.9 ml of N/40 $Na_2S_2O_3$. The second sample was incubated for 5 days and in the D.O. determination, 204.4 ml solution required 2.1 ml of N/40 $Na_2S_2O_3$. Calculate B.O.D. of the polluted water.
65. Outline the significance of B.O.D. measurement.
66. Why has the 5-day incubation period been selected for BOD determination?
67. What should be the composition of a good 'Dilution Water' for BOD determination?

3.4.3 *Determination of Chemical Oxygen Demand (C.O.D.) of a waste water sample*

The COD is usually defined as the amount of oxygen used while oxidizing the organic matter content of a sample with a strong chemical oxidant under acidic conditions. The organic matter in the sample is related to the oxygen required (COD) in accordance with equation:

$$C_xH_yO_z + \left(x + \frac{y}{4} - \frac{z}{2}\right) O_2 \longrightarrow xCO_2 + \frac{y}{2} H_2O \qquad (3.64)$$

Since in the COD determination, the organic matter (both biologically oxidizable like glucose and biologically inert like cellulose) is completely oxidised to CO_2 and H_2O, the COD values are greater than BOD values (which represent the amount of oxygen that bacteria need for stabilising biologically oxidizable matter).

Significance

For samples from a specific source, COD can be related to BOD. The COD test is therefore widely used for measuring the pollutional strength of domestic and industrial wastes. Used in conjunction with the BOD test, it indicates the magnitude of the biologically resistant organic matter. The major advantage of the COD test is that the determination is completed in 3 hours, compared to the 5 days required for the BOD determination, and therefore steps can be taken to correct errors on the day they occur.

Reagents Required

1. Standard postassium dichromate solution (N/4)
2. Standard ferrous ammonium sulphate solution (FAS N/4)
3. Silver sulphate-sulphuric acid reagent
4. Ferroin indicator solution
5. Mercuric sulphate ($HgSO_4$).

Theory

A suitable aliquot of the sample is boiled with a known excess of $K_2Cr_2O_7$ in presence of conc. H_2SO_4. The organic matter is oxidised to CO_2 and H_2O:

$$C_xH_yO_z + \left[\frac{4x + y - 2z}{6}\right] Cr_2O_7^{2-} + \frac{4}{3}[4x + y - 2z]\ H^+ \xrightarrow{\text{Heat}}$$

$$xCO_2 + \left[\frac{4x + y - 2z}{3}\right] Cr^{3+} + \left[\frac{16x + 7y - 8z}{6}\right] H_2O \quad (3.65)$$

The excess of dichromate left unused is titrated with FAS standard solution using ferroin as indicator:

$$6Fe^{2+} + Cr_2O_7^{2-} + 14H^+ \longrightarrow 6Fe^{3+} + 2Cr^{3+} + 7H_2O \quad (3.66)$$

The amount of $K_2Cr_2O_7$ consumed corresponds to COD.

Procedure

To 50 ml of the sample taken in a 500-ml reflux flask, add 1 g $HgSO_4$ and some broken porcelain pieces or glass beads. Immerse the flask in cold water and slowly and 75 ml of $Ag_2SO_4 - H_2SO_4$ reagent while continuously shaking the flask. Add 25 ml of N/4 $K_2Cr_2O_7$ solution. Mix the contents of the flask thoroughly. Fix the water condenser and reflux for 2 hours. Wash the condenser with distilled water into the flask. Cool and dilute to about 300 ml. Add 2 to 3 drops of Ferroin indicator and titrate against N/4 FAS until the first sharp colour-change from blue green to reddish brown. Record the volume of FAS used as A ml.

Blank Titration

Take a volume of distilled water, equal to that of the sample, in another reflux flask. Add the same amounts of reagents, reflux for 2 hours and titrate in the same way as in the sample titration. Record the volume of FAS used as B ml. COD of the sample corresponds to (B-A) ml of standard (N/4) FAS.

Precautions

The addition of $Ag_2SO_4 - H_2SO_4$ reagent to the sample is an exothermic process and may lead to loss of material. To avoid this, it is important to mix the reflux mixture slowly with shaking and the flask should be cooled during mixing. It is better to fix the reflux condenser in the flask and then add the major portion of the $Ag_2SO_4 - H_2SO_4$ reagent through the open end of the condenser.

Observations and Calculations

Volume of sample taken = 50 ml

Volume of N/4 FAS used in the sample (test) titration = A ml

Volume of N/4 FAS used in blank titration = B ml

Therefore, volume of N/4 FAS equivalent to $K_2Cr_2O_7$ used in satisfying COD = (B – A) ml

$$\underset{\text{(sample)}}{N_1V_1} = \underset{\text{(FAS)}}{N_2V_2}$$

$$N_1 \times 50 = \frac{1}{4} \times (B - A)$$

or $$N_1 = \frac{(B - A)}{4 \times 50}$$

Therefore, C.O.D. $$= \frac{(B - A)}{4 \times 50} \times 8 \text{ g/l}$$

$$= \frac{(B - A) \times 8 \times 1000}{4 \times 50} \text{ mg/l}$$

$$= (B - A) \times 40 \text{ mg/l}$$

Exercises

68. What are the major sources of error in the above method?
69. How are the above-mentioned errors minimised ?

3.5 Residual Chlorine and Chlorine Demand

Water is the chief vehicle of transmission for some diseases like dysentry, cholera, typhoid, etc. (called waterborne diseases), caused by pathogenic bacteria. Growth of other living organisms such as algae and fungi in water may be extremely harmful when such water is used for certain industrial purposes. Chlorination is an important disinfection process which serves to destroy or deactivate the disease producing and other harmful organisms.

Chlorination may produce adverse effects also. Phenols, if present in water, are converted into mono-, di-, or trichloro-phenols which impart unacceptable taste and odour to water. Some forms of aquatic life are adversely affected by chlorination of waters containing ammonia or amines. Humic substances which are present in most water supplies are converted into several chlorinated products, the most harmful of which is the carcinogenic chloroform ($CHCl_3$). In spite of these adverse effects, chlorination has not been abandoned because of the immense benefits arising from it which far outweigh the risks involved.

Chlorine is applied to water in its elemental form (Cl_2 gas or a concentrated solution of Cl_2 in water) or as hypochlorite (NaOCl or bleaching powder). Chlorine reacts with water to produce hypochlorous acid which further dissociates to give H^+ and OCl^-:

$$Cl_2 + H_2O \longrightarrow HOCl + H^{++} Cl^- \qquad (3.67)$$

$$HOCl \longrightarrow H^+ + OCl^- \qquad (3.68)$$

The relative amounts of Cl_2, HOCl and Cl^- depend on the temperature and pH. At the pH of the most natural waters, HOCl and Cl^- predominate. Ammonia, formed by dissociation of NH_4^+ ion impurity present in water, is slowly attacked by Cl_2 or HOCl to form small amounts of mono-, di- and trichloramines:

$$NH_4^+ \longrightarrow NH_3 + H^+ \; [K_{dissociation} \approx 10^{-10}] \qquad (3.69)$$

$$NH_3 + HOCl \longrightarrow \underset{\text{Monochloramine}}{NH_2Cl} + 8H_2O \qquad (3.70)$$

$$NH_2Cl + HOCl \longrightarrow \underset{\text{Dichloramine}}{NHCl_2} + 8H_2O \qquad (3.71)$$

$$NHCl_2 + HOCl \longrightarrow \underset{\text{Trichloramine}}{NCl_3} + 8H_2O, \quad (3.72)$$

the mono- and dichloramines being important disinfectants. With higher concentrations of Cl_2 (1.5 mole Cl_2 per mole of NH_3), N_2 or NO_3^- are formed depending upon pH, temperature and reaction time:

$$2NH_3 + 3Cl_3 \longrightarrow N_2 + 6H^+ + 6Cl^- \quad (3.73)$$

$$NH_3 + 4Cl_2 + 3H_2O \longrightarrow NO_3^- + 9H^+ + 8Cl^- \quad (3.74)$$

Chlorination beyond this stage (called Break-point chlorination) increases the amount of Cl_2, HOCl and OCl^- which are referred to as free chlorine residuals. While the free chlorine residuals are quickly dissipated in the distribution system, the chloramines (referred to as combined chlorine residuals) are stable. This affords longer contact time, thus lowering the concentration of the disinfectant required for equivalent destruction of bacteria. Chloramines act as chlorine reserves killing bacteria left during the initial treatment and guard against any inadvertent bacterial contamination at a later stage. Their presence is therefore particularly desired when water is to be transported over long distances. When a chlorinated water is devoid of combined chlorine residuals, a small amount of NH_3 or ammonium salt is added after chlorination.

An overdose of chlorine not only increases the cost of disinfection but imparts an unpleasant taste to drinking water and is also injurious to health. For efficient disinfection of domestic water supplies, swimming pools and treatment of waste waters, measurement of both free and combined chlorine residuals is essential.

3.5.1 *Determination of Total Chlorine Residuals (Iodometric Method)*

Reagents Required

1. Standard sodium thiosulphate solution (N/100).
2. Glacial acetic acid.
3. Potassium iodide.
4. Freshly prepared starch solution.

Theory

When KI is added to water at pH 3 – 4, both free and combined chlorine residuals react to oxidise iodide to free iodine:

$$Cl_2 + 2I^- \longrightarrow I_2 + 2Cl^- \quad (3.75)$$

$$OCl^- + 2I^- \longrightarrow I_2 + Cl^- + H_2O \quad (3.76)$$

$$HOCl + 2I^- + H^+ \longrightarrow I_2 + Cl^- + H_2O \quad (3.77)$$

$$NH_2Cl + 2I^- + H^+ \longrightarrow I_2 + Cl^- + NH_3 \quad (3.78)$$

$$NHCl_2 + 4I^- + 2H^+ \longrightarrow 2I_2 + 2Cl^- + NH_3 \quad (3.79)$$

The liberated iodine is titrated with a standard solution of $Na_2S_2O_3$, using starch solution as indicator near the end-point:

$$I_2 + 2S_2O_3^{2-} \longrightarrow 2I^- + S_4O_6^{2-} \qquad (3.80)$$

Procedure

Pipet out 100 ml of the water sample into an iodine titration flask containing 1 g of KI and 2 ml of glacial acid. Stopper the flask and shake well to mix the solution. Rinse the stopper and the sides of the flask with distilled water (using a wash bottle) into the flask. Add $\frac{N}{100}$ $Na_2S_2O_3$ solution from a burette slowly with constant shaking until the colour of the solution becomes very light yellow or straw yellow. Add 1 ml of freshly prepared starch solution and titrate further until the blue colour disappears. Record the volume of $Na_2S_2O_3$ solution used as A ml. Take concordant readings.

Precautions

1. The titration should be completed rapidly in order to avoid atmospheric oxidation of iodide.
2. First disappearance of blue colour should be taken as the end-point.

Observations and Calculations

Volume of water sample taken for each titration = 100 ml

Concordant volume of $\frac{N}{100}$ $Na_2S_2O_3$ used = A ml

$$\underset{\text{(Sample)}}{N_1V_1} = \underset{(Na_2S_2O_3)}{N_2V_2}$$

$$N_1 \times 100 = \frac{1}{100} \times A$$

Therefore, N_1 (Normality of sample w.r.t. total chlorine residuals) $= \frac{1}{100} \times A \times \frac{1}{100}$

Total chlorine residuals $= \frac{1}{100} \times A \times \frac{1}{100} \times 35.5$ g/l

$$= \frac{1}{100} \times A \times \frac{1}{100} \times 35.5 \times 10^3 \text{ mg/l}$$

$$= 3.55 \times A \text{ ppm}$$

Exercises

70. Why and how is pH adjusted between 3 and 4?
71. What are the common impurities that interfere in the determination?

3.5.2 *Determination of Free Chlorine Residuals and Combined Chlorine Residuals (Chloramine and dichloramine) in a treated water (DPD-Ferrous Titrimetric Method)*

Reagents Required

1. Standard ferrous ammonium sulphate solution (N/100 FAS)

2. Phosphate buffer (pH 6.2 – 6.5)
3. Postassium iodide
4. DPD indicator solution

Theory

(a) In the pH range 6.2 – 6.5, free chlorine residuals oxidise N, N-diethyl-p-diphenylenediamine (DPD) quantitatively to produce a red colour which can be titrated to a sharp colourless end-point by standard ferrous ammonium sulphate (FAS) solution.

(b) If a small crystal of KI is added at this stage, monochloramine oxidises DPD to produce equivalent red colour which is immediately titrated with FAS.

(c) Addition of more KI (about 1 g) produces more red colour due to the slow oxidation of DPD by dichloramine which is titrated with FAS after waiting for about 2 minutes.

Procedure

Add 100 ml of the sample to a titration flask containing 5 ml of phosphate buffer solution (pH 6.2 – 6.5) and 5 ml of DPD indicator solution. Mix and titrate rapidly with standard FAS until the red colour disappears. The volume of FAS (A ml) used corresponds to the amount of free chlorine residuals in the sample.

Add 1 crystal of KI to the above solution. Mix and rapidly titrate against FAS until the red colour is again discharged. The volume of FAS (B ml) consumed is equivalent to monochloramine present in the sample.

Add 1 g of KI to the above solution and mix well. Allow to stand for 2–3 minutes and titrate against FAS. The volume consumed (C ml) corresponds to the amount of dichloramine in the sample.

Precaution

1. After the addition of 1 crystal of KI, the solution should be immediately titrated with no loss of time as otherwise dichloramine may also react.

Observations and Calculations

(a) Free chlorine residuals:

Volume of sample taken = 100 ml

Volume of $\frac{N}{100}$ FAS used = A ml

$$\underset{\text{(Sample)}}{N_1V_1} = \underset{\text{(FAS)}}{N_2V_2}$$

$$N_1 \times 100 = \frac{1}{100} \times A$$

or

$$N_1 = \frac{1}{100} \times A \times \frac{1}{100}$$

Amount of free chlorine residuals $= \frac{1}{100} \times A \times \frac{1}{100} \times 35.5$ g/l

$$= \frac{A \times 35.5}{100 \times 100} \times 10^3 \text{ mg/l}$$

$$= 3.55 \times A \text{ ppm}$$

(b) Combined chlorine residuals present as monochloramine $= 3.55 \times B$ ppm

(c) Combined chlorine residuals present as dichloramine $= 3.55 \times C$ ppm

Exercises

72. Why is the pH control so important in this determination?

73. EDTA is added in the preparation of DPD indicator solution. What is its purpose?

3.5.3 *Chlorine Demand*

Chlorine reacts with a wide variety of inorganic and organic materials present in water. The demand of these impurities for chlorine has to be satisfied before chlorine becomes available for disinfection. The difference between the amount of chlorine applied to a water sample and residual chlorine is reported as *Chlorine Demand.* It depends on the nature of impurities present in the sample, the amount of chlorine applied, the amount of residual chlorine desired, contact time, pH and temperature. Chlorine demand can be determined by treating a series of aliquots of the water sample with known but varying dosages of chlorine or hypochlorite (such as bleaching powder) and determining the residual chlorine at the end of the contact period. This will give the dosage required for any desired residual at the temperature of the experiment.

3.5.4 *Determination of the amount of bleaching powder required to disinfect a water sample by Horrock's Test*

Reagents Required

1. Bleaching powder
2. Potassium iodide
3. Glacial acetic acid
4. Freshly prepared starch solution.

Theory

Increasing amounts of a solution of known concentration of bleaching powder are added to the same known volumes of the sample taken in a number of containers. After allowing sufficient time for disinfection to take place, a few crystals of KI and glacial acetic acid are added to each of the containers. This leads to liberation of iodine, in the containers in which disinfection is complete and there is residual chlorine, which is detected by the production of blue colour on addition of starch solution.

$$OCl^- + 2H^+ + 2I^- \longrightarrow I_2 + H_2O + Cl^- \quad (3.81)$$

$$I_2 + \text{Starch solution} \longrightarrow \text{Blue Colour}$$

The method is approximate as no standard solutions are used and the residual chlorine is not measured quantitatively. However, the method is very quick and when conducted with care, it can lead to efficient disinfection. The procedure has been so designed that even a layman can conduct it easily, and make water safe for drinking.

Procedure

Take seven glass containers or bottles marked at 100-ml level. Transfer one level tea-spoonful (about 1 g) of bleaching powder to one of the containers, add a little distilled water and stir with a glass rod until it is converted into a paste. Add more distilled water to fill the container up to the mark and stir vigorously. Leave the solution to let the suspension settle down. Fill the remaining containers with water sample up to the mark. Number them as 1, 2, 3, 4, 5 & 6, with a glass marking pencil, and arrange in a row. Using a medicine dropper (delivering 20 drops per ml), transfer 1, 2, 3, 4, 5 & 6 drops of the prepared solution of bleaching powder to the containers numbered 1,2,3, etc., respectively. Stir thoroughly the contents of each container. Allow to stand for half an hour (contact period). Add to each container a few crystals of KI and 1 ml of glacial acetic acid followed by 1 ml of freshly prepared starch solution. Stir the contents gently with the rod and look for the development of blue colour. Note the container showing blue colour with minimum number of drops of the bleaching powder solution. If none of the containers shows a blue colour, throw the contents. Wash the containers thoroughly with distilled water and perform the test with 7, 8, 9, 10, 11 & 12 drops of the bleaching powder solution. Record the contact period and temperature.

Precautions

(1) While preparing the paste, care should be taken to break with a glass rod (preferably with a flattened tip) any lumps of the bleaching powder left.

(2) The tip of the dropper should be clean and the drops should be added slowly so that they are fully formed.

(3) Contact period for all the samples should be constant (the same), i.e. bleaching powder solution or KI and acetic acid should be added, as far as possible, simultaneously to all the samples.

Note: The test can be made more accurate by taking a known weight of the bleaching powder for preparing the solution.

Observations and Calculations

Volume of water sample taken in each container = 100 ml

Concentration of bleaching powder solution used = 1 level spoonful (~ 1 g)/100 ml

Contact period = ... minutes

Temperature of the experiment = ... °C

Sr.No. of container	No. of drops of B.P. solution added	Appearance of blue colour	Minimum No. of drops producing blue colour
1	1	No	
2	2	No	
3	3	–	
4	4	–	(n, say)
5	5	Yes	
6	6	Yes	
–	–	–	
–	–	–	

1 level spoonful of B.P. ≅ 100 ml of B.P. solution

$= 100 \times 20$ drops

Volume of sample disinfected by n drops $= 100$ ml

" 1 drop $= \frac{100}{n}$ ml

" 100×20 drops $= \frac{100 \times 2000}{n}$ ml

$= \frac{200}{n}$ litres

B.P. required for $\frac{200}{n}$ liters of the sample = 1 level spoonful (1 ~ g)

" 1 " $= \frac{1 \times n}{200}$ "

" 10^3 " $= \frac{1 \times n}{200} \times 10^3$ "

$= 5$ n level spoonful

$= 5$ n g.

Exercises

74. What is meant by Chlorine Demand of a water sample?
75. Name a few chlorine-reactable materials that may be present in water as impurities.

76. Describe the nature of the reactions which these substances undergo with chlorine.

77. Specify the optimum chlorine dosage, contact time and pH range for effective disinfection.

78. What is meant by stabilising the chlorine ?

79. What is Break-point Chlorination ? What is its significance ?

80. What do you understand by Prechlorination and Post-chlorination ?

3.6 Turbidity and Coagulation

The term turbidity is applied to finely divided particles suspended in water, due to which light passing into the sample is absorbed and scattered. Visual depth in such water is restricted. The turbidity of a water sample is related inversely to its 'clarity'. It is simply an optical property of the particles suspended in water and not a direct assessment of their amount present, since it is related to the size distribution of the particles. Turbidity in water may be of organic or inorganic origin. It is caused by suspended particles of clay, loam, sand, silt, decomposing food material, contamination with municipal or industrial waste water, organic coloured compounds, algae and other micro-organisms.

Significance

Consumers of municipal water supplies are always reluctant to accept even very slightly turbid waters as they often consider it to be polluted by domestic waste water—the cause of many water-borne diseases. Permissible turbidity for drinking water is between 5–10 units and preferably it should be less than 5. Turbidity makes filtration very difficult. To get good quality water, filters of special design have to be used, thereby increasing the cost. A high turbidity value of the filtered water points to faulty filter operation. When turbidity is due to contamination with sewage, disinfection is less effective because some of the pathogenic bacteria are enclosed in the sewage particles and escape the killing action of the disinfectant.

Measurement of turbidity is useful in deciding whether water from a particular source can be used without chemical coagulation. Turbidity measurement, before and after coagulation by different chemicals, is used to select the most effective and economical substance and to determine its optimum dose.

In the removal of suspended solids from domestic and industrial wastes of frequently changing composition, turbidity measurement, being much quicker than suspended solids determination, is especially useful to adjust the optimum dosages of chemicals required.

Turbidity Standards

Since turbidity is caused by a wide variety of materials, the following arbitrary standard has been chosen to compare turbidity of different waters : 1 unit turbidity is the turbidity produced by 1 part of fuller's earth in a million parts of distilled water (1 unit = 1 mg SiO_2/1, and the size of particles of SiO_2 used must lie within certain specific limits—the particles should pass through a 200-mesh sieve).

Standard suspensions of pure SiO_2 were used originally to calibrate Jackson Candle Turbidimeter, which is now used for routine work and no standards are required. In this method, interference caused to the passage of light is measured visually and the results are reported as Jackson Turbidity Units (JTU).

The instrumental method which, instead, is based on the measurement of the intensity of light scattered by turbidity using Nephelometry, gives different results which are reported in Nephelometric Turbidity Units (NTU).

3.6.1 *Determination of Turbidity between 2501000 JTU by Jackson Turbidimeter*

Procedure

Shake the sample well and transfer 10–15 ml of it into a cool Jackson tube (made from colourless glass, with a flat, polished, optical glass bottom and calibrated for direct turbidity readings) and fix it in position in turbidimeter. Light the candle or bulb at the base. Pour slowly, through a funnel, additional portions of the sample into the Jackson tube and after each addition, view the lighted candle (or bulb) from above. Continue addition of the sample until the contour (outline) of the flame or bulb is no longer visible. Extinguish the flame (or switch off the bulb). Take out the Jackson tube and record turbidity in JTU units. Take out a small amount of the sample; place the tube in position, light the candle (or bulb) and adding very small portions of the removed sample at a time, approach the end-point very carefully. Take a number of readings and report the mean value.

Precautions

(1) The Jackson tube should be kept clean, both inside and outside.

(2) Scratching of the tube should be avoided.

(3) The charred portion of the wick (if candle is used) should be trimmed and the distance of the flame from the bottom of the tube should be kept constant.

3.6.2 *Determination of Turbidity between 25–100 JTU with Bottle Standards*

Procedure

Take a number of glass bottles of the same size, shape and colour characteristics. Place the sample in one bottle and a number of standard suspensions of varying turbidities (made by suitably diluting, with distilled water, a natural turbid water standardised with Jackson Turbidimeter) in other bottles. Look through the sides of the bottles (sample and standards) at the same object (ruled paper or printed matter) and compare the distinctness with which the object can be seen. The turbidity of the sample is equal to the turbidity of the standard that most closely produces the same interference to visual perception as the sample.

Precautions

(1) As the number and size of particles changes with time, the standard suspensions must be frequently replaced.

(2) The standards should be well shaken before use.

(3) All the bottles (containing the sample and the standards) should be illuminated, well and equally, with a source of light so placed that no rays reach the eye directly.

Exercises

81. Distinguish between Jackson Turbidimetry and Nephelometric Turbidity.

82. What is meant by Coefficient of Fineness? What is its significance?

83. What is the Platinum Wire Method of Turbidity measurement? What is its 'specific' use?

84. Name the chemical added to the turbidity standards to prevent the growth of bacteria or algae.

3.6.3 *Coagulation*

Finely divided particles of colloidal dimensions, suspended in water, settle at a very slow rate and readily pass through the pores of the filter. To make sedimentation and filtration practicable, the size of the particles has to be increased. This is achieved by coagulation followed by flocculation. *Coagulation* is the process of adding chemicals to destabilise particles in a colloidal dispersion while *Flocculation* is their subsequent enmeshment, by gentle and prolonged mixing, to form discrete, visible, suspended particles that settle under gravity alone and are easily filterable.

When aluminium and iron salts are added to water, multi-positive hydroxo and polynuclear species of compounds are formed. [Model example of dinuclear species:

$$[Al(H_2O)_6]^{3+} + H_2O \rightleftharpoons [Al(H_2O)_5OH]^{2+} + (H_3O)^+$$

$$[Al(H_2O)_5\,OH]^{2+} \overset{H_2O}{\rightleftharpoons} [Al(H_2O)_4(OH)_2]^+ + (H_3O)^+$$

$$2\,[Al(H_2O)_4(OH)_2]^+ \overset{2\,(H_3O)^+}{\rightleftharpoons} (H_2O)_4\,Al\overset{\overset{H}{|}}{\underset{\underset{H}{|}}{<\!\!\!\!{}^{O}_{O}\!\!\!\!>}}Al\,(H_2O)_4]^{4+} + 4H_2O$$

The overall reaction being,

$$2\,[Al\,(H_2O)_6]^{3+} \rightleftharpoons [(H_2O)_4\,Al\overset{\overset{H}{|}}{\underset{\underset{H}{|}}{<\!\!\!\!{}^{O}_{O}\!\!\!\!>}}Al\,(H_2O)_4]^{4+} + 2(H_3O)^+, \quad (3.82)$$

the solution becomes acidic].

$[Al_8(OH)_{20}]^{4+}$ has been suggested to be the most effective species in coagulation. The coagulant species are then readily adsorbed on the surface of the suspended particles (particularly true for hydrophobic clay particles) — electrical attraction enhancing deposition. This causes decrease in the electrical charges and so the particles are destabilised and are in a position to agglomerate. Mixing or turbulence promotes collisions which result in lasting union. The progressively increasing mass of flocculent material produced by the hydrolysis of the coagulant may enmesh the turbidity particles and act like a 'sweep' as it settles.

In case of hydrophillic colloids (organic material — as in sewage) containing polar groups like hydroxyl, carboxyl or phosphate, destabilisation is largely due to chemical combination with the hydrolysis products of the coagulant.

3.6.4 *To determine the minimum dose of a coagulant required to coagulate a given sample by Jar Test and to compare the effectiveness of aluminium sulphate and ferric sulphate as coagulants for a given sample at room temperature*

Significance

A certain minimum dose of the coagulant must be added for effective coagulation. Additional doses are required for increased turbidity but the relation is not linear. While a smaller dose is required for very high turbidities with variable particle size, very low turbidities are difficult to remove. In addition, the process is influenced by a number of inter-related factors such as pH, colour, mineral content and composition, temperature, the duration and degree of agitation and the nature of the coagulant used. Therefore, the optinum dose and conditions cannot be predicted on the basis of the results of physical and chemical analysis of water and must be determined experimentally. Coagulation is widely used in lime-soda softening and in removal of colloidal silica from waters meant for domestic consumption or for many industrial uses.

Measurement of turbidity before and after the coagulation process and the amount of coagulant left (not removed) are used to control and evaluate the efficiency of coagulation. $FeCl_3$, $Al_2(SO_4)_3$ and $Fe_2(SO_4)_3$ produce acid on hydrolysis (reaction 3.82). Thus, if the sample is acidic, either alkali should be added in the form of carbonate or lime, or $NaAlO_2$ which produces alkali on hydrolysis ($NaAlO_2 + 2H_2O \rightleftharpoons Al(OH)_3 + NaOH$) should be used as coagulant.

The visible result of coagulation is the formation of a deposit in the form of porous gelatinous flakes that settle at the bottom of the vessel. If this is not observed, no coagulation has taken place.

Reagents Required

1. Aluminium sulphate solution (1.5%)
2. Ferric sulphate solution (0.5%)

Procedure

1. Take 8 beakers of 500-ml capacity each.
2. Transfer, with a measuring cylinder, 250 ml of the given water sample into each beaker.
3. Divide the beakers into two sets and mark them, with a glass marking pencil, as I_1, I_2, I_3, I_4 and II_1, II_2, II_3 and II_4.
4. Attach the first set of sample beakers to the stirring device (a laboratory flocculator).

5. Add alum solution (15g $Al_2(SO_4)_3 \cdot 18H_2O$/litre) with a graduated micro-pipet in amounts of 0.25 ml, 0.5 ml, 0.75 ml and 1.0 ml into the beakers I_1, I_2, I_3 and I_4 respectively.
6. Stir the samples rapidly for about half a minute followed by 15-minute gentle stirring at a rate of about 40 rpm.
7. Remove the beakers from the stirring device and let stand for settling the flocs.
8. Observe the flocs after every 10 minutes and record the characteristic (as slight, flocculent, considerable or heavy) in a tabular form.
9. If no coagulation takes place within an hour, repeat the experiment with dosages of 1.25 ml, 1.50 ml, 1.75 ml and 2.0 ml of alum solution.
10. Repeat the whole process taking 0.25 ml, 0.5 ml, 0.75 ml and 1.0 ml of ferric sulphate solution (5g $Fe_2(SO_4)_3$/litre), instead of alum solution, into the second set of four beakers and record the observations in the table.
11. Select the minimum dosage of the better coagulant.

Observations and Calculations

Volume of sample taken = 250 ml

Concentration of alum solution used = 15 g/l

Concentration of $Fe_2(SO_4)_3$ solution used = 5 g/l

Sl.No. of beaker	Coagulant used	Coagulant dose (ml)	Characteristic of floc	Selected minimum dose
I_1		0.25 ml		
I_2	Aluminium sulphate	0.50 ml		A ml
I_3		0.75 ml		
I_4		1.00 ml		
I_5		–		
I_6		–		
–		–		
II_1		0.25 ml		
II_2	Ferric sulphate	0.25 ml		B ml
II_3		0.75 ml		
II_4		1.00		
II_5		–		
–		–		

Quantity of aluminium sulphate required $= A\ ml/250\ ml$

$$= \frac{A \times 15}{1000}\ g/250\ ml$$

$$= 15 \times A\ mg/250\ ml$$

$$= 60\ A\ mg/l$$

Quantity of $Fe_2(SO_4)_3$ required $= B\ ml/250\ ml$

$$= \frac{B \times 5}{1000}\ g/250\ ml$$

$$= 5 \times B\ mg/250\ ml$$

$$= 20\ B\ mg/l$$

Precautions

As far as possible, the time and rate of stirring should be maintained constant for all the samples.

Exercises

85. Specify the advantages of using a coagulant.
86. What is a flocculant aid? Give one example.
87. What do you understand by 'Liquid Alum'?

3.7 Solids

The impurities present in a water sample, whether suspended or dissolved, that have a negligible vapour pressure at 105°C may be termed as 'Solids'. Solids may adversely affect the quality of water in a number of ways. While waters containing suspended solids are not at all acceptable for drinking purposes, it is desirable that dissolved solids may not be more than 500 mg/l. Higher concentrations may cause laxative or sometimes the reverse effect (constipation) upon travelling people whose bodies are not adjusted to such waters, but there appears to be no ill-effect on residents regularly using such waters.

Suspended solids are also objected to for bathing or recreational (swimming) purposes. Higher levels of suspended solids in lakes or streams inhibit the penetration of sunlight and promote conditions favourable for growth of pathogenic (disease causing) bacteria.

Both suspended and dissolved impurities cause priming and wet steaming and deposite sludge and scale in the boiler.

3.7.1. *Classification*

(A) *Total Solids*, reported in mg/l, is the amount of the residue left in the vessel after evaporation of a sample and its subsequent drying in an oven at a specified temperature. Total solids determination is the easiest test conducted on industrial wastes to have an idea of the extent of dissolved inorganic salts.

(B) *Total Suspended Solids* represent, in mg/l, portion of the 'total solids' retained by a filter, and

(C) *Total Dissolved Solids*, is the portion that passes through the filter.

The magnitude of the suspended solids and dissolved solids thus depends on factors like the type of the filter and its pore-size, the physical nature and the particle size of the suspended impurities, etc.

In drinking water, most of the matter is in the dissolved form. The amount of suspended solids increases with the extent of pollution and is maximum in sludges, where the dissolved fraction becomes less important.

(D) *Fixed Solids and Volatile Solids* : 'Fixed Solids' is the term applied to the residue of total, suspended or dissolved solids after ignition for a specified time at a specified temperature. The weight loss on ignition is called 'Volatile Solids'. If the ignition is conducted at a temperature at which

(i) the organic matter is completely converted into CO_2 and H_2O, and

(ii) there is minimum loss of inorganic substances due to decomposition and volatilization,

then the loss in weight is approximately equal to the organic content of the solid fraction of waste-water, activated sludge or industrial wastes.

The ignition is therefore usually conducted at 550° ± 50° C and the only sources of error are

(a) decomposition of thermally unstable $MgCO_3$:

$$MgCO_3 \xrightarrow{350^\circ C} MgO + CO_2 \tag{3.83}$$

and (b) volatilisation of ammonium salts not removed during drying. However, in the test on suspended solids, the inorganic dissolved salts are removed during filtration and so no error is caused. The volatile matter content of suspended solids is usually more than three-fourths whereas that of dissolved solids is very very small.

The results of the test are used to control aeration in the activated sludge process and form the basis for the design and operation of sludge digestion, vacuum-filter and incineration units in Sewage Treatment and Disposal.

(E) *Settleable Solids* is the material settling out of suspension. It depends on the method used and the time allowed for settling. It represents the suspended solids that are coarse and have specific gravity higher than that of water. The term is usually applied to sewage and indicates the amount of sludge that may be obtained in the sedimentation tank. The test is widely used in deciding the necessity of a sedimentation unit in the Sewage Disposal System. The measurement is also useful in predicting the physical behaviour of industrial and municipal wastes being drained into natural bodies of water.

3.7.2 *Determination of 'Total Solids' (dried at 103–105°C) in a water sample*

Procedure

Transfer a suitable aliquot (corresponding to about 0.1 g of dissolved matter) of the well-mixed sample to a dried (at 103–105°C for 1 hour, dry at 550 ± 50°C if volatile solids are to be measured) and preweighed nickel (or platinum) crucible and evaporate to dryness on a water bath. Dry in an electric oven at 103 to 105°C for 1 hour, cool in a desiccator and weigh. Repeat the process of drying, cooling and weighing until the weight loss is less than 4% of the previous weight.

Observations and Calculations

Let the volume of sample taken = V ml

Weight of empty crucible (dried at 103–105°C) = w_1 g

Weight of crucible + residue (dried at 103 –105°C) = w_2 g

Therefore, weight of solids in V ml of the sample = $(w_2 - w_1)$ g

and 'Total Solids' $= \dfrac{w_2 - w_1}{V} \times 10^3$ g/l

$$= \frac{w_2 - w_1}{V} \times 10^3 \times 10^3 \text{ mg/l}$$

$$= \frac{w_2 - w_1}{V} \times 10^6 \text{ ppm}$$

Precautions

(1) To keep the crucible clean, (i) it should be supported on a glass or porcelain ring with a flange on the water bath during evaporation. (ii) The outer surface of the crucible should be wiped before placing it in the oven for drying.

(2) If the determination is to be completed in a short span of time, the volume of the sample should be reduced by gentle boiling on a hot plate before transferring the crucible to water bath. However, care should be taken to avoid spattering.

3.7.3 *Determination of 'Total Suspended Solids' dried at 103–105°C*

Procedure

Mix the sample well. Filter under suction a known volume through a glass-fiber filter disc previously washed and dried at 103–105°C to constant weight (550 ± 50°C for volatile content of suspended solids). Wash with distilled water. Continue suction to completely drain the water. Remove the filter disc, place it on a flat aluminium sheet for support and dry in an oven at 103–105°C for 1 hour. Cool in a desiccator and weigh. Repeat the drying process to get constant weight.

Observations and Calculations

Let the volume of the sample filtered = V ml

Weight of the dried (103–105°C) glass filter disc = w_1 g

Weight of filter disc + suspended solids, dried at 103-105 °C = w_2 g

Therefore , weight of suspended solids in V ml of the sample = $(w_2 - w_1)$ g and 'Total Solids' $= \dfrac{w_2 - w_1}{V} \times 10^3$ g/l

$= \dfrac{w_2 - w_1}{V} \times 10^6$ ppm

3.7.4 *Determination of 'Total Dissolved Solids' dried at 180°C*

Drying at 180°C removes most of water of crystallization

[e.g., $MgSO_4 \cdot 7H_2O \longrightarrow MgSO_4 \cdot H_2O + 6H_2O$ (3.84)]

and some CO_2 due to conversion of bicarbonates into carbonates and their partial decomposition to oxides. Evaporation followed by drying at 180°C usually gives results closer to those obtained from complete analysis for mineral species than the results obtained by drying at 103–105°C.

Procedure

Transfer the filtrate[1] obtained in experiment 3.7.3 to a dried [180°C for 1 hour; dry at 550 ± 50°C for volatile content of dissolved solids] and weighed platinum crucible and evaporate to dryness on a steam bath. Dry at 180°C for 1 hour in an oven, cool in a desiccator and weigh. Repeat the drying process until there is no further loss in weight.

Note 1: If a glass fibre filter disc is not available, a Gooch crucible with a filter-mat prepared from glass fibre or asbestos fibre may be used for filtration.

Observations and Calculations

Let the volume of the sample filtered = V ml

Weight of dried (at 180°C) empty crucible = w_1 g

Weight of crucible + residue, dried at 180°C = w_2 g

Therefore, weight of dissolved solids in V ml of the sample = $(w_2 - w_1)$ g

and 'Total Dissolved Solids' $= \dfrac{w_2 - w_1}{V} \times 10^3$ g/l

$= \dfrac{w_2 - w_1}{V} \times 10^6$ ppm

3.7.5 *Determination of 'Fixed' and 'Volatile' Solids*

Procedure

Ignite the residue obtained in experiments 3.7.2, 3.7.3 or 3.7.4 in a muffle furnace at a temperature of 550 ± 50°C. Cool the crucible or the filter disc, as the case may be, in a desiccator and weigh. Finish the ignition to constant weight.

Precautions

(1) The temperature should be strictly controlled as higher temperatures lead to the decomposition of inorganic matter.

(2) Before placing the sample in the muffle furnace, destroy all inflammable material by controlled firing of the sample with bunsen flame. This will eliminate mechanical losses due to decomposition.

Observations and Calculations

Let the volume of sample taken $= V$ ml

Weight of dried filter disc or crucible $= w_1$ g

Weight of filter disc or crucible + residue (dried) $= w_2$ g

Weight of filter disc or crucible + residue, after ignition $= w_3$ g

Weight of total solids, suspended solids or dissolved solids, as the case may be $= (w_2 - w_1)$ g

Weight of 'Fixed Solids' left after ignition $= (w_3 - w_1)$ g

Weight of 'Volatile Solids' (weight loss on ignition)

$$= (w_2 - w_1) - (w_3 - w_1)$$

$$= (w_2 - w_3) \text{ g}$$

Therefore, 'Fixed Solids' (of total, suspended or dissolved matter, as the case may be) $= \dfrac{w_3 - w_1}{w_2 - w_1} \times 100\%$

'Volatile Solids' $= \dfrac{w_2 - w_3}{w_2 - w_1} \times 100\%$

3.7.6 *Determination of 'Settleable Solids' by Imhoff Cone*

Procedure

Mix the sample well and pour it into Imhoff cone up to 1-litre mark. Leave it undisturbed for 45 minutes; gently stir the sides of the cone with a glass rod and let the material settle for another 15 minutes. Record the volume of the material that settles and report the result in ml per litre.

Precaution

The volume of any liquid pockets present between large particles should be estimated and substracted from the volume of settled solids. Any floating material completely separated from the settled material should not be included in the settleable solids.

Observations and Result

Volume of the sample taken = 1 litre

Volume of solids settled $= V$ ml

'Settleable Solids' $= V$ ml/litre.

4

LUBRICATING OILS, GREASES AND EMULSIONS

The resistance to motion that operates when one solid surface is moved tangentially with respect to another solid surface, with which it is in contact, is known as Frictional Resistance, and results in considerable loss of energy and damage to the contacting surfaces.

But friction is not all that useless. An innumerable number of processes of everyday life—lighting a match stick, walking, starting or stopping of a vehicle, gripping objects by hand, and what not—are dependent for their effectiveness on the presence of friction in large enough amounts. When friction is less, slippery conditions exist such as walking on sand, climbing up an oiled staff, etc.

In addition to the Van der Waals forces of attraction (molecular forces of cohesion) operating between the molecules of the contacting surfaces, resistance to motion is caused by interlocking of minute projections—asperities or peaks, may be of the order of 1000 Å (quite large when considered on molecular scale) — existing on the surfaces of even the smoothest objects. As one body moves past the other, some of the peaks are broken off, leading to wear and tear of the surface. Friction generates heat (highly localised at the points of contact between the rubbing surfaces) and coupled with high pressure developed even under small loads (as the actual points of contact are very small as compared to the apparent contact area between the objects) causes fusion of the material at the peaks and accounts for the formation of welded junctions. If the relative motion of the bodies is to be maintained, additional force is required to break these welded junctions which in turn generates more heat. The generation of heat is therefore a self–accelerating process and may eventually lead to large scale seizure and bring the moving bodies to a grinding halt.

The losses due to frictional resistance form a substantial part of the total energy consumption of mankind. In order to conserve the depleting energy reserves, it is essential that the frictional resistance be brought down to the optimum. Friction can be reduced by

(i) decreasing the roughness of the surfaces involved,

(ii) proper design of the moving parts,

(iii) using surfaces of low coefficient of friction, or

(iv) interposition of a substance of low shear strength between the moving surfaces. The substance that is used for this purpose is called a *Lubricant* and the process of reducing friction by use of a lubricant is known as *Lubrication.*

In addition to reducing energy losses and wear and tear of the moving surfaces, a lubricant is usually called upon to perform one or more of the following inter–related functions:

(a) To act as a heat dissipating medium (coolant)

(b) To form a seal (to prevent leakage, to keep out dirt)

(c) To act as a medium for power transmittal (hydraulic)

(d) To control corrosion (prevent rust)

(e) To dampen shock (dissipation of mechanical energy through fluid friction— dash pots, gears, etc.)

(f) To act as detergent or remove contaminants (flushing action to remove sludges from IC Engines)

(g) To act as electric insulator (transformers, switch gears).

Depending upon the operating conditions and the lubricant characteristics, a lubricant may form a fluid film (~1000 Å thick) in between the moving surfaces thereby removing all points of contact between them (Hydrodynamic or Fluid Film Lubrication). Solid friction is completely substituted by fluid friction and the coefficient of friction (0.001–0.01) is governed by viscosity of the fluid.

Under conditions of high load, low speed, or when the lubricant has a low viscosity, it may be squeezed out of position. Lubrication is then maintained by a very thin film (only a few molecules thick) of the lubricant adsorbed on the surface (Boundary Lubrication). The moving surfaces are not completely separated but their points of contact are considerably decreased. The coefficient of friction depends on the sticking characteristics (oiliness) of the lubricant and is much higher (0.05–0.15) than in the case of hydrodynamic lubrication.

When the pressure and temperature are enormously high, even the boundary film may break. Lubrication under such conditions is maintained by Extreme Pressure Additives — certain substances added to the lubricant which, under the conditions of extreme pressure and temperature, react with the solid surface in question producing compounds having low shear strength, thus reducing the friction.

A wide variety of materials—air, liquids (animal and vegetable oils, mineral oils, synthetic oils), emulsions (oil-in-water and water-in-oil), suspensions (aquadag and oildag), semi-solids (greases) and solids (graphite, molybdenum disulphide, etc.) — have been employed for the lubrication purposes.

4.1 Oils

'Fatty' oils of vegetable and animal origin such as olive oil, palm oil, rapeseed oil, linseed oil, cottonseed oil, castor oil, tallow oil, lard oil, sperm oil, seal and whale oils were the first to be used. They are usually the mixtures of mixed glycerides of saturated and unsaturated long chain acids and can be represented by

$$\begin{array}{l} CH_2OOC{\cdot}R\check{z} \\ | \\ CHOOC{\cdot}R' \\ | \\ CH_2OOC{\cdot}R'' \end{array}$$

These oils are called 'Fixed' oils as they cannot be distilled without decomposition. They have good oiliness but

(a) are costly

(b) easily undergo oxidation, get polymerised and thickened, and

(c) have a tendency to get hydrolysed in presence of water or moisture.

So they are now used only to a limited extent in the manufacture of greases and as additives to improve the oiliness of other lubricants. They have largely been replaced with cheap, more easily and abundantly available and more stable mineral oils obtained by refining of petroleum. Chemically, mineral lubricating oils are mixtures of hydrocarbons ($C_{12} - C_{50}$), and may contain only a small amount of oxygen, present as impurity, while this element is the main constituent of fatty oils.

Selection of a lubricant for a particular job is a highly professional task and depends on such a wide variety of factors as the design of machine, the operating conditions (temperature, pressure, speed of movement, the duration for which the machine is to be used continuously, the nature of the environment), mode of application of the lubricant and its physical and chemical characteristics. It is now universally realised that the physical and chemical properties such as viscosity, specific gravity, flash point, volatility, acid value, saponification value, iodine value, etc., as measured in the laboratory, do not indicate exactly how a lubricant will behave under the operating conditions. These tests are therefore mainly used in the classification of and specifications for the type of lubricant required. They are also helpful in comparing the composition and evaluating the potential of the commercially available products. For a reasonably good prediction, the properties measured in the laboratory should be correlated to the service conditions expected to be met with in actual use. It is therefore desirable that variation of these characteristics with temperature, load, etc., and the extent and nature of possible contaminants should be studied. The suitability has finally to be established under the working conditions.

Usefulness of tests increases if test results can be compared directly with those obtained by others. This will be possible if the tests are performed in accordance with Universally accepted common specifications. The majority of the tests have been standardised by British Institute of Petroleum (IP) and by American Society for Testing and Materials (ASTM). A few of these tests are outlined below.

4.2 Viscosity and Viscosity Index

Significance

Viscosity is a measure of flowability at definite temperatures. The flow properties of oils influence the rate of production of an oil-well, the transport of crude oil and refined products in pipelines and the performance of an oil as a lubricant in a machine.

Viscosity is the single most important property of the lubricating oils which determines their performance under the operating conditions. A lubricating oil should have sufficient viscosity to enable it to stay in position. On machine parts moving at slow speeds under high pressures, a heavy oil (high viscosity) should be used as it better resists being squeezed out from between the rubbing parts. Light oils (low viscosity) can be used, however, when lower pressures and higher speeds are encountered (since high speed permits a good oil wedge to form) and should be preferred as they do not impose as much drag on high speed parts as heavy oils do. In fact, under hydrodynamic lubrication, solid friction is completely substituted by fluid friction and so the frictional resistance encountered depends directly on the viscosity of the oil. Therefore, for minimum friction, the thinnest (least viscous) oil that will stay in position should be used.

Change of viscosity with temperature: Viscosity Index

With rise in temperature, forces of cohesion between the molecules of a fluid are weakened, resulting in a decrease in viscosity. When the same lubricant has to function satisfactorily at widely varying temperatures (as is encountered in hydraulic systems, crankcases of internal combustion engines, automatic transmissions and gear cases and portable air compressors), the variation of its viscosity with temperature must be negligibly small. Otherwise, the lubricant may become very thin at higher temperatures (usually above 200°C during take off or landing of an aircraft) and may be squeezed out of position; or it may become highly viscous at very low temperatures (at times about – 50°C, when the oil is pumped into aircraft engines in colder regions of the world) and may even cease to flow.

The variation of viscosity with temperature is either indicated by Viscosity–Temperature Curves (V–T Curves) or measured on an arbitrary scale known as *Viscosity Index (V.I.).*

V.I. represents the average decrease in viscosity of an oil per degree rise in temperature between 100°F and 210°F. The viscosity (at 100°F) of the oil whose V.I. is to be calculated is compared with that of two standard oils, having the same viscosity at 210°F as the oil under test. One of these reference oils is chosen from a standard set made from Pennsylvania crude (consisting mainly of paraffins and showing a relatively small decrease in viscosity with rise in temperature) and arbitrarily assigned a V.I. of 100; the other is chosen from a standard set made from Gulf–Coast crude (consisting mainly of alicyclic hydrocarbons or naphthenes and exhibiting a relatively large variation in viscosity with temperature) and arbitrarily assigned a V.I. of 0 (zero). Mathematically,

$$\text{Viscosity Index} = \frac{L - U}{L - H} \times 100 \qquad (4.1)$$

where U = viscosity at 100°F of the oil whose V.I. is to be determined

L = viscosity at 100°F of an oil of zero V.I. having the same viscosity at 210°F as the oil whose V.I. is to be determined

and H = viscosity at 100°F of an oil of 100 V.I. having the same viscosity at 210°F as the oil whose V.I. is to be calculated.

An oil whose viscosity changes rapidly with change in temperature has a low V.I. (the slope of V–T curve is high) while the one whose viscosity changes only slightly has a high V.I. (the V–T curve is flatter). Addition of linear polymers increases V.I. and oils with V.I. higher than 100 have been prepared. Methods to calculate values of V.I. higher than 100 are also available.

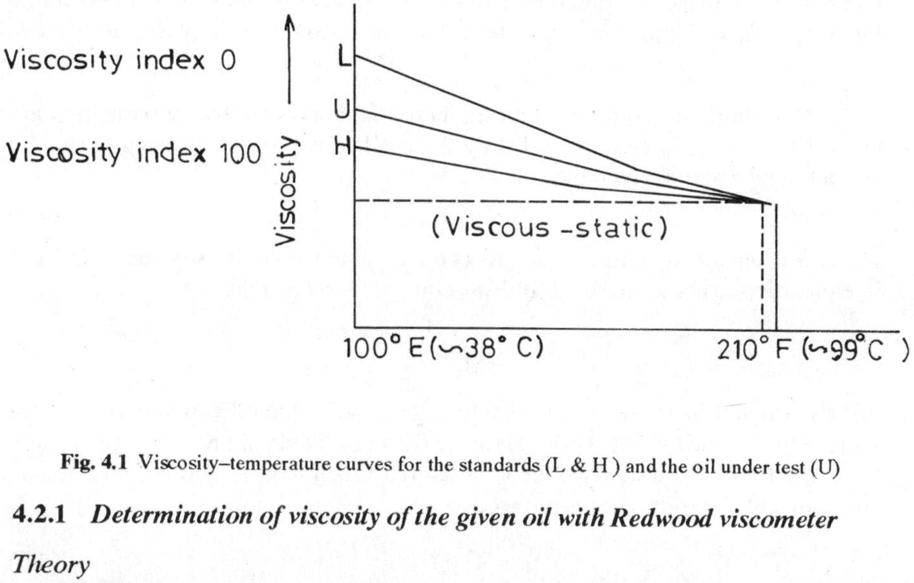

Fig. 4.1 Viscosity–temperature curves for the standards (L & H) and the oil under test (U)

4.2.1 *Determination of viscosity of the given oil with Redwood viscometer*

Theory

The Absolute or Dynamic viscosity is the tangential force per unit area required to move one horizontal plane of the fluid at unit velocity with respect to another maintained at a unit distance apart by the fluid. When the force is 1 dyne/cm^2, the distance between the layers or planes 1 cm and the velocity gradient 1 cm/sec, the viscosity is 1 poise. This being a large unit, absolute viscosity is more commonly expressed in centipoise $\left(= \frac{1}{100} \text{ poise}\right)$. The viscosity, also known as coefficient of viscosity, η, of water at 20°C is 1.002 centipoise.

When a liquid is made to flow through a capillary tube by a pressure gradient, and the time for a given volume to flow along the tube is measured, then at a constant temperature, the coefficient of viscosity, η, is given by Poiseuille's equation

$$\eta = \frac{\pi P r^4 t}{8 V L} \tag{4.2}$$

where P = pressure difference between the ends of the tube

L = length of the tube

r = radius of the tube

t = time in seconds

V = volume of the liquid flowing through the tube in t seconds.

When the 'Inlet correction' and the 'Kinetic Energy Correction' are incorporated, the above equation takes the form

$$\eta = \frac{\pi P r^4 t}{8 V (L + k_1 r)} - \frac{V \rho k_2}{8 \pi (L + k_1 r) t} \tag{4.3}$$

where ρ is the density of the liquid and k_1 and k_2 are constants which depend on the form of the apparatus and are determined experimentally ($k_1 = 0 - 1.6$, $k_2 = 1 - 1.5$).

For accurate measurements on mineral oils, glass capillary viscometers are used. The driving force is provided by the head of the test oil flowing vertically down the tube and is given by

$$P = h \rho g, \tag{4.4}$$

where h is the mean height of the oil. The tube dimensions are so chosen that $k_1 r$ in equation (4.3) becomes negligible as compared to L, so that

$$\eta = \frac{\pi h \rho g r^4 t}{8 V L} - \frac{V \rho k_2}{8 \pi L t} \tag{4.5}$$

For a particular instrument (Viscometer), the head of the oil is maintained at the same height h, and the time t in secs. for the flow of a fixed volume of the oil through the capillary of fixed dimensions is measured. Then h, g, r, V and L all become constant and equation (4.5) is reduced to

$$\eta = C \rho t - \frac{B \rho}{t} \tag{4.6}$$

Dividing both sides by ρ, we get

$$\frac{\eta}{\rho} = C t - \frac{B}{t} \tag{4.7}$$

$$\cong C t \tag{4.8}$$

when t is very large. The time t in seconds measured experimentally is reported as Relative Viscosity and its magnitude depends on the viscometer used and the volume of the liquid that flows through the tube of the viscometer. Therefore, while reporting the relative viscosities of oils, the viscometer used must be specified.

The quantity η/ρ, the ratio of absolute viscosity to density of the liquid is known as Kinematic Viscosity. It is denoted by ν and is expressed in centistokes when η is expressed in centipoise. The viscometers are usually engraved with the constants B and C such that when t is in seconds, ν is in centistokes. The kinematic viscosities are thus independent of the viscometer used.

If measurements are made on two different liquids using the same viscometer, from equation (4.8) we get

$$\frac{\eta_1}{\rho_1} = C t_1 \tag{4.9}$$

and

$$\frac{\eta_2}{\rho_2} = C t_2 \tag{4.10}$$

where η_1, ρ_1 and t_1 are the absolute viscosity, density and the time of flow of a fixed volume, respectively, for the first liquid and η_2, ρ_2 and t_2 the corresponding values for the second liquid. Dividing equation (4.9) with (4.10) and rearranging, we get

$$\eta_1 = \eta_2 \frac{\rho_1 t_1}{\rho_2 t_2} \tag{4.11}$$

Thus knowing η_2, the absolute viscosity of one liquid, ρ_1 and ρ_2, the densities of the two liquids at the temperature of the experiment and measuring t_1 and t_2, the absolute viscosity of the other liquid can be calculated.

Redwood Viscometers

Redwood viscometers are available in two sizes — No. 1 (Universal) and No. 2 (Admirality). The two viscometers, RW1 and RW2, are identical in principle, shape and method of testing a sample. The difference lies in the dimensions of the discharge capillary (tube/jet/orifice):

Viscometer	*Dimensions of Jet*	
	Diameter	Length
RW1	1.62 mm	10 mm
RW2	3.8 mm	50 mm

The rate of discharge of the oil through RW2 is nearly 10 times the discharge through RW1 and so the RW2 receiving–flask is designed with a wider mouth.

RW1 is commonly used for light or thin oils (e.g., kerosene, mustard oil, etc.). For highly viscous liquids (e.g., fuel oil, mobile oil, glycerol, etc.), when the flow time with RW1 may exceed 2,000 seconds, RW2 is used.

Procedure

Level the viscometer with the help of levelling screws. Fill the outer bath with water and connect to the electric mains. Clean the oil cup and the discharge jet with xylol followed by passing a small amount of the test oil through the jet, using a plunger. Place the ball valve on the agate jet to close it and pour the test oil into the cup to such a level that the metal indicator (pointer) fixed on the wall of the oil cup just dips in the oil. Insert a thermometer and a stirrer and cover with lid. Keep stirring the water in the bath and the oil in the cup and adjust the bath temperature until the oil attains the desired constant temperature. Place a clean and dry Kohlrausch flask (marked to 50-ml capacity) immediately below and directly in line with the discharge jet (Figure 4.2). Remove the ball valve with one hand, simultaneously starting the stop watch with the other. Oil from the jet flows into the flask. Stop the timer when the lower meniscus of the oil reaches the 50-ml mark on the neck of the receiving flask. Record the time elapsed in seconds. Repeat the experiment to take a number of readings and report the mean value in seconds (Relative viscosity) mentioning the viscometer used and the test temperature.

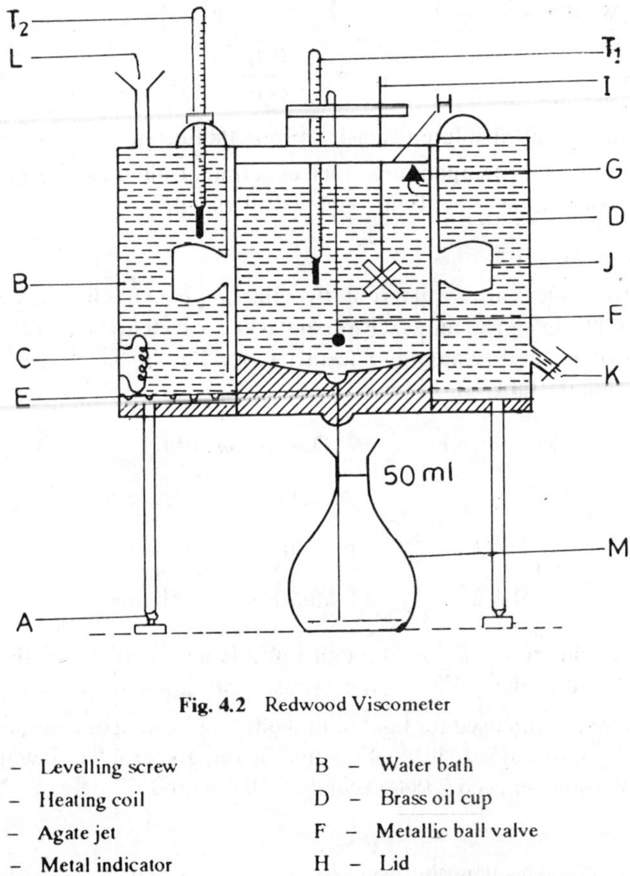

Fig. 4.2 Redwood Viscometer

A	– Levelling screw	B	– Water bath
C	– Heating coil	D	– Brass oil cup
E	– Agate jet	F	– Metallic ball valve
G	– Metal indicator	H	– Lid
I	– Oil stirrer	J	– Water bath stirrer blade
K	– Water outlet	L	– Water inlet
M	– Kohlrausch flask	T_1	– Test Thermometer
T_2	– Bath thermometer		

Precautions

(1) Before testing, the oil should be filtered through muslin cloth or a 100-mesh wire-strainer to remove solid particles that may otherwise clog the jet. Moisture also clogs the jet and if present, the sample should be filtered through lintless filter paper.

(2) The receiving flask should be placed in such a way that the oil stream from the orifice (jet) strikes the neck of the receiving flask and does not cause foaming.

(3) To prevent the oil from overflowing, the jet valve should be closed immediately after stopping the timer.

(4) After each reading, oil should be completely drained out of the receiving flask. The flask should preferably be washed with xylol and dried before repeating the test.

(5) The test may be expedited by preheating the oil sample, before straining, in an aluminium cup (to not more than 3°F higher than the temperature of the test.)

Exercises

88. What is meant by 'Inlet Correction'?
89. What is the minimum Efflux Time (Flow time or Discharge time) recommended for (Redwood) Viscometers?
90. The viscosity of an unknown oil was measured with Saybolt Universal Viscometer (S.U.V.) at temperatures 100°F and 210°F and was found to be 600 seconds and 60 seconds, respectively. Calculate the Viscosity Index (V.I.) of the oil using the following table:

	Name of the oil	V.I.	S.U.V. seconds	
			at 100°F	at 210°F
1.	Pennsylvanian oil	100	420	60
2.	Gulf oil	0	780	60

91. How does the viscosity of a liquid vary with rise in temperature?
92. What is meant by 'all weather lubricants'?
93. Give some examples of lubricants that have high Viscosity Index.
94. How is the Viscosity Index of a lubricating oil improved?
95. What is meant by the term 'Viscous–Static'?
96. How can a viscous–static lubricant be prepared?

4.3 Cloud and Pour Points

Lubricating oils derived from petroleum usually contain dissolved paraffin wax and other asphaltic or resinous impurities, their amounts depending on the efficiency of dewaxing and refining process used. These impurities tend to separate out of the oil at lower temperatures. When a petroleum oil is chilled under specified conditions, the temperature at which paraffin wax or other solidifiable materials, normally dissolved in oil, begin to separate out from solution in the form of minute crystals, causing the oil to become less transparent, cloudy or hazy in appearance, is known as the Cloud Point of the oil.

If the cooling is continued further, the amount of the separating material increases and a stage is reached when the oil solidifies and stops flowing. The lowest temperature at which an oil will flow or pour under prescribed conditions, when it is cooled undisturbed at a fixed rate, is called its Pour Point.

The cloud point determination is limited only to transparent oils; otherwise there may be slight variation in results due to the human error involved, since the decrease in transparency is to be visually observed.

Significance

Cloud point of a lubricant to be applied by a capillary feed system or wicking arrangement must be low so that the oil flow does not stop due to deposition of crystals of wax in the capillary or wick interstices. Cloud point is helpful in identifying the temperatures at which wax separation may clog the filter screens in the fuel intake system of diesel engines. Oils of naphthenic type, which are almost wax-free, show very low cloud points and this fact may be useful in identifying the source of the oil.

Pour Point is more important since in the lubrication of any machine subjected to low temperature, the lubricant must flow freely, especially during the start-up period. A high pour point may lead to solidification of the lubricant that may cause jamming of the machine. Pour point also establishes the lowest temperature at which an oil can be transferred by pouring or below which, because of extremely poor mobility of the oil, lubrication by gravity flow process is less reliable.

4.3.1 *Determination of cloud point of an oil*

Bring the oil sample to be tested to a temperature at least 25°F above the expected cloud point. If the sample contains moisture, dry it by shaking with a little anhydrous sodium sulphate followed by filtration through dry lintless filter paper. Pour the clear oil into the test jar upto the etched mark (or into a flat-bottom glass tube, about 3 cm in diameter and 12 cm high, to a depth of about 5 cm). Tightly close the jar with a cork carrying a thermometer with its bulb touching the bottom of the jar (Figure 4.3). Insert the test jar inside a holding jacket (made of glass or copper) which is immersed in a freezing mixture suitable for obtaining the desired temperature. After every 2°F fall in temperature of the oil, take out the test jar from the jacket, inspect for cloudiness and immediately replace it in the jacket. Record the temperature at which such inspection first reveals a distinct cloudiness in the oil near the bottom of the jar and report it as the cloud point.

4.3.2 *Determination of pour point of an oil*

Proceed as in the cloud point test with the difference that the thermometer bulb is just completely immersed in the oil, instead of touching the bottom (Figure 4.3).

Starting at a temperature about 20°F above the expected pour point, take out the jar after every 5°F fall in temperature and tilt it just enough to see any movement of the oil. Immediately replace the jar in the jacket. At a point where the oil in the jar shows no movement on tilting, hold the jar in a horizontal position for 5 seconds. If there is some movement of the oil, replace the jar immediately in the jacket and repeat the test for flow after further 5°F fall in temperature. Continue the test until no movement of the oil is observed when the test jar is held in a horizontal position for exactly 5 seconds. Record the reading on the test thermometer as Solid Point. Add 5° to this temperature to get the pour point.

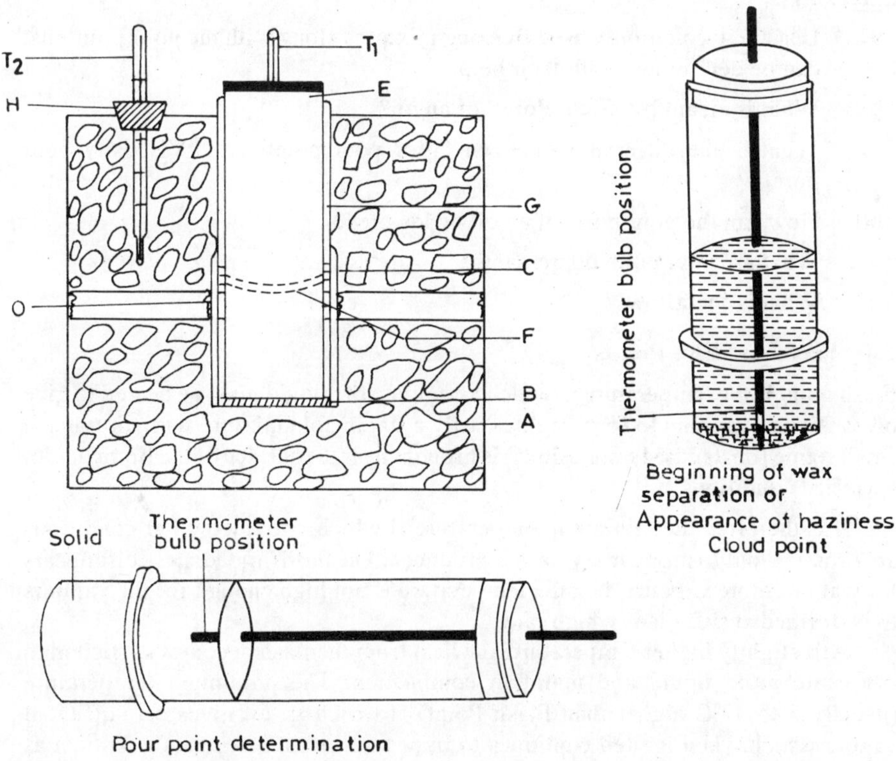

Fig. 4.3 Cloud point and pour point apparatus

A –	Freezing mixture	E –	Test jar
B –	Cork disc	F –	Ring gasket
C –	Glass or Copper jacket	G –	Air gap
D –	Support for holding the jacket	H –	Cork
T_1 –	Test thermometer	T_2 –	Bath thermometer

Precautions

(1) The test jar should not touch the jacket. This is achieved by placing a cork disc at the bottom of the jacket and using a ring gasket around the test jar.

(2) The complete operation of the removal and replacement of the test jar should be completed within 3 seconds.

(3) When the pour point is very low, a number of freezing mixtures with decreasing lower temperature should be used.

(4) When the separation of wax crystals starts, great care should be taken not to disturb the mass of the oil. Even the thermometer should not move in the oil. Any disturbance will delay solidification and so lower results will be obtained.

Exercises

97. List a few commonly used freezing mixtures along with the pour point that can be determined with their help.

98. What is meant by 'Solid Point' of an oil?

99. What is the difference between 'Wax pour point' and 'Viscosity pour point'?

100. How can the pour point of an oil be lowered?

101. What are pour point depressants?

102. What is 'paraflow'?

4.4 Flash and Fire Points

Flash point is the temperature to which a combustible liquid must be heated to give off sufficient vapours to form momentarily a flammable mixture with air when a small flame (of standard dimensions) is brought near the surface of the liquid under specified conditions.

It is therefore the minimum temperature at which, provided other conditions are satisfied, only a momentary flash is produced. The flash immediately disappears for want of more vapours, i.e., the temperature is not high enough for the vapours to be formed at sufficiently high rate.

At a slightly higher temperature, the heat from the flash becomes sufficient to evaporate more liquid and maintain combustion. This minimum temperature (usually 5 to 40°C higher than Flash Point) at which an oil gives off sufficient vapour which when ignited continues to burn for at least 5 seconds is known as FIRE POINT of the oil.

A fire may develop if a set of the following three conditions (known as FIRE TRIANGLE) is simultaneously satisfied:

(a) A source of ignition (Ignition sources are abundant—anything from static electricity, flame, an arc, a single spark, a cigarette, a match to ash from a pipe, and many more).

(b) Oxygen (abundantly present in air).

(c) Combustible vapours within the Explosive Range—a combustible vapour does not burn even when ignited in presence of air (O_2) if its concentration in the vapour–air mixture is (i) less than this range ('too lean') or (ii) more than this range ('too rich').

[This latter fact is utilised in making hydrocarbon mixtures less flammable (for use as Industrial Cleaners) by the addition of (i) a non-combustible liquid such as CCl_4 (not widely used because of high toxicity), or

(ii) less-combustible chlorinated substances such as trichloroethylene and methylchloroform, as they bring down the concentration of the combustible vapour below the explosive range and make them 'too lean' to ignite.]

Flammable liquids are volatile and speed of evaporation (Evaporation Rate) increases appreciably when the liquid is heated. The vapours being heavier than air, go on accumulating at the lowest level, such as pits under the tanks, where their concentration reaches within the explosive range and explode and burn in the presence of an ignition source. This can be prevented by mechanical ventilation of any pits at the bottom of oil tanks—this increases the rate of diffusion of vapours and the explosive range is not approached.

Significance

Though unrelated to the lubricating property of the oils, a knowledge of Flash and Fire Point is helpful in providing safeguards against fire hazards during their storage, transportation, handling and use. Their practical importance for transformer oils is obvious. They are also of immense importance for illuminating oils—to ensure safety, flash point of illuminating oils (e.g., kerosene) should be reasonably above the average maximum atmospheric temperature of a country.

Although flash point and fire point are not sufficient as the sole indices of a fire hazard, they have been used for comparative ratings of the fire hazard potential of different flammable liquids and for labelling their containers accordingly—liquids having flash points less than 140°F are called FLAMMABLE LIQUIDS and those with flash points above 140°F are called COMBUSTIBLE LIQUIDS.

Fire-insurance companies are concerned about the flash points of the oils stocked by their Policy Holders. Also, Railways are normally concerned about the flash points of the oils they carry.

The volatility of a liquid markedly increases when it is heated to or above its flash point. So a low flash point indicates appreciable evaporation losses and possibility of formation of gas-locks in fuel pipes of SPARK Ignition Engines.

Flash and Fire points have also been used to detect solvent contamination and to determine the approximate extent of dilution of the Lubricating Oils.

Flash point is higher for the oils of paraffin base than those of naphthenic base. This test is therefore a rough guide as to the base of an oil.

The test is also useful to refiners in controlling the manufacturing process.

4.4.1 *Determination of Flash Point by Abel's Flash Point Apparatus*

The cylindrical oil cup (made of brass) and the lid (also made of brass and provided with a paddle stirrer, an opening for a thermometer, an arrangement for applying a small test flame and three small openings—one for the application of test flame and the other two for the entry of air into the oil–covered by a sliding shutter) are thoroughly cleaned and dried. The cup is filled with the oil sample to a level such that the tip of the metallic pointer fixed on the side of the cup just dips in the oil. The lid is tightly fixed on the cup, the thermometer is inserted and the shutter is closed. The cup is now supported by its flange over a copper air–jacket which is enclosed by cylindrical vessel made of copper (Figure 4.4). Through a funnel, the outer vessel is filled with warm water at 130°F (for oils having flash point up to 90°F) or with cold water (for oils flashing from 90°F to 120°F) which is heated electrically. The test flame is lighted and adjusted to the size of a white bead mounted on the cover. The sample is heated at a rate of 5°F per minute and the stirring paddle is turned at a speed of approximately one revolution per second. When the temperature of the oil reaches within 15°F of the probable flash point, first application of the test flame is made by pulling the sliding shutter outwards when the test flame drips into the central opening in the lid and comes in contact with the ascending vapour–air mixture. Subsequently, the test flame is applied at

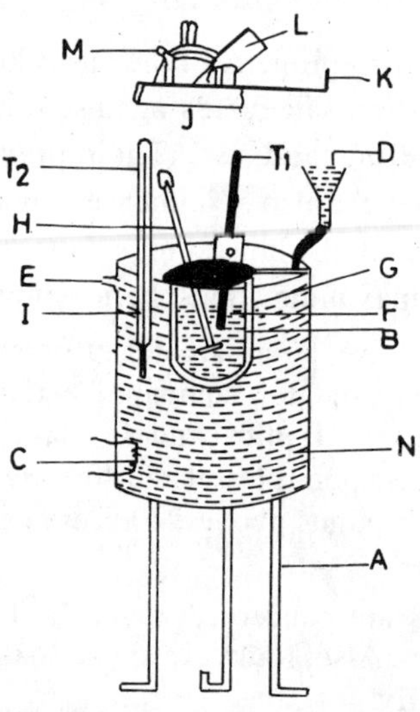

Fig. 4.4 Abel's flash point apparatus

A –	Tripod stand	B –	Copper air jacket
C –	Heating coil	D –	Water inlet
E –	Water outlet	F –	Air gap
G –	Brass oil cup (with flange)	H –	Stirrer
I –	Pointer	J –	Lid
K –	Sliding shutter	L –	Test flame
M –	White bead	N –	Water bath
T_1 –	Test thermometer	T_2 –	Bath thermometer

every 2°F rise of temperature. When the application of the test flame first produces a distinct blue flash in the interior of the oil cup, the temperature on the test thermometer is recorded (let it be t_1°F) and heating is discontinued. The temperature of the oil continues to rise for sometime. The oil is then allowed to cool down (if the cooling rate is too slow, some cold water may be added to the bath through the funnel). When the temperature comes down to within 10°F of t_1, test flame is again applied at every 2°F fall of temperature. The lowest temperature at which flash is produced is recorded. Let it be t_2°F.

$$\text{The flash point of the oil sample is given by} = \frac{(t_1 + t_2)}{2} \text{ °F} \tag{4.12}$$

Precautions

(1) As moisture affects the flash point, all parts of the cup and its accessories should be dried before placing oil in the cup.

(2) No oil should remain between the sliding and fixed plates forming the cover of the cup. If necessary, these should be separated and cleaned. Care should also be taken to prevent wetting of the cup above the pointer tip.

(3) With very low flashing oils, the sample (sometimes the oil cup itself) may be cooled in melting ice before filling.

(4) Always a fresh portion of the oil sample should be used. A second determination on the same portion of the oil shows a higher flash point.

(5) The thermometer bulb should dip into the oil.

(6) For applying the test flame, the slide should be drawn open slowly and closed quickly.

(7) Stirring should be discontinued during the application of the test flame.

Exercises

103. What is the purpose of the air jacket surrounding the oil cup?
104. What is Spontaneous Ignition Temperature (S.I.T.) of a liquid?
105. What are the factors that affect the flash and fire points of oils?
106. How does the presence of water affect the flash point of an oil?
107. How is free water removed from an oil?
108. What is meant by a 'Freaky' flash?
109. Why fatty oils do not have sharp or distinct flash points?
110. What is meant by Flash Point 'Closed' and Flash Point 'Open'? Which one is reproducible and why?
111. Describe the main difference between the Abel's closed-cup and Pensky–Marten's closed-cup methods for flash point determination.

4.5 Aniline Point

Since like dissolve like, aniline is readily soluble in those lubricants which are rich in aromatic and naphthenic compounds (all containing ring structures). In lubricants richer in paraffins, dissolution takes place at higher temperatures. The tendency of a lubricant to mix with aniline is expressed in terms of Aniline Point of the sample.

Aniline Point (also known as Standard Aniline Point), with respect to petroleum oils, is the lowest temperature at which the oil is completely miscible with an equal volume of freshly distilled aniline. Alternatively, Aniline Point is the minimum equilibrium solution temperature for equal volumes of aniline and the lubricant sample.

Significance

Aniline point of any lubricant is a measure of its aromatic content. A lubricant with a low aniline point (therefore having high aromatic content) will tend to attack (solvate and swell) the rubber seals, used in the system to prevent leakage. The sensitivity of a rubber to a lubricating oil depends on the characteristics of the rubber formulation (composition), and for a rubber seal of fixed composition, the severity of attack increases with decrease in Aniline Point.

The best way to ascertain the probable action of a lubricating oil on rubber is, of course, to immerse the particular rubber in the oil and observe any swelling or softening; but this is a time consuming exercise and may involve weeks or months. The aniline point of an oil, which can be experimentally measured within hours, therefore, can be used as advance information as to the possible deterioration of rubber sealing, gasket and packing material in the presence of the oil.

A lubrication of high aniline point is recommended for systems in which rubber seals are being used. Also the lubricants of almost same aniline points should be used in a system, as any variation may change the volume of rubber seals and leakage might take place.

Mixed Aniline Point

There may be certain lubricants (with very high aromatic content) which when mixed with equal volume of aniline may remain completely miscible and separation into different phases may not be observed even at the time of solidification. For determining such low aniline points, 1 volume of sample is mixed with 2 volumes of aniline and 1 volume of a suitable diluent (*n*-hexane or *n*-heptane). Addition of the diluent lowers the miscibility of aniline with the sample and so with decrease in temperature, separation of phases can be easily observed. The equilibrium solution temperature observed under these conditions is known as Mixed Aniline Point, which can be used in the same way as the Standard Aniline Point.

4.5.1 *Determination of Aniline Point of an oil*

Equal volumes of the sample and aniline are heated to effect complete dissolution and then cooled under controlled conditions. The temperature at which the two phases separate, as indicated by the sudden appearance of cloudiness throughout the medium, is reported as aniline point of the sample.

Procedure

The whole apparatus is cleaned and dried at 100–110°C. 5–10 ml of pure aniline (dried over KOH pellets, filtered and distilled afresh) and an exactly equal volume of the sample (dried by shaking with anhydrous Na_2SO_4 and filtered) are transferred to a test tube (2.5 × 15 cm) made of heat resistant glass. The tube is fitted with a cork, holding an electrically operated wire or glass rod stirrer and a thermometer (Aniline Point Thermometer of appropriate range) with its bulb about 5 mm above the bottom of the tube. The tube is inserted into an outer air jacket (4 × 17.5 cm) also made of heat resistant glass (Figure 4.5). The aniline and sample mixture is stirred to get a homogeneous solution. If miscibility is not complete at room temperature, the jacket (holding the tube) is immersed in a hot bath. Stirring is continued and the temperature of the bath is raised until the solution is complete. The jacket is then withdrawn from the hot bath and, while stirring gently, the temperature is allowed to fall at a rate below 1°C per minute (a cold bath may be used, if necessary). The temperature at which the outline of the thermometer bulb

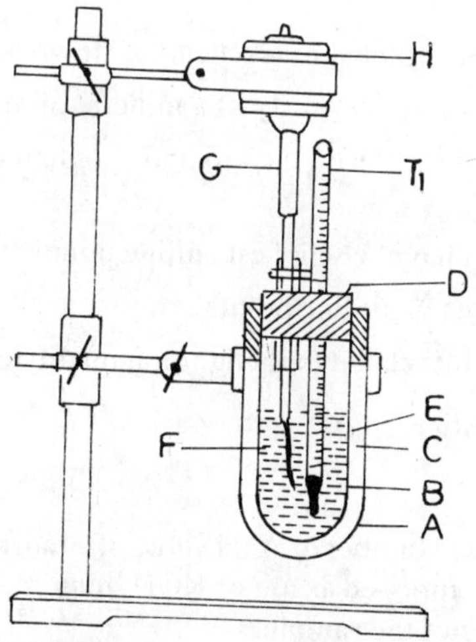

Fig. 4.5 Aniline point apparatus

A – Outer air jacket
B – Test tube
C – Air gap
D – Corks
E – Equal volumes of oil and aniline
F – Glass stirrer with auger tip
G – Rubber tubing
H – Variable speed motor
T_1 – Aniline point thermometer

is obscured, due to cloudiness or haziness in the solution, is reported as the aniline point of the sample. This is 1–2°C below the temperature at which the turbidity is first observed.

Precautions

(1) The whole apparatus and all the reagents must be perfectly dry as the presence of even traces of moisture gives erroneous high results.

(2) Aniline being hygroscopic, water should not be used even in hot and cold baths. Instead a nonaqueous, non-volatile, transparent liquid should be used.

(3) In case the expected aniline point is below the dew point of the atmosphere, the space above the aniline-sample mixture in the tube should be filled with dry N_2.

(4) Aniline being highly toxic, it should not be sucked into the pipet by mouth. A pipet provided with a rubber suction bulb or an aspirator should be used.

(5) Stirring should be done at such a rate as to avoid any splashing of the liquid or formation of air bubbles.

Exercises

112. What is the purpose of the air jacket enclosing the test tube?

113. What is the effect of the viscosity of a mineral oil on its action on rubber?

114. What is the relation between the aromatic content of a lubricating oil and its aniline point?

115. Which type of oils have the highest aniline points?

116. What are the ranges of aniline point thermometers?

117. How is aniline point related to the ignition quality of a diesel fuel?

4.6 Neutralization Number

Definitions

1. Total Acidity, Acid Number or Acid value of a lubricating oil is the amount of titrating base, expressed as mg of KOH, required to neutralize all acidic constituents of 1 g of the sample.
2. [Total] Basicity, Base Value (Alkali Value) or Base Number (Alkali Number) of a lubricating oil is the amount of titrating acid, expressed as mg of equivalent KOH, required to neutralize all the basic constituents of one gram of the sample.
3. Inorganic Acidity, Strong Acid Number or Strong Acid Value represents the mg of KOH used to neutralize the mineral acid content of 1 g of the sample. The difference of Total Acidity and Inorganic Acidity is equal to Organic Acidity.
4. Strong Base Number or Value is the amount of acid, expressed as mg of equivalent KOH, used to neutralize the strong basic constituents of 1 g of the sample.

The terms Neutralization Number or Neutralization Value may be used to represent any of the above values.

Sources of Acidity and Significance

Fatty oils consist mostly of glyceryl or other esters of higher fatty acids — in some cases, notable amounts of free acids themselves are present. The amount of free acid present depends on the source from which the oil is taken. The acid content or value of fatty oils increases with time due to hydrolysis with moisture and is therefore a rough indicator of the age of the oil, i.e., it gives an idea of how old a fatty oil is. The deterioration in the flavour of the edible oils with time is due to the increase in the free acid concentration formed by hydrolysis and oxidation.

The presence of free mineral acids in lubricating oils is very rare and is a pointer to external contamination.

New, unblended petroleum oils should have very low neutralization values usually ranging from 0.02 to 0.1. Values higher than this indicate faulty refining, *i.e.*, acids and bases used during the refining process have not been completely removed. Blended or compounded oils may have higher values of neutralization number because of the presence of additives such as oiliness carriers (fatty acids, fatty oils), oxidation and corrosion inhibitors (phenols, aniline), etc. The test is therefore used to maintain specification control on new lots of the lubricant.

As the oil is used, the neutralization number may increase due to contamination (e.g., SO_2 from combustion of S in the fuel, CO_2 from combustion or that present in atmosphere) and/or oxidation of the oil. The oxidation of the oil results in the formation of oil soluble alcohols, ketones, acids and peroxides (which may polymerise to give insoluble resins) thereby increasing the acid number, viscosity and darkening the oil colour. The rate and extent of oxidation of the oil during use depends on temperature, length of exposure to air or oxygen, amount of moisture, catalysts present (formed by the action of oxidation products on the metal surface), type of oil and the inhibitors used.

Periodic determination of Acid Number, therefore, can be used to indicate the progress of oxidation of the lubricant and for systems operating consistently under exactly the same conditions without external contamination, a record of such determinations is helpful in deciding the stage, with a considerable degree of accuracy, where the lubricant needs replacement.

Although the neutralization number gives the amount of acid or base present in the lubricating oil, it gives no information about their source and corrosive nature.

4.6.1 *Determination of total acid number of an oil*

Reagents Required

1. Standard alcoholic potassium hydroxide solution (N/100)
2. A suitable titration solvent (p.7)
3. Phenolphthalein indicator solution
4. *p*-Naphtholbenzoin indicator solution

Theory

A known weight of the oil sample is dissolved in a suitable solvent and titrated with a standard alcoholic solution of KOH to a definite end-point.

$$H^+ + OH^- \longrightarrow H_2O \quad (4.13)$$

$$R-COOH + KOH \longrightarrow RCOOK + H_2O \quad (4.14)$$

Procedure

Weigh accurately the beaker containing the oil sample and a dropper. Transfer 150 drops (about 5 g) of the oil to a titration flask and take the final weight of the beaker, oil and dropper. The difference gives the weight of the oil sample. Add with a measuring cylinder 100 ml of a suitable titration solvent. [Select from 1 to 6 (p.7), a solvent that completely dissolves the sample]. Shake vigorously to dissolve the oil (in case of alcohol as solvent, heat the reaction mixture to boiling on a water bath). Add 20–25 drops of *p*-naphtholbenzoin solution or 10–12 drops of phenolphthalein indicator and titrate against N/100 alcoholic KOH solution until a sharp colour change from orange to green or bluish green (in case of p–naphtholbenzoin indicator) or until appearance of pink colour (in case of phenolphthalein indicator). Record the volume of alkali used as A ml. Run a blank on the reagents used and record the volume of alkali consumed as B ml.

4.6.2 *Determination of strong acid number or inorganic acidity of an oil*

Reagents Required

1. Standard potassium hydroxide solution (N/100)
2. Methyl orange indicator solution.

Procedure

Transfer 25–50 g of accurately weighed oil sample to a 50-ml separating funnel. Add 100–150 ml hot distilled water, shake vigorously and allow to stand for the separation of oil and water layers. Drain water layer into a 500–ml conical flask. Wash oil layer with two to three 50-ml fractions of hot distilled water and collect the washings in the same conical flask. Add 3–4 drops of methyl orange indicator to the contents of the conical flask and titrate against N/100 KOH solution until the colour changes from red to yellow. Record the volume of alkali used as A ml. Run a blank on the same volume of hot water and record the volume of alkali used as B ml.

Observations and Calculations

Initial weight of beaker, oil and dropper = W_1 g

Final weight of beaker, oil and dropper = W_2 g

Weight of oil sample taken = $(W_1 - W_2)$ g

Volume of N/100 KOH used in the test = A ml

Volume of N/100 KOH used in the blank = B ml

Therefore volume of N/100 KOH used against acid in $(W_1 - W_2)$ g of oil $= (A - B)$ ml

$$\text{Acid value (or strong acid value, as the case may be)} = \frac{\text{ml of KOH used} \times \text{Normality} \times 56}{\text{Weight of sample}}$$

$$= \frac{(A - B)}{(W_1 - W_2)} \times \frac{1}{100} \times 56$$

$$= \frac{(A - B)}{(W_1 - W_2)} \times 0.56$$

Precautions

1. The solvent used should be freshly distilled.
2. The alcoholic solution of KOH must be standardised just before the test.
3. While carrying out the test on 'Fatty Oils', the reaction mixture should not be heated.
4. In case the pink colour fades repeatedly after subsequent additions of alkali, the titration should be completed rapidly and the first appearance of pink colour should be taken as the end-point.
5. Phenolphthalein is a satisfactory indicator for pale oils but when the test sample is highly coloured (red or black), a smaller weight of the sample should be taken and *p*-naphtholbenzoin indicator should be used or better the end-point should be determined potentiometrically.

Exercises

118. When phenolphthalein is used as indicator, the pink colour at the end–point sometimes fades away after a few seconds. What is the possible cause of this?

119. In determining the acid values of 'fatty oils', what will happen if the reaction mixture is heated?

120. What inference is drawn if addition of *p*-naphtholbenzoin indicator to the oil sample, dissolved in a suitable solvent, produces a green or bluish green colour? How is the titration performed?

4.7 Saponification Value or Number or Koettsdoerfer Number

Saponification is the hydrolysis of an ester with NaOH or KOH to give alcohol and sodium or potassium salt of the acid. The term originated from the alkaline hydrolysis of fatty oils which led to formation of soaps.

Saponification Number or Value of an oil is the number of mg of KOH required to saponify fatty material present in 1 g of the oil.

Mineral oils, being mixtures of hydrocarbons, do not react with KOH and so are not saponifiable. Vegetable and animal oils, however, are mixtures of glyceryl esters of fatty acids and so require large amounts of alkali to get hydrolysed. Their saponification values are consequently very high and each fatty oil has its own characteristic value.

Significance

The measurement of saponification value, in the absence of impurities that can react with alkali (such as inorganic or organic acids), can be used

(a) to distinguish between fatty oils and mineral oils.

(b) to identify a given fatty oil (for fatty oils, saponification value is a measure of molecular weight).

(c) to determine, approximately, the extent of adulteration (if any) in a given oil.

(d) to determine the extent of compounding (fats and oils added to improve oiliness) in a lubricant. When the type of fatty ingredient in a compounded oil is known, its exact amount is given by

$$\text{Percentage of fatty oil} = \frac{C}{F} \times 100$$

where C = saponification value of the compounded oil (or lubricant)

F = saponification value of the fatty oil

Though increases in saponification values of lubricants during use are of the same order as those in acid values, they have not been correlated to the extent of deterioration of the oil.

4.7.1 *Determination of saponification value of an oil*

Reagents Required

1. Standard hydrochloric acid (N/2)
2. Alcoholic potassium hydroxide solution (N/2)
3. Ethyl alcohol or ethyl methyl ketone
4. Phenolphthalein indicator solution

Theory

A known weight of the sample is mixed with a known excess of standard alcoholic KOH solution and refluxed.

$$\begin{array}{lcll} CH_2OOC{\cdot}R & & CH_2OH & \\ | & & | & \\ CHOOC{\cdot}R' + 3KOH & \longrightarrow & CHOH & + \underbrace{RCOOK + R'COOK + R''COOH}_{\text{Potassium salts of fatty acids (soaps)}} \\ | & & | & \\ CH_2OOC{\cdot}R'' & & CH_2OH & \end{array} \quad (4.15)$$

(where R, R′ and R″ are alkyl groups)

The unreacted KOH is titrated back with standard acid using phenolphthalein as indicator.

$$OH^- + H^+ \longrightarrow H_2O \tag{4.16}$$

Procedure

Transfer about 5 g of accurately weighed oil sample to 500-ml alkali-resistant conical flask. Add 50 ml of N/2 alcoholic KOH and 50 ml of alcohol or ethyl methyl ketone (to act as a solvent) through a pipet. Add the same amounts of solvent and N/2 alcoholic KOH to another flask for Blank determination. Fit the two flasks with air condensers (a long narrow glass tube) and reflux the contents on a water bath for a minimum period of 1 hour. Cool the contents slightly, disconnect the condenser and rinse it with a small amount of distilled water into the flask. Add 10–12 drops of phenolphthalein indicator and titrate the contents of two flasks with standard solution of HCl (N/2) until the pink colour has just disappeared.

Observations and Calculations

Weight of the sample taken = W g

Volume of alcoholic KOH added to both the flasks = 50 ml

Volume of solvent added to both the flasks = 50 ml

Volume of N/2 HCl used in the sample determination = A ml

Volume of N/2 HCl used in the Blank determination = B ml

Then the volume of N/2 HCl equivalent to KOH used in saponifying W g of the sample = $(B - A)$ ml

$$\text{mg of KOH present in } (B - A) \text{ ml of N/2 KOH} = (B - A) \times \frac{1}{2} \times 56$$

$$\text{Saponification value of the sample} = \frac{(B - A)}{W} \times 28$$

Precautions

1. If the room temperature is appreciably high (as in summer months) and there is a possibility of loss of vapour to the atmosphere, the air condenser should be replaced with a water condenser and boiling should be done at a slow rate.
2. To hasten the process of saponification, the sample flask should be occasionally shaken during the reflux operation.
3. Ethyl methyl ketone as a solvent should be preferred as it raises the boiling point and thus hastens saponification.

4.8 Iodine Value

Oils and fats, both animal and vegetable, are mixtures of mixed glycerides represented by the general formula

$$\begin{array}{l} CH_2OOCR \\ | \\ CHOOCR' \\ | \\ CH_2OOCR'' \end{array}$$

and are formed from glycerol and monobasic acids. The acids present in combination with glycerol have all even number of carbon atoms (C_4 to C_{22}) and may be saturated and/or unsaturated. Commonly occurring acids are:

	Acid	Important source
1.	$C_{15}H_{31}COOH$ Palmitic acid	Butter, palm oil
2.	$C_{17}H_{35}COOH$ Stearic acid	Fats
3.	$CH_3{\cdot}(CH_2)_7{\cdot}CH{=}CH{\cdot}(CH_2)_7\ COOH$ Olcic acid	Olive oil, palm oil, peanut oil, soyabean oil
4.	$CH_3{\cdot}(CH_2)_4{\cdot}CH{=}CH{\cdot}CH_2{\cdot}$ $CH = CH{\cdot}(CH_2)_7 \cdot COOH$ Linoleic acid	Cottonseed oil, linseed oil, soyabean oil, corn oil
5.	$CH_3{\cdot}CH_2{\cdot}CH{=}CH{\cdot}CH_2{\cdot}CH{=}CH{\cdot}$ $CH2CH{=}CH(CH_2)_7{\cdot}COOH$ Linolenic Acid	Linseed oil
6.	$CH_3{\cdot}(CH_2)_3{\cdot}CH{=}CH{\cdot}CH{-}$ $CH{\cdot}CH{=}CH{\cdot}(CH_2)_4{\cdot}CO{\cdot}(CH_2)_2{\cdot}$ $COOH$ Eleostearic acid	Tung oil

The combinations having predominance of short-chain saturated fatty acids or long chain fatty acids with a considerable degree of unsaturation are liquids at ordinary temperatures and are called oils; while others, that are solids at ordinary temperatures, are known as fats.

The degree of unsaturation of oils and fats is reported in terms of their Iodine Number or Iodine Value (I.V.) which is the number of grams of Iodine equivalent to the amount of Iodine Monochloride (ICl) absorbed by 100 g of the oil.

Significance

Each oil has its own characteristic iodine value (varying between narrow limits), and so the determination of iodine value can be utilised to detect and even to determine (though approximately) the extent of contamination in any specific oil.

Iodine values have been used to classify oils into

(i) Drying oils—Linseed oil, tung oil, perilla oil, etc., having iodine value (I.V.) > 150

(ii) Semi-drying oils—Soyabean oil, dehydrated castor oil, etc., with iodine value between 100 and 150

(iii) Non-drying oils—Castor oil, coconut oil, olive oil, etc., with iodine values <100.

Unsaturation confers drying characteristics on oils—when a thin layer of an oil with a high degree of unsaturation is spread on a smooth surface and exposed to air and light, it takes up oxygen at the double bond sites, and gets oxidised and polymerised

[through the intermediate formation of peroxides,

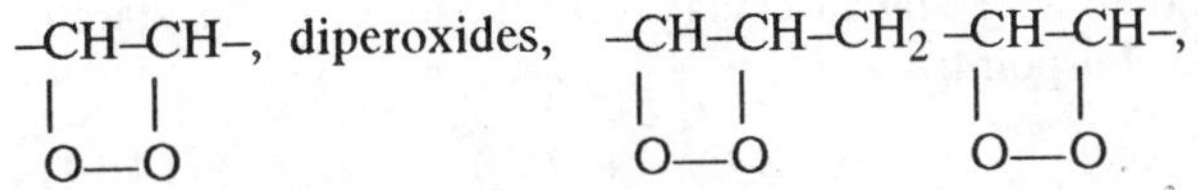

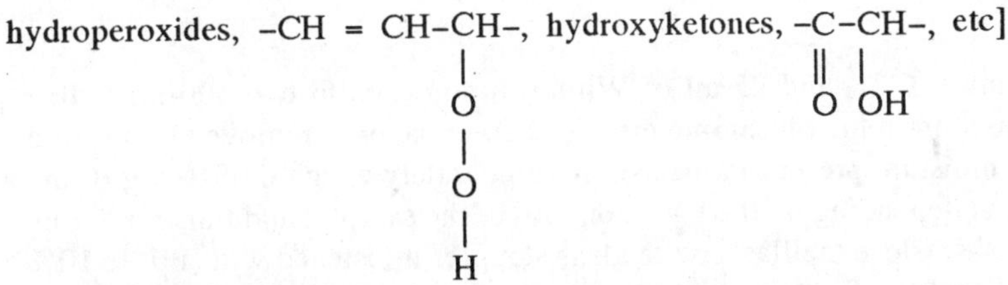

into a hard, coherent, tough, infusible, cross-linked, resin-like film. The rate of drying and the nature of the film formed depends (in addition to other factors like temperature, presence of driers, etc.) on the degree of unsaturation or I.V. of the oil used.

Thus an oil of high I.V. is most suitable for the manufacture of paints and varnishes but for an oil to be used as a lubricant, its I.V. should be as low as possible so that during use it is not deteriorated to any appreciable extent due to oxidation and polymerisation.

4.8.1 *Determination of Iodine Value of an oil*

Reagents Required

1. Standard sodium thiosulphate solution (N/10)
2. Wij's solution
3. Potassium iodide solution (10%)
4. Carbon tetrachloride or chloroform
5. Freshly prepared starch solution.

Theory

A known weight of the oil is dissolved in CCl_4 or $CHCl_3$ and treated with a known excess of Wij's solution (solution of ICl in glacial acetic acid). One molecule of ICl adds on each double bond:

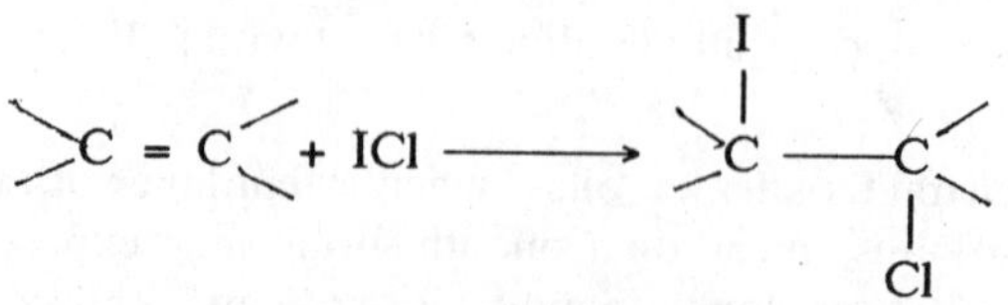 (4.17)

After the reaction is complete, KI solution is added which is oxidised to I_2 by the unreacted ICl:

$$ICl + I^- \longrightarrow I_2 + Cl^- \qquad (4.18)$$

The liberated iodine is titrated with standard $Na_2S_2O_3$ solution using starch solution as indicator near the end-point:

$$2S_2O_3^{2-} + I_2 \longrightarrow S_4O_6^{2-} + 2I^- \qquad (4.19)$$

From the amount of iodine (in the form of ICl) added and that left unconsumed, as measured iodometrically, the iodine value of the sample can be calculated.

Procedure

Transfer 25 ml of CCl_4 and 25 ml of Wij's solution each to two 500-ml iodine titration flasks. Strain the oil sample through a filter paper to remove all the solid material and moisture present in the sample. Accurately weigh 0.05–0.5 g of the sample [weight depending on the I.V. (200–50) of the sample] and transfer to one of the two flasks. Close the flasks with glass stoppers moistened with a little 10% KI solution. Swirl the flasks to mix the contents intimately and keep in dark for about 1 hour at a temperature below 30°C. Add 20 ml of KI solution to both the flasks, washing down the stopper, and dilute with about 100 ml of distilled water. Titrate the liberated iodine with N/10 $Na_2S_2O_3$ solution. When the colour of the solution turns light yellow, add about 1 ml of freshly prepared starch solution and continue the titration until the first disappearance of blue colour. Record the volume of N/10 $Na_2S_2O_3$ solution used in sample titration as A ml and that in blank titration (without oil) as B ml.

Precautions

(1) Wij's solution should be added with the help of burette or a vacupet. This solution should never be sucked into the pipet with mouth.

(2) The sample and the glassware used must be completely dry.

(3) The amount of the sample taken should be such that the Wij's solution added is about 100% in excess of the expected requirement.

(4) As far as practicable, the addition of solutions (Wij's solution or KI solution) to the two flasks should be simultaneous.

(5) During titration, the flasks should be frequently stoppered and shaken vigorously to ensure complete titration of the iodine present in CCl_4 layer.

Observations and Calculations

Weight of the oil sample taken = W g

Volume of N/10 $Na_2S_2O_3$ used to titrate excess iodine in sample determination = A ml

Volume of N/10 $Na_2S_2O_3$ used to titrate total iodine in blank titration = B ml

Therefore, volume of N/10 $Na_2S_2O_3$ equivalent to iodine (or ICl) consumed by W g of the Oil = $(B - A)$ ml

Iodine present in (B – A) ml of N/10 iodine solution

$$= (B - A) \times 127 \times \frac{1}{10} \times \frac{1}{10^3} \text{ g}$$

$$\text{Therefore, Iodine value} = \frac{(B - A)}{W} \times 127 \times \frac{1}{10} \times \frac{1}{10^3} \times 100$$

$$= \frac{(B - A)}{W} \times 1.27$$

Exercises

121. What is the effect of the presence of moisture in the sample, acetic acid or glassware, on the test?
122. Why is the reaction mixture kept in dark?
123. Why is the glass stopper moistened with KI soultion?

4.9 Density and Specific Gravity

The density (ρ) of a substance is the mass of its unit volume, and is expressed in g/ml. At 4°C, 1 ml of distilled water weights 1 g and so its density at this temperature is unity which is taken as standard. The density at any other temperature $t°$ is expressed in comparison to this standard and represented as $\rho_{4°}^{t°}$. The specific gravity or Relative Density ($\rho_{t°}^{t°}$) is the ratio of the mass in air of a given volume of the substance at a stated temperature $t°$ to the mass in air of an equal volume of distilled water at the same temperature. The density of the substance ($\rho_{4°}^{t°}$) in absolute units, at temperature $t°$, can be obtained from relative density ($\rho_{t°}^{t°}$) by multiplying it with the density of water at that temperature.

Significance

Though density or specific gravity of an oil is not important in assessing its performance as a lubricant, their determination is essential for measuring viscosity (p. 87) and Surface Tension (p. 215) of the lubricating oils. The density or specific gravity of materials must be known wherever weight–volume relationship is of concern, such as in marketing, shipment or storage of petroleum products. These values also give a rough idea of

(a) the type of the crude (naphthene base type crudes have the lowest density while the paraffin base type have the highest density), and

(b) the relative amounts of gasoline and kerosene in the crude.

API Gravity

Specific gravity values involve several decimal points which are not convenient to use. Therefore, those associated with petroleum industry usually prefer to use API gravity, an arbitrary scale (established by American Petroleum Institute) expressed in whole number degrees. API gravity is a derivative of the conventional gravity and is defined as

$$\text{Degrees API scale} = \frac{141.5}{\text{sp. gr. } 60/60\text{ F}} - 131.5$$

where sp. gr. 60/60 F represents the ratio of the weights of equal volumes of the oil and distilled water, both at 60°F. On this scale, pure water has an API gravity of 10. Liquids lighter than water have values higher than 10 while those heavier than water have values lower than 10. API and sp.gr. scales thus run in opposite directions — the denser the oil, the lower the API gravity.

4.9.1 *Determination of sp. gr. and density of a liquid say alcohol, kerosene, benzene, etc., by*

(a) Hydrometer

(b) Westphal balance.

Theory

When greater accuracy is required or when only small quantities are available, liquid densities are best obtained from the mass required to fill a vessel of accurately defined volume, e.g., a specific gravity bottle (p.217) or a pycnometer. With careful technique, the densities measured with pycnometers are precise to five significant figures.

For not too viscous liquids available in sufficient amounts, density is more conveniently determined by buoyancy devices namely Westphal balance or a Hydrometer. These devices are based on Archimedes principle according to which the buoyant effect (the upthrust acting on an object immersed in a liquid) is directly proportional to the weight of the liquid displaced.

(a) Hydrometer

A hydrometer is a float caliberated (with water) to indicate specific gravity (usually on API scale) of a liquid by the extent (depth) to which it submerges into the liquid. Though it permits very rapid measurements, a hydrometer is usually employed only where approximate specific gravity is acceptable.

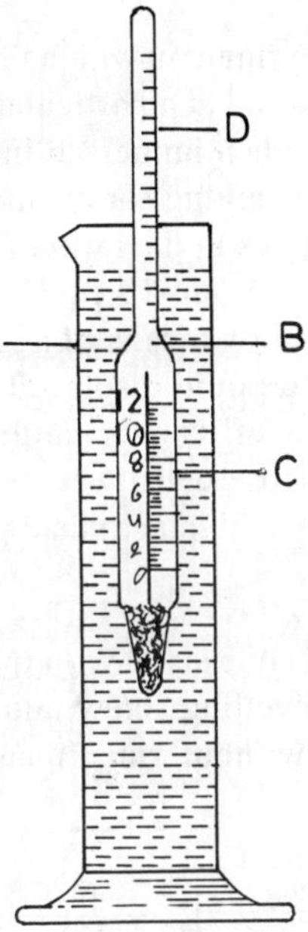

Fig. 4.6 Thermo-hydrometer

A - Hydrometer jar B - Thermohydrometer
C - Thermoscale D - Gravity scale

Procedure

Pour the liquid sample under test into the hydrometer jar slowly, so as to avoid air bubbles. Suspend in the sample a thoroughly cleaned and dry thermo-hydrometer (a hydrometer with a thermometer enclosed in its body, Fig. 4.6). When the hydrometer has come to rest, depress it slightly into the liquid and then release. When the hydrometer has again become stationary, note simultaneously the readings on the gravity scale and the thermo-scale of the hydrometer.

Precautions

(1) Any air bubbles collecting at the surface of the liquid should be removed by touching them with a piece of clean filter paper.

(2) There should be no air bubble sticking to the surface of the hydrometer.

(3) There should be sufficient sample in the jar to allow the hydrometer float freely.

(4) The hydrometer should not touch the walls of the jar.

(5) Before noting the reading, sufficient time should be allowed for the hydrometer to become completely stationary.

(6) If a thermohydrometer is not being used or when the sample is coloured, the temperature of the sample should be recorded separately with a thermometer.

(b) *Westphal balance*

A westphal balance is an instrument in which the upthrust on a small sinker (a glass plummet) immersed in the liquid at a particular temperature is compared with the upthrust on the same sinker when immersed in water up to the same extent and at the same temperature. The weighing (or balancing) is made by means of a series of U-shaped riders (with masses in the ratio of 1.0000 : 0.1000 : 0.0100 : 0.0010) and a notched beam (graduated into 10 equal parts) with a hook at the end from which the sinker is suspended by means of a fine platinum wire. When the beam is horizontal, the mass of the liquid displaced is exactly equal to the sum of the rider moments which, in an efficiently calibrated (usually at 60°F) Westphal balance, directly gives the relative density of the liquid. The results are only slightly less accurate (precise to 3rd or 4th place) than pycnometric measurements.

Procedure

Mount the beam on the adjustable stand with the plummet suspended (in air) from its hook. By means of the levelling screw and the threaded counterpoise (screw weight), balance the beam (without riders) until the indicator points D_1 and D_2

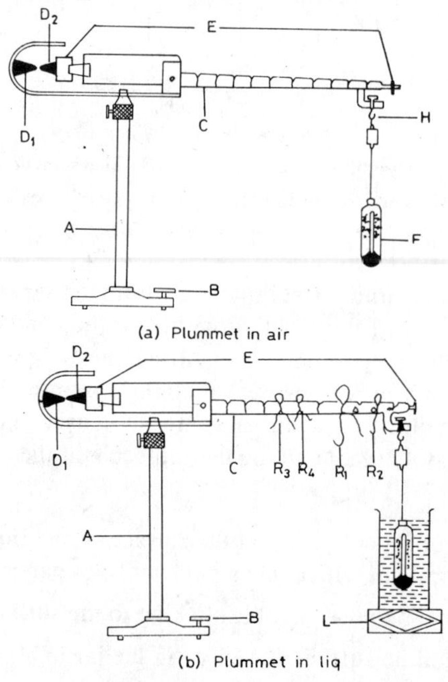

Fig. 4.7 Westphal Balance

A - Adjustable stand	B - Levelling screw
C - Notched beam	D_1, D_2 - Indicator points
E - Threaded counterpoise	F - Glass plummet
H - Hook	L - Lab jack
R_1 - Unit Rider	R_2 - 0.1000 Rider
R_3 - 0.0100 Rider	R_4 - 0.0010 Rider

(Fig. 4.7(a)) are opposite to each other. Now immerse the plummet in the liquid sample taken in a cylinder (placed on a lab-jack or an adjustable stand) and balance the beam again, this time by placing the graded riders on the notched beam. Simultaneously, adjust the height of the cyclinder so that the plummet is just completely immersed in the liquid (Fig. 4.7(b)). Also record the temperature on the thermoscale enclosed in the plummet.

Sum of the rider moments give the sp. gr. of the liquid relative to water at the calibration temperature (usually 60°F) which is marked on the plummet or the beam. For example, if the unit (large) rider is placed on the 7th notch and the 0.1000, 0.0100 and 0.0010 riders on notches 9, 4 and 5, respectively (as shown in Fig. 4.7 (b)), then

$$\text{Sp. gr.} = \frac{7}{10} \times 1.0 + \frac{9}{10} \times 0.1 + \frac{4}{10} \times 0.01 + \frac{5}{10} \times 0.001 = 0.7945$$

To convert this value into density, multiply by the density of water at the calibration temperature.

For more accurate work, or when the plummet has been chipped or the test temperature is more than 20°F off the calibration temperature, the westphal balance is first adjusted with water:

Assemble the balance with its glass plummet immersed in water. Place the unit (large) rider on mark 10, or hang it from the hook if only 9 divisions are shown on the beam. Level the beam by means of screw weight and note the temperature and the depth to which the plummet is submerged below the surface of water.

Now remove the plummet and dry or rinse it with the liquid under test if miscible with water. Immerse the plummet in the liquid under test and level the beam by means of riders. The position of the riders gives the relative density of the liquid at the temperature of the experiment.

Precautions

(1) The wire used for suspending the plummet should be extremely fine.

(2) There should be no air bubble sticking to the surface of the plummet or the portion of the wire dipping in the liquid.

(3) The plummet should not touch the sides of the cylinder.

(4) The temperature of the measurement and the depth of submersion of the plummet in the liquid should, as far as possible, be the same as in case of water.

(5) If a lab-jack or an adjustable stand for the cylinder is not available, a dropper may be used to adjust the level of the liquid in the cylinder.

(6) If the plummet does not enclose a thermometer, or when the sample is coloured, a separate thermometer should be used to note the test temperature.

Exercises

124. What are pycnometers?
125. Is it possible to place more than one rider at the same mark on the beam?
126. What is meant by rider moment?
127. Why are the masses of riders chosen to differ by a factor of 10?
128. Predict the position of the unit rider(s) when the density of the liquid under test is (a) less than 1 (b) greater than 1.
129. How is the sp. gravity of (a) highly volatile, and (b) highly viscous liquids determined by Westphal balance?
130. How is the specific gravity measured at the experimental temperature converted to 60/60°F value?
131. An oil storage tank is to be designed to hold 1000 litres. What expansion capacity should be provided in the system if the temperature is expected to rise by 100°F?

4.10 Emulsions

Emulsions are dispersions of small droplets of one liquid in another liquid, neither liquid being soluble in the other. They are electrically stabilised by some type of surface active substances know as emulsifiers. When the amount of the emulsifier is small, the emulsion is relatively unstable ('loose' emulsion) and the dispersed droplets coalesce fairly easily to give large drops ('cracking' of emulsion). When the concentration of the emulsifier is large, the dispersed droplets are finer and the emulsion is hard to break ('tight' or stable emulsion).

Emulsions are used as lubricants for certain specific jobs:

(a) Oil-in-water (o/w) emulsions are used as coolants in metal working, grinding, boring, turning, cutting, etc., and as lubricants for large diesel-motor pistons.

(b) Water-in-oil (w/o) emulsions (also called Invert Emulsions) are used mainly as fire-resistant hydraulic fluids and as compressor and pneumatic tool lubricants.

Depending upon the prospective service, the emulsifying tendencies of an oil can be a benefit or a drawback. Emulsification is *Advantageous* when an oil is to be used in presence of water, as in case of steam cylinder oils where wet steam is involved and lubrication of the cyclinder walls and valve seats is maintained by emulsified oil. Therefore, oils which most readily form stable emulsions with water are added to straight mineral cyclinder oils to increase their tendency to emulsify and to stablize the emulsions formed.

Formation of emulsions in most operations may be *Harmful* as emusification increases the viscosity and consequently the coefficient of friction. Emulsions have a tendency to collect dirt, grit particles and other foreign matter which may cause abrasion and wearing out of the lubricated parts. Emulsions may form sludges, insulate coolers, clog oil-lines and filters and may even cause corrosion of the metal surface by carrying salts from water through the lubricating system and by developing an e.m.f. between the bearing and the journal. Emulsions are also ideal medium for microbiological growth. Since the emulsion formed must be removed, a considerable amount of oil is lost.

These problems can be minimised by using a lubricant that would not emulsify during use or form only a loose emulsion that readily breaks (Rapid water separation reduces the Resting Time required for lubricating oils). In steam turbines, condensation of steam leads to the formation of water-in-oil emulsion whose Demulsibility (the ability of a lubricant to separate from its emulsion with water) depends on the composition of the oil, the impurities left in the lubricant during refining, on the presence of additives, oxidation products of the lubricant during its use and other foreign contaminants. The tendency to demulsify decreases with the time the lubricant has been in use.

The demulsibility of a lubricant is reported in terms of the time in which an emulsion produced from 40 ml of the sample and 40 ml water (at 130°F; 180°F for lubricants with viscosities more than 450 SUS) cracks (breaks) to give two distinct layers, when left undisturbed. In case of incomplete demulsification at the end of 1 hour, a 3-ml cuff or emulsion is neglected; if it is more, the volume in ml of the oil, water and emulsion are reported.

Steam Emulsion Number (S.E.N.) determines emulsion characteristics resulting from steam condensing in oil. It is the time in seconds in which oil and water separate in distinct layers from an emulsion (at 200–203°F) produced by bubbling steam into 20 ml of oil until the volume of emulsion becomes 40 ml.

S.E.N. is useful in defining the demulsibility of a new oil. Exceptionally high values during service indicate foreign contamination or malfunctioning of some component.

4.10.1 *Water Emulsion Test*

In the Water Emulsion Test (United States Steel Method), 40 ml of the oil sample is taken in a 100-ml graduated cylinder which is immersed in a constant temperature bath maintained at 130°F or 180°F. 40 ml of distilled water heated to the temperature of the bath are added slowly to the oil so as to complete the addition in one minute. The mixture is then emulsified by stirring for 5 minutes with a paddle rotating at 1500 revolutions per minute. After switching off, the paddle is slowly raised out of the cylinder. Simultaneously, a stop watch is started. The cylinder is then allowed to remain in the bath at the test temperature. At suitable time intervals, the cylinder is taken out only to inspect and is reimmersed in the bath immediately after each reading. The results of the experiment are recorded in the following tabular form:

Time in minutes	1	2	3	4	5	6	7	8	9	10	15	20	30	40	50	60
ml of the oil layer																
ml of cuff or emulsion left (intermediate layer)																

Precautions

(1) The graduated test cylinder should be immersed in the constant temperature bath at least up to 85-ml mark.

(2) After raising the paddle out of the cylinder, it should be held in that position so as to allow as much of emulsion as possible to return to the cylinder.

4.10.2 *Steam Emulsion Number*

In the ASTM Test, 20 ml of the oil to be tested is taken in the standard graduated test tube (A) which is suspended in a beaker containing water (B-Emulsifying bath) at an initial temperature of about 70°F. Steam from the generator (C) is passed through a delivery tube (D) into the oil (Figure 4.8). The steam jet violently agitates the oil which becomes intimately mixed with water (formed by the condensation of steam) and an emulsion is formed. The latent heat of condensation of steam raises the temperature of the emulsion. Steam is admitted at such a rate (adjusted with the help of pinch cock E_3) as to raise the temperature of the emulsion to about 190°F in about 1 minute. Steaming is continued till the volume of emulsion in the test tube becomes 40+3 ml. At this stage, the delivery tube and the thermometer along with the cork are quickly withdrawn from the test tube which is rapidly transferred to the other beaker containing water (F-Separating bath), maintained at a temperature of 200-203°F by bubbling steam from the generator. Immediately as

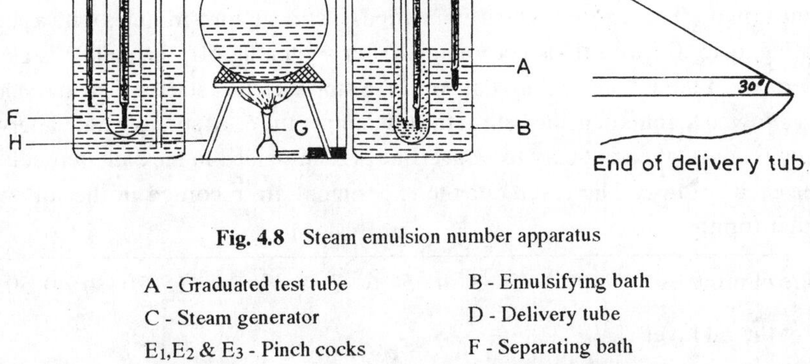

Fig. 4.8 Steam emulsion number apparatus

A - Graduated test tube
B - Emulsifying bath
C - Steam generator
D - Delivery tube
E_1,E_2 & E_3 - Pinch cocks
F - Separating bath
G - Burner
H - Tripod stand
I - Clamp stand
T_1,T_2 & T_3 - Thermometers

the delivery tube is withdrawn, a stop watch is started to measure the time. The observations are recorded in the tabular from (as in Water Emulsion Test 4.10.1). The time in which the emulsion breaks is reported as Steam Emulsion Number of the oil.

Precautions

(1) Before immersing the steam delivery tube into the test tube, steam should be passed through the delivery tube until condensation in the tube ceases.

(2) The steaming rate should be so adjusted as to double the volume of the contents of the tube in 4–6 minutes.

(3) The end of the delivery tube dipping into the oil should be cut at an acute angle and should rest at the bottom of the test tube so that no portion of the oil escapes agitation.

(4) A narrow vertical groove should be made in the side of the cork to permit the uncondensed steam to escape.

Note: The importance of these static tests in industry has considerably decreased. They have been replaced with a Dynamic Demulsibility Test which measures the ability of an oil to separate from water under actual circulating conditions and correlates better with practice.

Exercises

132. Name some specific lubricants which must have high water separation tendency, i.e. low S.E.N.

133. What steps should be taken in order that a lubricant may possess good demulsifying characteristics?

134. What are 'Soluble Oils'?

135. What are defoamants? Give some examples.

4.11 Greases

A grease is essentially a semi-solid to solid combination of a thickening or gelling agent (soap or a mixture of soaps) and a liquid lubricant (petroleum oil, fatty oil or synthetic oil). It may also contain other ingradients which are added to impart special characteristics.

Whereas oils move of their own accord, pressure has to be applied to greases to make them flow. Thus greases give a higher coefficient of friction than lubricating oils; yet they are preferred to oils under the following conditions:

(1) Where the oil is squeezed out due to heavier loads, low speeds, or does not remain in place due to intermittent jerks.

(2) Where the design of the machine is such that the parts to be lubricated can in no way retain the liquid lubricant, e.g., open gears.

(3) Where the lubricant is also required to act as a seal against the entrance of dirt or moisture.

(4) Where dripping or splashing of oils cannot be tolerated, e.g., machines preparing edible articles, paper, textile, etc.

(5) Where an oil is seldom added, e.g., electric motor bearings.

(6) Where frequent lubrication is either inconvenient or uneconomical (automobile wheel bearings).

4.11.1 *Consistency or mechanical stability*

Some greases which, on being in service, retain their original hardness or softness for a long duration are said to be more consistent than those which more readily lose their hardness on being worked. Consistency, as applied to a lubricating grease, is thus its ability to resist a change in its stability or degree of stiffness likely to be produced by an operating mechanical shear (whether in laboratory or in service). Consistency is thus a characteristic of plasticity (as viscosity is of fluidity) and represents the ability of a grease to resist deformation under the application of force.

Penetration Number, Value, Index or Yield Value

Penetration number with respect to a lubricating grease is the depth (in tenths of an mm) that a cone of standard dimensions penetrates vertically into the sample under test, under prescribed conditions of weight (150 g), temperature (25°C) and time (5 seconds).

Significance of Penetration Test

Penetration value of a grease is a measure of its degree of stiffness (i.e., whether it is hard or soft). The hardness or stiffness of a grease may vary from that of a heavy viscous liquid to that of a stiff solid mass (e.g., a cake of soap). The lower the penetration value, the harder the grease — which means pumping of grease is difficult and coefficient of friction is more. However, the grease can better withstand higher loads at lower speeds. A higher penetration number means a softer grease, lower resistance to flow, increased pumpability but unable to withstand higher loads and more attention will have to be paid to prevent leakage.

Measurement of penetration number of 'unworked' (as marketed — a grease that has received only the minimum handling in transfer from sample-can to test-apparatus and which has not been subjected to the action of a grease worker) and 'worked' (subjected to standard shear rate with a manual or mechanically-operated worker) samples can be utilised to assess the consistency of a grease, and also to study its gel-fluid reversibility (thixotropic nature). It is also used as a control test for product uniformity and serves as a basis for dividing greases into different grades.

4.11.2 *Determination of penetration number of a grease*

Procedure

Using a corrosion-resistant steel spatula, the sample is packed into the grease cup taking care to exclude all air. The cup is then placed in a bath maintained at 25°C for about one hour. The surface of the grease is then smoothened and brought at level with the edge of the cup by scrapping off excess of grease with the blade of the spatula.

The grease cup is now placed on the levelled table of the heavy base of the penetrometer (Figure. 4.9). The standard cone (made of brass and fitted with detachable hardened-steel tip) is cleaned and fixed to the holder shaft. The holder is then slowly and very carefully lowered so that the cone tip just touches the grease surface. The dial rod is now pressed down so that it comes into contact with the holder shaft, fixed vertically below, and the reading of the indicator on the circular dial is recorded. By pressing the push button, the shaft is released and the cone falls without measurable friction. Simultaneously, a stop watch is started. The cone is allowed to penetrate into the grease for exactly 5 seconds and then the push button is released. The dial rod is now pressed down until it is stopped by the holder shaft, and the indicator reading on the circular dial is again recorded. The difference in

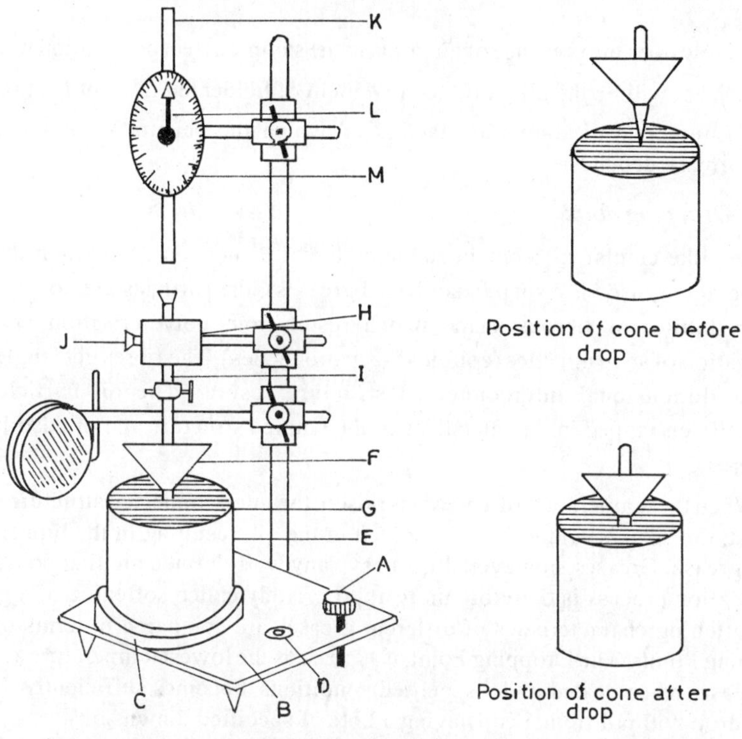

Fig 4.9 Penetrometer

A - Levelling screw B - Heavy base
C - Table D - Spirit level
E - Grease cup F - Standard brass cone
G - Hardened-steel tip H - Holder
I - Shaft J - Push button
K - Dial rod L - Indicator
M - Circular dial

the two readings directly gives the distance, in tenths of an mm, through which the cone has penetrated into the grease. The test is repeated 2 to 3 times and the average is reported as the penetration number of the grease.

Precautions

(1) For perfect positioning of the cone in contact with the grease surface, a mirror should be used to remove parallax.

(2) As far as possible, all air voids in the sample should be eliminated.

(3) On pressing the push button, the cone must have a free fall, i.e., the shaft should move in the holder without any measurable friction.

(4) Before taking subsequent readings, the cone should be wiped perfectly clean and the surface of grease should be smoothened.

Exercises

136. How does the presence of air voids in the sample affect penetration number?

137. What will be the effect if the movement of holder shaft is not frictionless?

138. Enumerate the important factors on which the penetration number of a grease depends.

4.11.3 *Dropping point*

Greases, like emulsions, contain stable colloidal dispersions, but when thickner (gelling agent) used is a soap (soap-based greases), the particles are not spherical but crystallites of fibrous shape which result from polymerisation or linear aggregation of soap micelles (colloidal soap molecules). The fibres get tangled and a three-dimensional interconnected structure results. The oil particles get physically entrapped in the interstices of the tangled structure and/or adsorbed on the fibres.

When the temperature of a grease is raised, the inter-connected structure ceases to exist; the soap is said to have dissolved in the oil, resulting in the liquification of the grease. Greases, however, do not have any well-defined melting points. The liquification process is therefore more appropriately called softening of a grease. The softening characteristics of different greases are compared in terms of their Dropping Points. The Dropping Point of a grease is the lowest temperature at which the grease, when heated under specified conditions, becomes sufficiently fluid so that a drop will fall from a cup having a hole of specified dimensions.

The dropping point is a qualitative indicator of heat resistance or thermal stability of a grease, which depends on the nature of the soap and the fatty oil present and, to some extent, on the viscosity of the mineral oil (added to adjust the consistency of the grease). Lime soap greases generally have lower dropping points than greases of sodium, lithium or mixed base. The dropping point of some greases, particularly those of aluminium soaps, appreciably decreases upon ageing. Greases containing saturated fatty acids have higher thermal stability. The dropping point can be increased by adding non-soap thickners such as carbon black, bentonite, collodial silica, etc.

Significance

Dropping point provides a practical limit of temperature above which a grease cannot be used as a lubricant in the semi-solid state. The test is useful as a means of identifying greases. In the manufacture of greases, periodic determination of dropping point is used to ascertain the uniformity of the finished product.

4.11.4 *Determination of dropping point of a grease*

Procedure

A chromium-plated brass cup, with an opening of standard dimensions at its bottom, is taken and its larger opening is pressed into sample grease until the cup is filled and a small amount of grease extrudes out of the smaller opening. Excess of grease is removed with a spatula. A wedge of grease, shaped into an inverted cone by revolving the cup against the standard polished metal rod (provided for this purpose along with the apparatus) inserted through the length of the cup, is removed such that only a smooth uniform thin film of grease remains on the walls of the cup.

The cup is now lowered into a pyrex glass test tube with a rim or indentations near the bottom to support the cup (Figure 4.10). A thermometer is then inserted through a cork into the tube to such a depth that the tip of the bulb is just above the bottom of the grease cup. The assembly is then lowered into a heating bath (a glass tub containing a transoparent oil and fitted with a thermometer and a stirrer).

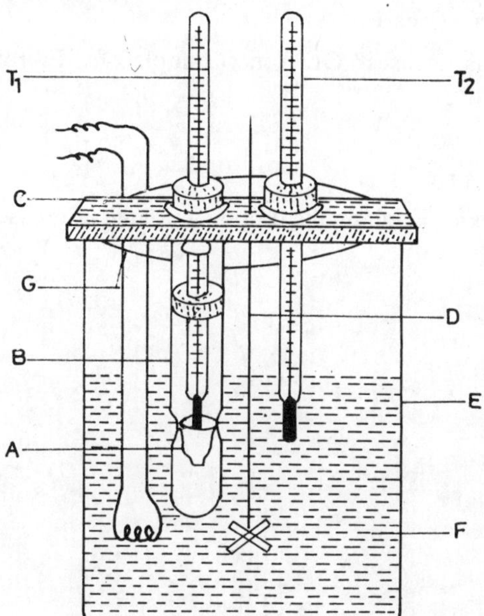

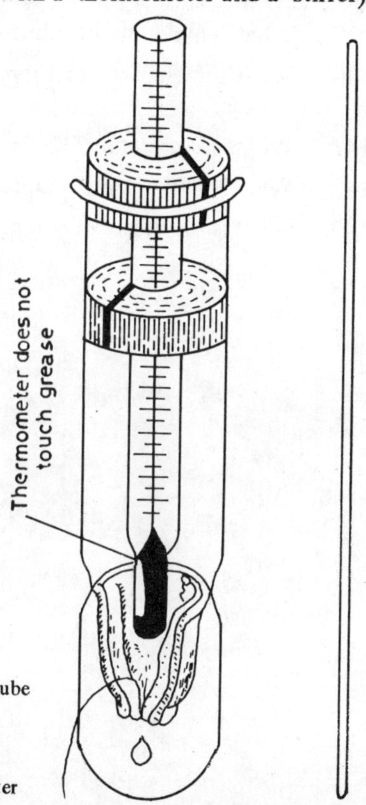

Fig 4.10 Dropping point apparatus

A - Chromium plated brass cup
B - Pyrex glass test tube
C - Cork
D - Stirrer
E - Glass tub
F - Transparent oil
G - Heating coil
T_1 - Test thermometer
T_2 - Bath thermometer

The temperature of the heating bath is then slowly raised (electrically) while the stirrer is working. As the temperature rises, the grease softens and slowly exudes from the orifice of the cup. When a drop of grease gets detached from the bottom of the cup and falls into the tube, the temperatures shown by the two thermometers are recorded. The average of the temperatures is reported as the Dropping Point of the grease sample.

Precautions

(1) The grease film on the walls of the cup should be as thoroughly uniform as possible.

(2) The thermometer bulb should not touch the grease.

(3) The bulbs of the test thermometer and bath thermometer should be, as far as possible, at the same level.

(4) Initial heating may be fast (at a rate of about 5°C per minute) but when the temperature of the bath is about 15°C below the expected dropping point, the heating rate should be so reduced that the two thermometers show a temperature difference of less than 2°C.

(5) The temperature should be recorded when the grease drop completely separates from the bottom of the cup, i.e., if the drop has a tailing end, it should break completely.

Exercises

139. What is meant by bleeding of a Grease?

140. What are Smooth or Fibrous Greases? Give one example of a Fibrous Grease.

141. What is Petrolatum?

142. What is meant by a complex soap?

143. What is the effect of complexing agent on the dropping point of a grease?

Observations and Calculations

Let the weight of empty capsule or crucible $= A$ g

Weight of capsule (or crucible) + sample, before drying $= B$ g

Weight of capsule (or crucible) + residue, after drying $= C$ g

Inherent Moisture Content $= \dfrac{B - C}{B - A} \times 100\%$

The sum of the weight loss on air-drying and inherent moisture gives Total Moisture Content. If a coke sample is dried to constant weight between temperature 105–200°C, without any preliminary crushing, the loss in weight on percentage bases gives the total moisture content within an error of 0.5%.

5.3 Ash Content

The useless and non-combustible inorganic matter that is left after burning off the organic matter in coal is known as Ash.

Free or Extraneous Ash arises from

(i) Ankerites and cleats (usually composed of $CaCO_3$ & $FeCO_3$ along with some $MgCO_3$) deposited, inside the cracks and along joint planes in the matured coals, from mud or water that percolates through the coal beds.

(ii) Sedimentary materials (slate, clay, pyrites, etc.) deposited simultaneously with the coal-forming material and present in definite layers in the coal, and

(iii) Sand and clay mixed with coal from the roof and floor of the mine.

This portion of ash can be removed by selective mining and cleansing process (hand picking, washing, etc.)

Inherent Ash (oxides of Na, Mg, Ca, K & Si) arises from the mineral matter originally present in the cellulosic plant debris from which coal was formed. It rarely amounts to more than 2%.

Adventitious (Non-essential) mineral matter (clayey or silicious) present in mud or water in contact with the decomposing vegetation (during the anaerobic decay state) gets intimately mixed with coal or is present as dirt bands alternate with purer coal layers. The amount of adventitious ash varies widely and may go upto 10%. The inherent and adventitious mineral matter cannot be removed by cleansing operations and together constitute what is know as 'Fixed Ash'.

Significance

Ash reduces the heating value of coal. There is a heat loss of about 1.5% for each 1% ash present in coal. It increases transportation, handling and storage costs. The furnaces may have to be shut down for the removal and disposal of ash and increased labour costs are involved.

The chemical composition of ash, which is determined by the usual methods of analysis, differs from coal to coal and usually varies between the following limits:

Constituent	%	Constituent	%
SiO_2	40–60	MgO	0.5–5
Al_2O_3	20–40	TiO_2	0–3
Fe_2O_3	2–25	$Na_2O + K_2O$	1–6
CaO	1–15	SO_3	0.2–15

As coal (or coke formed from it) is used as a reducing agent in several metallurgical operations, the composition of the ash affects the slag and metal composition and its characteristics, and is the most important factor governing the selection of a proper flux.

The fusion temperature of ash is of considerable importance and is closely related to its composition. In general, it has been found that ash containing Al_2O_3 and SiO_2 in the ratio in which they are present in pure dehydrated clay $Al_2O_3 \cdot 2SiO_2$ (45.8% Al_2O_3 + 54.2% SiO_2 by weight) is the most infusible (called refractory ash), and the fusion temperature decreases with increasing proportion of basic oxides like Na_2O, K_2O, MgO, CaO and FeO.

If the ash has low softening temperature, it fuses when coal is burnt on grates, and lumps of ash (called 'clinker') get deposited on the fire bars. This blocks the grate holes and spaces between coal particles; passage of air becomes irregular and restricted, and uneven heating takes place. This leads to increased clinkering. Clinker removal from grates is very difficult and laborious. Some coal particles may get embedded in the clinkers and escape combustion, thus leading to loss of fuel. Melted fly ash may be deposited on boiler tubes restricting heat transfer.

Molten slag from low melting ash penetrates into the pores of the refractory lining of a boiler or the furnace. Difference in the coefficients of expansion and contraction of the slag and the refractory material causes Spalling of the refractory lining thereby reducing its life.

Presence of some high melting ash is, however, necessary to protect the grates (in steam generation, producer gas and water gas manufacture) from direct contact with incandescent coal, which might cause oxidation of the grate bars.

5.3.1 *Determination of ash content of a coal/coke sample*

Procedure

Place the porcelain/silica capsule or platinum crucible containing dried coal (from the inherent moisture test), without cover, in a muffle furnace at a low temperature and gradually raise the temperature to 700°C. Stir the residue with a platinum or nichrome wire (to hasten ignition) and ignite for half an hour at a temperature between 700 and 750°C. Moisten the residue with a drop or two of alcohol. If black particles appear (due to the presence of unburnt carbon), continue ignition for another 15 minutes. Cool in a desiccator and weigh. Repeat the process of ignition, cooling and weighing until the difference in weights between two successive weighings is less than 1 mg.

For tests on coke, heat the capsule or crucible containing dried coke in a muffle furnace or over a burner to redness. Finish the ignition to constant weight (± 1 mg) at a temperature between 900 and 950°C.

Precautions

(1) To avoid mechanical loss due to rapid expulsion of volatile matter, the temperature should be raised at a slow rate.

(2) For coals containing high percentage of calcite and pyrite, reproducible results are not obtained because of the varying amounts of sulphate–sulphur being retained in the ash. To overcome this difficulty, the sample is heated at a lower temperature (~ 500°C) for a longer duration of time which ensures complete oxidation and expulsion of pyrite sulphur before the decomposition of calcite takes place when the temperature is raised to 750°C.

Observations and Calculations

Weight of empty and previously ignited capsule or crucible = A g

Weight of capsule + air dried sample = B g

Weight of capsule + oven dried sample = C g

Weight of capsule + residue, after ignition = D g

Ash content (Air dry basis) $= \dfrac{D-A}{B-A} \times 100\%$

Ash content (Oven dry basis) $= \dfrac{D-A}{C-A} \times 100\%$

5.4 Volatile Matter

When coal is heated gradually in the absence of air, moisture and occluded gases escape first. As the temperature rises, there is evolution of H_2S and some unsaturated hydrocarbons. At about 350°C, active decomposition of coal begins and continues with rise in temperature, accompanied by evolution of large quantities of gas and tarry vapours consisting mainly of hydrocarbons. Around 700°C, the amount of hydrogen in the evolved gases increases. Hydrogen is the major constituent of the gases liberated at higher temperature.

The volatile matter expelled thus is not present as such in coal but results from thermal decomposition of the coal material. The total amount of the volatile matter evolved and its composition depends, in addition to the quality of coal, on the temperature, rate of heating and the time for which heating is continued. For comparable results, therefore, conditions specified by ASTM are generally followed.

Volatile matter is the percentage loss in weight of coal (minus % of moisture) when it is heated in the absence of air for exactly 7 minutes at 950 ± 20°C, in a crucible of standard dimensions.

Significance

The volatile matter of coals may consist of combustible gases (like CO, H_2, CH_4 and other hydrocarbons) and incombustible gases (like CO_2 and N_2). The presence of incombustible gases is always undesirable as they do not have any heating value. Volatile matter of coals may be as high as 50% and when such coals are burnt, large proportion of the fuel distils over as gas or vapour which burns with the production of flame. Coals with high volatile matter will therefore burn with long smoky flames, the smoke causing pollution of the environment. The calorific intensity of such coals is less since the heat produced is distributed over large space. Such coals are therefore not suitable for production of steam where intense heat is required. Coals with high volatile matter are. however, suitable for manufacture of coal gas, particularly when the recovery of by-products is desired.

For complete combustion of the large volatile matter. secondary air will be required and the volume of the furnace will have to be increased to prevent the escape of unburnt vapours which otherwise will lead to loss of heat. The furnace design therefore depends on the volatile matter.

Whereas it is very easy to ignite a coal with a high volatile matter, burning of low volatile coals necessarily requires forced draft; its intensity increasing with decreasing volatile matter.

Caking quality of coal varies with its volatile matter. Coking power decreases with increase or decrease of volatile matter and coals having less than 14% volatile matter have no caking tendency.

5.4.1 *Determination of volatile matter of a coal sample*

Procedure

Weigh exactly about 1 g of the sample in a previously dried (at 950 ± 20°C) and weighed platinum crucible (diameter 2.5–3.5 cm, height 3–3.5 cm, capacity 10–20 ml) with a closely fitting lid which has a vent for the escape of volatile matter. Spread the sample evenly, close with the lid and place in a muffle furnace maintained at 950 ± 20°C and shut the door. After heating for exactly 7 minutes, take out the crucible and first bring down its temperature rapidly (to avoid oxidation of the contents) by placing it on a cold iron plate and then transfer the warm crucible to a desiccator to bring it to room temperature. Take the final weight of the crucible and the contents.

Precaution

If 'Sparking' (mechanical loss due to suddenly heating the sample) is observed, the sample-containing crucible should first be heated to raise the temperature to 600 ± 50°C in 6 minutes, followed by heating for exactly 6 more minutes at 950 ± 20°C.

Observations and Calculations

Let the weight of empty crucible = A g

Weight of crucible + sample, before heating = B g

Weight of crucible + sample, after heating $= C$ g

% Volatile matter $= \left[\dfrac{B-C}{B-A} \times 100 - \%\text{ moisture}\right]$

5.5 Fixed Carbon

The residue, remaining after the volatile matter of coal has been expelled, contains the mineral matter originally present and the non-volatile or 'Fixed' carbon. The fixed carbon in coal is thus calculated as follows:

$$\%\text{ Fixed carbon} = 100 - (\%\text{ moisture} + \%\text{ ash} + \%\text{ volatile matter}) \quad (5.2)$$

'Fixed' carbon varies inversely with volatile matter. It represents the amount of coal that will burn in the solid state with primary air forced or drawn through the hot bed of coal, thus having a bearing on furnace design. High 'Fixed' carbon ascertains high calorific value. The following empirical relationship (Goutel formula) holds true for high ranking coals:

$$Q = 82 \times FC + \alpha \times VM \quad (5.3)$$

Where Q = calorific value of coal (cals/g)

FC = % of fixed carbon

VM = % of volatile matter

and α = a factor that depends on the nature of volatile matter and its value decreases with rise in volatile matter.

Exercise

144. (i) 2.5 g of a coal sample was taken in a silica crucible weighing 19.35g. After heating in an electric oven at 105-110°C for 1 hour, the crucible weighed 21.765 g. The residue was then ignited at 700-750°C to a constant weight when the crucible weighed 19.595 g.

(ii) In another experiment, 2.5 g of the same sample was heated in a platinum crucible (weighing 19.345 g), Covered with a vented lid, at a temperature of 950 ± 20°C for exactly 7 minutes. After cooling, the crucible weighed 20.873 g.

(iii) Calculate the percentage of fixed carbon and the calorific value of coal using Goutel's formula, value of α being 80.

5.6 Ultimate Analysis

It is the determination of carbon, hydrogen, nitrogen, sulphur, phosphorus, ash and oxygen (by difference) in coal. The analysis is exact and helpful:

(1) In calculating the calorific value of coal by Dulong's formula

Gross or Higher Calorific Value H.C.V. $= \dfrac{1}{100}\left[8080\,C + 34460\left(H - \dfrac{O}{8}\right) + 2240\,S\right]$ kcal/kg

(5.4)

and $$\text{Lower Calorific Value (L.C.V.)} = \text{H.C.V.} - \frac{9}{100} H \times 587 \text{ kcal/kg} \quad (5.5)$$

where *C*, *H*, *O* and *S* represent the percentage of respective elements.

(2) In calculating the amount of air required for perfect combustion of the coal and

(3) In ascertaining whether a particular coal can be used in metallurgical operations or not.

5.7 Carbon and Hydrogen

Carbon being the major constituent of coal, its amount determines, to a large extent, the calorific value of the coal. Also, the carbon content increases with the degree of alteration in the natural coalification process (wood ~50%, peat 50–60%, lignites 60–75%, bituminous 75–92%, anthracite 92–98%). The percentage of carbon is therefore an important factor in assessing the rank and age for classifying coals. Coals with high carbon content usually require a small combustion chamber although there is no direct relation between the two.

The percentage of hydrogen in coal (3–6%) depends to a large extent on the hydrogen content of the original vegetable matter (3.5–8%) and to a small extent on the degree of transformation. Hydrogen content has a large bearing on the calorific value of the coal because the heating value of hydrogen per unit weight (34460 k cal/kg) is more than four times the heating value of carbon (8080 k cal/kg). However a small part of this heat is lost in evaporating the water formed form the combustion of H. Hydrogen is mostly associated with volatile matter of coal and therefore has a bearing on the size of the combustion chamber and to some extent on the suitability of coal for a specific use or purpose.

Principle of Determination

Carbon and hydrogen are determined by burning a weighed quantity of coal, in presence of CuO, in a stream of oxygen free from moisture and CO_2. The carbon and hydrogen of the sample are converted into CO_2 and H_2O:

$$C + O_2 \longrightarrow CO_2 \quad (5.6)$$

$$2H + \frac{1}{2}O_2 \longrightarrow H_2O \quad (5.7)$$

After complete oxidation and purification from interfering substances, the products of combustion are passed first through a water absorbing unit and then through a CO_2 absorbing unit. The increase in weights of these units gives respectively the weights of H_2O and CO_2 formed from which the percentage of hydrogen and carbon present in the sample can be calculated:

$$\% \text{ Hydrogen} = \frac{2}{18} \times \frac{\text{Weight of } H_2O \text{ formed}}{\text{Weight of coal burnt}} \times 100 \quad (5.8)$$

$$\% \text{ Carbon} = \frac{12}{44} \times \frac{\text{Weight of } CO_2 \text{ formed}}{\text{Weight of coal burnt}} \times 100 \quad (5.9)$$

5.8 Sulphur in Coal

Sulphur content of coals generally varies from 0.5 to 3%, with some exceptions of higher amounts. It occurs principally in two forms:

(a) As iron pyrites (occasionally as gypsum, $CaSO_4 \cdot 2H_2O$, also), mixed along with coal, most of which can be removed by
   (i) hand picking from larger sizes of coal, and
   (ii) gravity separation by washing smaller sizes of coals with water.

(b) As organic sulphur compounds forming part of the coal itself. The organic sulphur usually
   (i) ranges from 0.5 to 0.6%
   (ii) is almost uniformly distributed throughout the coal material, and
   (iii) can not be removed by 'cleansing' operations.

Significance

Although sulphur does have some heating value (2240 kcal/kg), SO_2 and SO_3 formed during the combustion of sulphur-containing coals are extremely harmful.

$$4FeS_2 + 11O_2 \longrightarrow 2Fe_2O_3 + 8SO_2 \qquad (5.10)$$

$$CaSO_4 \xrightarrow{\text{heat}} CaO + SO_3 \qquad (5.11)$$

$CaSO_4 + SiO_2$ (usually present in coals in small amounts)

$$\xrightarrow{\text{heat}} CaSiO_3 + SO_3 \qquad (5.12)$$

The gases cause environmental pollution and in moist atmosphere produce acids (H_2SO_3 and H_2SO_4) which cause corrosion of the metallic equipment around.

We can have an idea of the pollutional hazards from the fact that sulphur oxides equivalent to 3 tons of concentrated H_2SO_4 are discharged daily into atmosphere by a small Thermal Power Plant burning per day 100 tons of coal that contains only 1% sulphur.

When sulphur-containing coals are carbonised, 20–30% sulphur passes into coal gas, in the form of H_2S, CS_2 and thiophene, from which only H_2S is removed before the gas is supplied for domestic purposes.

Most of the 70–80% sulphur retained in coke (present partly as FeS, CaS or elemental sulphur) is oxidised to SO_2 when coke is burned and escapes to atmosphere. A part of FeS that escapes oxidation under mildly reducing conditions might fuse and lead to clinkering problem. When such cokes are used for metallurgical purposes, some sulphur might get mixed with metals and adversely affect their characteristics. Presence of sulphur in the coal used in ceramic industry adversely affects the product characteristics.

Rapid oxidation of iron pyrites present in coals exposed to air may lead to disintegration of the coal lumps thereby exposing fresh and easily oxidisable coal surface. Presence of iron pyrites is thus one of the factors responsible for spontaneous ignition of coal samples when exposed to air.

$$2FeS_2 + 7O_2 + 2H_2O \longrightarrow 2FeSO_4 + 2H_2SO_4 \qquad (5.13)$$

5.8.1 *Determination of sulphur content of a sample of coal*

Reagents Required

(1) Barium chloride solution (5%)

(2) Bromine solution (saturated)

(3) Concentrated hydrochloric acid

(4) Dilute hydrochloric acid

(5) Sodium hydroxide solution (5N)

(6) Methyl orange indicator solution

(7) Silver nitrate solution

(8) Eschka-Mixture — Thoroughly mix 2 parts by weight of sulphur-free MgO with 1 part by weight of sulphur-free anhydrous Na_2CO_3 and grind to pass through an ordinary flour sieve.

Theory

A known weight of powdered coal is fused with Eschka-mixture in presence of air. The sulphur in coal is oxidised to SO_2 which is absorbed by Eschka-mixture to produce the corresponding sulphites:

$$S + O_2 \longrightarrow SO_2 \qquad (5.14)$$

$$MgO + SO_2 \longrightarrow MgSO_3 \qquad (5.15)$$

$$Na_2CO_3 + SO_2 \longrightarrow Na_2SO_3 + CO_2 \qquad (5.16)$$

The sulphites are extracted with water and oxidised to sulphate using bromine water.

$$SO_3^{2-} + Br_2 + H_2O \longrightarrow SO_4^{2-} + 2H^+ + 2Br^- \qquad (5.17)$$

On adding barium chloride solution to the slightly acidified extract, the sulphate is precipitated as $BaSO_4$ which is filtered, washed, ignited and weighed.

$$SO_4^{2-} + Ba^{2+} \longrightarrow BaSO_4 \qquad (5.18)$$

From the weight of $BaSO_4$, the percentage of S in coal can be calculated as follows:

$$\% \, S = \frac{32}{233} \times \frac{\text{Weight of } BaSO_4}{\text{Weight of coal}} \times 100 \qquad (5.19)$$

Procedure

Transfer about 1 g of accurately weighed coal powder (60 mesh) to a silica, porcelain or platinum crucible. Add about 3 g of Eschka-mixture, mix thoroughly and cover with about 1 g more of Eschka-mixture. Place the crucible in a slightly slanting position in a muffle furnace at room temperature and heat gradually so as to raise the temperature to 850 ± 50°C in about 1 hour. Continue heating at this temperature for another 1 hour with occasional stirring with a stout nickel wire. When no black carbon particles are visible, allow the crucible to cool to room temperature. Remove the contents to a 200-ml beaker. Extract the residue completely using about 100 ml of hot distilled water. Heat for about half an hour and filter under suction. Wash the residue with hot water. Collect the filtrate and washings in a beaker, add 10–15 ml of saturated bromine solution and heat gently for a few minutes. Acidify with concentrated HCl and boil off the liberated bromine. Add two drops of methyl orange and just neutralise with NaOH solution. Add 1 ml of dilute HCl and heat the solution to boiling. Add slowly from a pipet, while stirring constantly with a glass rod, 10–15 ml. of hot 5 per cent solution of barium chloride. Cover the beaker with a clock glass, boil for 15 minutes and keep the solution hot, below its boiling temperature, by placing the beaker on a low temperature hot plate for 1 hour.

Filter, by decantation, through a Whatman filter paper No. 540/40. Transfer the precipitate to the filter paper and wash with small portions of hot distilled water until the washings are free from chloride (test a few drops of the filtrate with $AgNO_3$).

Hold the moist filter paper around the precipitate, loosely to avoid spattering, and place it in a clean platinum or silica crucible (previously ignited to a temperature of about 900°C, cooled in a desiccator and weighed). Cover the crucible with lid loosely to permit free access of air, place in a slightly inclined position in a cold muffle furnace and gradually raise the temperature to dry and smoke off the paper. Then raise the temperature to above 900°C and ignite to constant weight. Cool in a desiccator and weigh. Run a blank using the same amounts of reagents as in the regular determination.

Precautions

1. The reagents used must be sulphur-free
2. The sample should be heated with Eschka-mixture at a low heat in the beginning to avoid rapid expulsion of the volatile matter which may lead to
   (i) mechanical loss due to decrepitation, and
   (ii) incomplete absorption of SO_2 by Eschka-mixture.
3. A jet of hot water from a wash bottle may be used to aid the transfer of the precipitate to the filter paper. Any precipitate sticking to the walls of the beaker or to the glass rod may be loosened with a 'policeman'.
4. To avoid reduction of the $BaSO_4$ precipitate by carbon of the filter paper

$$BaSO_4 + 4C \xrightarrow{600^\circ} BaS + 4CO, \tag{5.20}$$

during ignition, the paper should be charred first at slow heat, without inflaming, and then the carbon should be slowly burnt off at a low temperature in free access of air.

Observations and Calculations

Let the weight of coal sample taken $= W$ g

Weight of $BaSO_4$ obtained from coal sample $= A$ g

Weight of $BaSO_4$ obtained in blank determination $= B$ g

% S in the sample
$$= \frac{32}{233} \times \frac{(A-B)}{W} \times 100 \qquad (5.19)$$

Exercises

145. Why is the precipitation of $BaSO_4$ carried out in a medium made only slightly acidic with HCl?

146. What is coprecipitation? How can the error due to coprecipitation of $Ba(NO_3)_2$ and $BaCl_2$ be avoided?

5.9 Nitrogen in Coal

Nitrogen content of coals generally varies between the limits 1–1.8% and comes from the proteins present in the parent vegetable matter. It has no calorific value and so its presence in coal is undesirable. If the combustion results in a very high temperature (as in Thermal Power Stations), a part of nitrogen is converted into its oxides which escape along with the flue gases and cause atmospheric pollution. During carbonisation, nitrogen of coal is converted to NH_3 which can be recovered from the volatile matter (coal gas) by scrubbing with water or dilute H_2SO_4.

5.9.1 *Determination of nitrogen in coal sample (Kjeldhal Gunning method)*

Reagents Required

1. Standard sulphuric acid (N/10)
2. Standard sodium hydroxide solution (N/10)
3. Sodium hydroxide solution (50%)
4. Concentrated sulphuric acid
5. Potassium sulphate
6. Anhydrous copper sulphate
7. Sucrose
8. Methyl red indicator solution
9. Red litmus paper.

Principle

A known weight of powdered coal is digested with concentrated H_2SO_4 in presence of K_2SO_4 (to raise the boiling point of H_2SO_4 and thus help in digestion) and $CuSO_4$, selenium powder or $CuSeO_3 \cdot 2H_2O$ (to act as catalyst). The nitrogen in coal gets converted into ammonium sulphate:

$$2N + 6H + H_2SO_4 \longrightarrow (NH_4)_2SO_4 \qquad (5.21)$$

As the solution becomes clear, it is treated with excess of NaOH or KOH and distilled when NH_3 gas is evolved:

$$(NH_4)_2SO_4 + 2\,KOH \longrightarrow 2NH_3 + K_2SO_4 + 2H_2O \qquad (5.22)$$

The ammonia so liberated is absorbed into a known volume (present in excess) of standard sulphuric acid

$$2NH_3 + H_2SO_4 \longrightarrow (NH_4)_2SO_4 \qquad (5.23)$$

and the unused acid (excess) is back titrated with standard NaOH solution using methyl red indicator:

$$H^+ + OH^- \longrightarrow H_2O \qquad (5.24)$$

The percentage of nitrogen in coal is calculated as follows:

$$\%N = \frac{\text{ml of acid used} \times \text{Normality of acid} \times 1.4}{\text{Weight of coal taken}} \qquad (5.25)$$

Procedure

Weigh accurately about 1 g of powdered coal and transfer it to a Kjeldahl's flask. Add about 10 g of K_2SO_4 and either 0.2 g of Se powder or 0.5 g of anhydrous $CuSO_4$ or 0.3 g of cupric selenite dihydrate ($CuSeO_3 \cdot 2H_2O$) followed by 25 ml concentrated H_2SO_4 along the sides of the Kjeldahl's flask to wash down any powder sticking to the neck or the sides of the flask. Close the flask with the loose glass stopper and shake to mix the contents. Place the flask in an inclined position and heat gently below the boiling point of the solution until the initial frothing ceases. Then raise the temperature to the boiling point of the solution and continue the digestion for at least two hours after the contents of the flask have become almost colourless.

Then cool the Kjeldahl's flask and transfer its contents to a round-bottom distillation flask. Wash the Kjeldahl's flask well with distilled water and transfer the washings to the distillation flask so that the total volume of the contents of the distillation flask are about 300 ml. Add a few chips of red litmus paper and fit the flask with a thistle funnel and a Kjeldahl's trap (to prevent splashing of the alkaline solution to the condenser during heating). Connect the Kjeldahl's trap to a condenser whose other end dips in 50 ml of N/10 H_2SO_4 taken in a conical flask. Through the thistle funnel, add concentrated NaOH solution until the contents of the flask are distinctly alkaline, as evidenced by the change in colour of the litmus paper from red to blue. Distil the contents of the flask for about one hour, maintaining the solution alkaline throughout. Discontinue boiling, and disconnect the flask, when the volume of the distillate in the conical flask becomes about 200 ml, and wash down the condenser with distilled water into the conical flask. Add 3-4 drops of methyl red indicator and titrate with N/10 NaOH until the colour changes from red to yellow and record the volume of acid used as *A* ml. Make a blank on 1 g of sucrose using the same amounts of other reagents as in the estimation on coal and record the volume of N/10 alkali used as *B* ml.

Precautions

(1) All the reagents used must be nitrogen-free.

(2) Before transferring the digested solution to the distillation flask, the Kjeldahl's flask should be cooled under tap. Also, dilution of the solution in distillation flask should be slow with shaking and cooling.

(3) To prevent bumping during distillation, a few porcelain chips or glass beads or about 1 g granulated Zn should be added.

(4) The solution in the distillation flask should be kept only moderately alkaline throughout. A considerable excess of NaOH often causes excessive frothing.

(5) When the solution froths excessively, a pea-sized piece of paraffin wax should be added to control it.

(6) To prevent back suction, the tube of the condenser, dipping in standard acid solution, should end in the shape of a funnel.

Observations and Calculations

Let the weight of coal taken $= W$ g

Volume of N/10 NaOH used in Blank (= amount of acid taken) $= B$ ml

Volume of N/10 NaOH used in sample determination (= amount of acid left unused) $= A$ ml

Therefore, amount of acid neutralised by NH_3

$$= (B - A) \text{ ml N/10 } H_2SO_4$$

$$\cong (B - A) \text{ ml N/10 } NH_3 \text{ solution}$$

$$\cong (B - A) \times \frac{1}{10} \times \frac{1}{1000} \times 17 \text{ g } NH_3$$

$$\cong (B - A) \times \frac{1}{10 \times 10^3} \times 17 \times \frac{14}{17} \text{ gN}$$

Therefore,
$$\%N = \frac{(B - A) \times 14}{10 \times 10^3} \times \frac{100}{W} = \frac{(B - A) \times 1.4}{W} \times \frac{1}{10} \qquad (5.25)$$

Exercises

147. Why is it necessary to add sufficiently large amount of concentrated H_2SO_4 during digestion?

148. What modification in the procedure is required if HgO or Hg is used as a catalyst during digestion with H_2SO_4?

149. How does addition of granulated zinc prevent bumping?

5.10 Oxygen Content of Coal

With the increase in the degree of transformation from wood to anthracite, the percentage of oxygen regularly decreases (wood = 35–43%; peat = 30–40%; lignites = 20–30%; bituminous = 5–20%; anthracite = 2–5%). The rank of the coal is therefore related to the oxygen content and decreases with increasing amount of oxygen. Oxygen in coal is supposed to be combined with hydrogen. An equivalent amount of hydrogen is therefore already burnt, i.e., not available for combustion, leading to a decrease in calorific value. It has been found that a 1% increase in oxygen content decreases the calorific value by about 1.7%. Increase in oxygen content of coals increases their tendency to retain inherent moisture. In case of bituminous coals, the caking power is known to decrease with increasing oxygen content. There is no direct method for determining the oxygen content of coal and has to be calculated by deducting the combined percentage of C, H, N, S and ash from 100.

$$\% \text{ Oxygen} = 100 - \text{percentage of } (C + H + N + S + \text{Ash}) \quad (5.26)$$

Exercises

150 (a) A sample of coal containing C,H,N,S and O was analysed as follows:

(i) 0.24 g of the sample on combustion gave 0.792 g of CO_2 and 0.0216 g of H_2O.

(ii) 1.4 g of the sample was Kjeldahlised and the ammonia liberated was absorbed in 50.0 ml of N/10 H_2SO_4. The resultant solution required 10.0 ml. of N/10 NaOH for complete neutralisation.

(iii) 3.2 g of the sample was analysed by Eschka method and gave 0.233 g of $BaSO_4$.

Using Dulong's formula, calculate the Gross and Net calorific value of the coal.

(b) Of the two types of Coal Analysis—Proximate and Ultimate– which one is more useful and why?

5.11 Combustion of a Carbonaceous Fuel – Flue Gas

When a fuel is burnt in air, the carbon and hydrogen of the fuel are converted into CO_2 and H_2O as per the following equation:

$$C_xH_yO_z + (x + y/4 - z/2)\,O_2 \longrightarrow x\,CO_2 + y/2 \cdot H_2O \quad (5.27)$$

and heat is liberated. The mixture of the gases which comes out of the furnace (combustion chamber) is known as Flue Gas. If the theoretical amount of air (containing exactly the same amount of O_2 as is required in accordance with the above reaction) is supplied and the combustion is complete, the flue gas consists of (in addition to water vapour) CO_2 and N_2. However, it is not possible to complete the combustion with the theoretical amount of air, and either a part of fuel escapes combustion or incomplete combustion of C to CO stage takes place. The flue gas, therefore, may contain small amounts of CO and/or hydrocarbons (C_xH_y). Escape of these gases (CO and C_xH_y) to the atmosphere causes environmental pollution. Also, a part of the total heat, that could be available from the fuel, is thus lost. On the other hand, if a large excess of air is used, the flue gas also contains this unused extra oxygen supplied. Also, a large excess of air added lowers the flame temperature (calorific intensity) and carries away a lot of sensible heat.

Hence for efficient utilization of heat available from a fuel, an optimum amount of air should be supplied. This amount cannot be theoretically computed. Other factors that affect efficient combustion are intimate mixing (or contact) of air and fuel and sufficient time to allow completion of combustion (governed by rate of combustion). Both these factors depend on the furnace design.

Whether efficient combustion is taking place or not is determined by the actual amounts of CO, O_2 or hydrocarbons present in the flue gas. Hence the analysis of the flue gas is of prime importance.

A large excess of CO in the flue gas indicates incomplete combustion, which means amount of air supplied is not sufficient. A large excess of O_2 in the flue gas indicates that air supply is much in excess. If both CO and O_2 are simultaneously present in the flue gas in appreciable amounts, combustion is incomplete, irregular and non-uniform, and improvement in the design of the combustion chamber, furnace or engine is required.

5.11.1 *Determination of the percentage of CO_2, CO, O_2 and N_2 in a flue gas (or automobile exhaust) by Orsat's apparatus*

Principle

A measured volume of the gas to be analysed is taken in a graduated burette which is connected through a capillary manifold to a series of pipets containing solutions of suitable absorbent reagents. The gas is successively forced through these solutions in a specific order. Each reagent removes a single constituent from the sample. The decrease in volume of the flue gas sample in each step gives the volume of the constituent removed in that step.

Solutions of Absorbent Reagents

(i) *Absorbent for CO_2* is a solution of 300 g of KOH pellets in one litre of distilled water prepared by stirring with a glass rod. The solution is stored in a rubber stoppered glass bottle and is decanted before use. When the gas sample is led into this solution, CO_2 content is retained by the solution as per the following reaction:

$$2KOH + CO_2 \longrightarrow K_2CO_3 + H_2O \qquad (5.28)$$

(ii) *Absorbent for O_2* is prepared by adding 80 g of pyrogallol crystals to 500 ml of KOH solution [prepared in (i)] in a glass-stoppered bottle kept in a cool place till dissolution is complete. When the gas sample is forced into this solution, the oxygen content is used up in oxidising pyrogallol.

(iii) *Absorbent for CO* is a solution of 75 g of Cu_2Cl_2 in 600 ml conc. HCl and 400 ml distilled water stored in a glass bottle containing copper turnings or wire. CO of the gas sample, when it is passed through this solution, is retained in the form of a complex:

$$Cu_2Cl_2 + 2CO + 4H_2O \longrightarrow 2Cu(CO)Cl{\cdot}2H_2O \qquad (5.29)$$

Sometimes the gas sample, after the major portion of its CO content has been absorbed by acid cuprous chloride, is passed through another pipet containing Cuprous Sulphate-β-Naphthol solution to remove the last trace of CO which combines with cuprous sulphate to form more stable $Cu_2SO_4{\cdot}2CO$ which does not give up CO as does $Cu(CO)Cl{\cdot}2H_2O$.

$$Cu_2SO_4 + 2CO \longrightarrow Cu_2SO_4{\cdot}2CO \qquad (5.30)$$

Procedure

(1) *Preparation of the apparatus*

After thoroughly cleaning the apparatus (Figure 5.1) apply grease to all the stoppers. Fill the jacket (A) around the burette (B) complete with water and the aspirator or the levelling bottle (C) upto two thirds with a confining liquid (a 25% NaCl solution acidified with dilute HCl to which a few drops of methyl red have been added). Take the three solutions, i.e., CO_2 absorbent, O_2 absorbent and CO absorbent in pipets p_1, p_2 and p_3 respectively. Close the stop cocks a_1, a_2, a_3 and open the three way stop cock D to atmosphere. Raise the aspirator (C) till the burette is filled with confining liquid. Now close cock D, open cock a_1 and lower C till the level of the solution in pipet p_1 is raised up to the fixed mark m_1. Close cock a_1. In a similar way, bring the level of the solutions in pipets p_2 and p_3 to fixed marks m_2 and m_3 respectively and close a_2 and a_3.

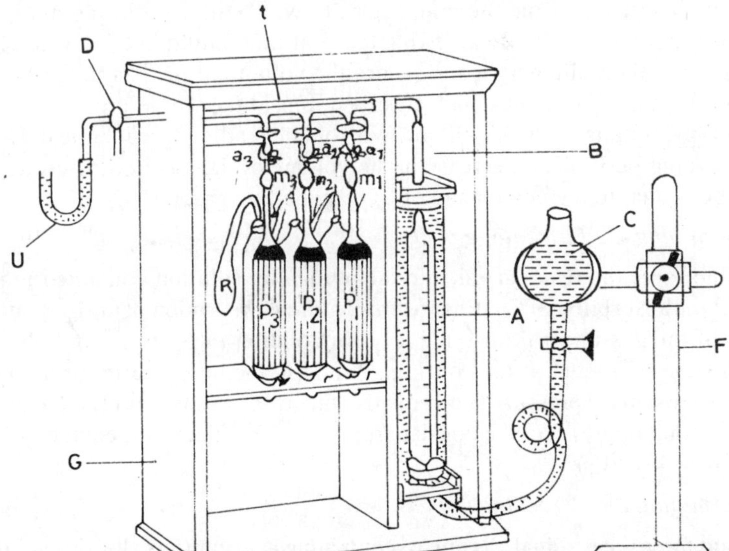

Fig. 5.1 Orsat apparatus

A - Jacket
B - Burette
C - Levelling bottle or Aspirator
D - Three way stop cock
E - Capillary manifold
F - Clamp stand
U - Tube containing anhydrous $CaCl_2$
G - Wooden case
R - Collapsible rubber bulb
a_1, a_2, a_3 stop cocks
p_1, p_2, p_3 pipets
m_1, m_2, m_3 fixed marks

(2) *Filling the Burette with the Sample*

Open cock D to atmosphere and raise C to expel air in the burette. Connect cock D to the flue gas supply (through U-tube containing anhydrous $CaCl_2$ secured by glass wool to retain any smoke particles and/or moisture present in the sample) and lower C to draw the gas into the burette. The gas gets mixed with the small amount of air present in the capillary manifold. Open cock D to atmosphere and expel the gas by raising C. Repeat the process of drawing in and expelling the gas twice or thrice. Finally, fill the burette completely with the gas, open D to atmosphere and raise the height of the aspirator C very carefully so as to equalise the levels of the liquid in the aspirator and in the burette in which it should stand at zero mark. Close the cock D. The burette now contains 100 ml of the gas under atmospheric pressure.

(3) *Determining the Percentage of CO_2*

Open a_1 and raise C to force the gas into p_1. A portion of the CO_2 content gets absorbed by KOH solution present in p_1. Lower C to take the gas back into the burette and again drive it into p_1. Repeat the process four to five times which should completely remove CO_2 from the sample gas. Lower C till the solution in pipet p_1 stands at the mark m_1 and close a_1. Bring the confining liquid in the aspirator and in the burette to the same height and record the volume of the gas in the burette. To be sure that all the CO_2 has been absorbed, repeat the whole process with the same pipet (p_1, containing KOH) till the two volume readings are identical. Let the constant reading be V_1. The decrease in the volume of the gas in the burette, i.e., V_1 ml is then equal to the percentage of CO_2 in the flue gas.

(4) *Determining the Percentage of O_2*

Open a_2 and force the residual gas into the pyrogallol solution contained in pipet p_2 to effect the absorption of O_2. Ensure the complete absorption of O_2 by a number of passes of the gas into p_2, raise the level of solution in p_2 to m_2 and close a_2. Record the constant level of the confining liquid in the burette after bringing the gas to the atmospheric pressure (i.e., equalise the levels of the liquid in the burette and the levelling bottle). Let the constant reading be V_2. Then percentage of O_2 in the flue gas = $(V_2 - V_1)$.

(5) *Determining the Percentage of CO*

Open a_3 and force the residual gas into p_3 containing acid cuprous chloride solution. Repeat as in (3) or (4) above with the difference that after each passage of the gas into the pipet p_3 and also before recording the volume of residual gas, pass the gas into the KOH pipet (p_1) to remove HCl vapours taken up from the acid cuprous chloride. Let the constant level, after complete absorption of CO, stand at V_3. Then percentage of CO in the flue gas = $(V_3 - V_2)$.

(6) *Determining the Percentage of N_2*

The gas left after absorption in p_1, p_2 and p_3 consists of N_2 which may contain small amounts of unburnt H_2 and/or hydrocarbons. Unless a detailed analysis is desired, these gases are reported as N_2.

$$\text{Percentage of } N_2 \text{ in the flue gas} = 100 - \left[V_1 + (V_2 - V_1) + (V_3 - V_2) \right]$$

$$= 100 - V_3$$

Precautions

1. Before filling the pipets with the respective solutions, the apparatus must be checked for leaks. (For this, close the cocks a_1, a_2, a_3 and open the cock D to atmosphere. Raise C to fill the burette with the confining liquid up to some definite level. Close D and slowly move the levelling bottle up and down. Any movement in the level of the liquid in the burette indicates a leak, which should be checked by properly greasing the stop cocks.
2. While expelling the air from the burette or forcing the gas into absorbent pipets by raising the level of the aspirator, care should be taken that no liquid enters the capillary manifold.
3. To avoid any contact between the atmospheric gases and the absorbent, the open end of the storage bulbs attached to the pipets should be closed with a collapsible rubber bulb.
4. While reading the volume of the gas in the burette, before or after absorption in a pipet, the aspirator must be held very close to the burette and the levels of the liquid in the burette and in the aspirator should be carefully adjusted to the same height (atmospheric pressure).

Exercises

151. What is the purpose of adding copper turnings or wire to the cuprous chloride solution?
152. Why are the pipets containing the absorbent solutions filled with glass tubes or beads?
153. What are the drawbacks of water being used as a confining liquid?
154. What is the purpose of adding a few drops of HCl and methyl red to the confining liquid?
155. Explain why it is necessary to follow a definite order for absorption of various constituents of a flue gas. List this order.
156. What is pyrogallol?
157. How does the presence of acidic gases like SO_2, H_2S, HCN, etc., in the flue gas affect the analysis?
158. How does the presence of H_2 or unburnt hydrocarbons affect the analysis?
159. What is the effect of temperature and pressure on the analysis?
160. Why is the graduated burette surrounded by a water-jacket?
161. Name a few modern instrumental methods of gas analysis and compare the usefulness of the Orsat apparatus against them.
162. Distinguish between a fuel gas and a flue gas. What is the approximate composition of a flue gas?

6

ORES AND ALLOYS

6.1 Iron Ore

Iron is easily attacked by humid atmosphere. So native iron in nature is very rare. It occurs principally as oxides :

(a) Heamatite or Red Iron Ore – Fe_2O_3

(b) Limonite or Brown Iron Stone – $Fe_2O_3H_2O$ or FeO (OH)

(c) Magnetite or Lode Stone – Fe_3O_4

Less abundant ores are sulphides and carbonate :

(d) Iron Pyrites – FeS_2

(e) Chalcopyrites – $CuFeS_2$

(f) Siderite or Sphatic Iron Ore – $FeCO_3$

The ores may be associated with small amounts of other metals like copper, cobalt, nickel, arsenic, molybdenum, chromium, etc. Some organic matter is also usually present. When the ores are to be used for making iron, the total iron content must be determined to ascertain the economic viability of the extraction process. The analysis involves the following steps :

(a) *Preparation of Solution of Iron Ore*

About 2 g (accurately weighed by difference method) of finely pulverised ore sample is transferred to an open crucible. It is first heated gently and then the sample is ignited by gradually raising the temperature to redness. The residue is cooled and using a platinum or nickel spatula, it is transferred to a porcelain basin. The silica crucible and spatula are washed with three 10 ml portions of conc. HCl and the washings are transferred to the porcelain basin. The basin is covered with a large funnel and heated on a hot plate for about half an hour. The resulting solution is diluted to about 50 ml with distilled water and allowed to cool when the undissolved matter settles down. The supernatant liquid is transferred by decantation to a 250-ml measuring flask. The residue is again heated with about 50 ml of 1:1 HCl for about half an hour and the liquid transferred to the measuring flask as before. The residue is washed with two 10-ml portions of 1:1 HCl, the washings transferred to the measuring flask and the volume is made up to the mark.

The decomposition of the iron ore with acid involves the following reactions :

$$Fe_2O_3 + 6H^+ \longrightarrow 2Fe^{3+} + 3H_2O \quad (6.1)$$

$$Fe_3O_4 + 8H^+ \longrightarrow 2Fe^{3+} + Fe^{2+} + 4H_2O \quad (6.2)$$

$$FeCO_3 + 2H^+ \longrightarrow Fe^{2+} + CO_2 + H_2O \quad (6.3)$$

$$FeS_2 + 2H^+ \longrightarrow Fe^{2+} + H_2S + S \quad (6.4)$$

(b) *Reduction of* Fe^{3+} *to* Fe^{2+}

Iron in a given solution is generally determined by titrating with a standard solution of $K_2Cr_2O_7$ or $KMnO_4$. Both methods determine the amount of Fe^{2+} ions in the solution. The ore solution prepared above contains some iron in the Fe^{3+} state also. So, in determining total iron, all the ferric should first be brought to the ferrous state. The following reducing methods may be used:

Method I—Reduction with Stannous Chloride

This method is used when the resulting solution can be titrated in presence of HCl, e.g., titration with $K_2Cr_2O_7$.

To a hot solution of iron ore, 3–4 N with respect to HCl, concentrated $SnCl_2$ solution is added in slight excess to ensure complete reduction of Fe^{3+} to Fe^{2+}:

$$Fe^{3+} + e \longrightarrow Fe^{2+} \times 2$$

$$Sn^{2+} \longrightarrow Sn^{4+} + 2e$$

$$2Fe^{3+} + Sn^{2+} \longrightarrow 2Fe^{2+} + Sn^{4+} \quad (6.5)$$

The solution is then cooled and the slight excess of $SnCl_2$ is removed by adding a saturated solution of $HgCl_2$.

$$Sn^{2+} \longrightarrow Sn^{4+} + 2e$$

$$2Hg^{2+} + 2Cl^- + 2e \longrightarrow Hg_2Cl_2$$

$$2Hg^{2+} + Sn^{2+} + 2Cl^- \longrightarrow Sn^{4+} + \underset{\text{(silky white ppt.)}}{Hg_2Cl_2} \quad (6.6)$$

A slight silky precipitate of Hg_2Cl_2 in the suspended form does not interfere to any significant extent.

6.1.1 *Determination of the amount of* Fe^{2+} *and total iron in the iron ore solution by* $K_2Cr_2O_7$

Reagents Required

1. Standard potassium dichromate solution (N/10)
2. Stannous chloride solution (5%)
3. Mercuric chloride solution (saturated)

4. Sodium diphenylamine sulphonate indicator solution
5. Syrupy phosphoric acid
6. Concentrated sulphuric acid
7. Concentrated hydrochloric acid

Theory

A known volume of the iron ore solution is titrated directly with standard $K_2Cr_2O_7$ solution in a medium acidified with H_2SO_4 (or HCl) and H_3PO_4 using sodium diphenylamine sulphonate as internal indicator :

$$Fe^{2+} \longrightarrow Fe^{3+} + e] \times 6$$

$$Cr_2O_7^{2-} + 14H^+ + 6e \longrightarrow 2Cr^{3+} + 7H_2O$$

$$Cr_2O_7^{2-} + 6Fe^{2+} + 14H^+ \longrightarrow 2Cr^{3+} + 6Fe^{3+} + 7H_2O \quad (6.7)$$

First permanent appearance of violet-blue colouration indicates the end-point and the volume of $K_2Cr_2O_7$ solution used corresponds to Fe^{2+} ions in solution.

For determining total iron, another suitable aliquot of the iron ore solution is first reduced by the $SnCl_2$ method and the total iron (which is now present as Fe^{2+}) is determined with standard $K_2Cr_2O_7$.

Procedure

(i) *Determination of Fe^{2+}*

With a pipet, transfer 50 ml of the iron ore solution to a titration flask. Add about 10 ml of dil. H_2SO_4 and 6–8 drops of sodium diphenylamine sulphonate indicator solution. Titrate with standard $K_2Cr_2O_7$ solution. When the colour of the solution changes to dark green, add 5 ml of syrupy phosphoric acid and continue titration until the colour changes to violet-blue. Take concordant readings.

(ii) *Determination of Total Iron*

With a pipet, transfer 20 ml of the iron ore solution to a titration flask. Add 5 ml of concentrated HCl and heat just to boiling. Add dropwise a 5% solution of $SnCl_2$ (in 1:1 HCl) with thorough shaking of the flask until the yellow colour due to $FeCl_3$ has given way to a faint green colour. Add 3–4 drops of $SnCl_2$ in excess to ensure complete reduction. Now cool the flask rapidly under tap and with a test tube add about 5 ml of a saturated solution of $HgCl_2$ in one lot and mix thoroughly. After 2 minutes, add 5 ml of syrupy phosphoric acid, 6–8 drops of the indicator solution and titrate rapidly with standard $K_2Cr_2O_7$ solution until there is a change of colour from deep green to violet blue. Take concordant readings.

Precautions

1. $SnCl_2$ solution should be added dropwise and should be thoroughly mixed after each addition.
2. $HgCl_2$ solution should be added to the cooled solution in one lot and not drop by drop.

3. The reaction between $SnCl_2$ and $HgCl_2$ being slow, 2–3 minutes should be allowed for its completion.
4. If on addition of $HgCl_2$,
   (a) no precipitate appears,
   (b) precipitate is heavy, crystalline and settles at the bottom of the flask, or
   (c) a grey or black colour of elemental Hg appears, the solution should be discarded and reduction should be repeated.

Observations and Calculations

Weight of iron ore taken = W g

Volume of iron ore solution prepared = 250 ml

Determination of Fe^{2+}

Volume of iron ore solution taken for each titration = 50 ml

Volume of N/10 $K_2Cr_2O_7$ used = A ml

$$\underset{(Fe^{2+}\text{ solution})}{N_1V_1} = \underset{(K_2Cr_2O_7)}{N_2V_2}$$

$$N_1 \times 50 = \frac{1}{10} \times A$$

$$\text{or} \quad N_1 = \frac{A}{10 \times 50}$$

Amount of Fe^{2+} per litre of iron ore solution $= \dfrac{A \times 56}{10 \times 50}$ g

$$\% \, Fe^{2+} \text{ in the ore} = \frac{A \times 56}{10 \times 50} \times \frac{250}{1000} \times \frac{100}{W}$$

$$= \frac{A \times 2.8}{W}$$

Determination of Total Iron

Volume of iron ore solution taken for each titration = 20 ml

Volume of N/10 $K_2Cr_2O_7$ used = B ml

$$\underset{(\text{Iron})}{N_1V_1} = \underset{(K_2Cr_2O_7)}{N_2V_2}$$

$$N_1 \times 20 = \frac{1}{10} \times B$$

$$\text{or} \quad N_1 = \frac{B}{10 \times 20}$$

Amount of iron per litre of the ore solution $= \dfrac{B \times 56}{10 \times 20}$ g

Therefore % iron in the ore $= \dfrac{B \times 56}{10 \times 20} \times \dfrac{250}{1000} \times \dfrac{100}{W}$

$$= \frac{B \times 7}{W}$$

Exercises

163. What is the importance of grinding in the analysis of iron ore?
164. What is the purpose of roasting the ore before its dissolution in acid?
165. Why is it necessary to add only a very small excess of $SnCl_2$ during the reduction of Fe^{3+}?
166. What happens if the $HgCl_2$ solution is added
    (a) when the solution is still hot,
    (b) drop by drop, or
    (c) it is not added in sufficient amount?
167. Why is the solution cooled immediately after the reduction is complete?
168. No precipitate is formed on adding $HgCl_2$ to an iron ore solution reduced with $SnCl_2$. What does it indicate?
169. What is the function of adding H_3PO_4 during the titration?
170. What is meant by Internal Indicator, External Indicator and Self Indicator? Give one example of each.
171. Name the external indicator used in the titration of Fe^{2+} ions with $K_2Cr_2O_7$. How is the indicator solution prepared?
172. Describe the detection of the end-point with this indicator.
173. What important precautions should be observed with the use of $K_3[Fe(CN)_6]$ as external indicator?
174. Why has the external indicator method been largely replaced with internal indicator methods?
175. What will happen if $K_3[Fe(CN)_6]$ is added to the titration flask?

Method II : *Reduction with Zn and dil. H_2SO_4*

This method is normally used when the resulting solution is to be titrated with $KMnO_4$. A known volume of the iron ore solution prepared in H_2SO_4 is taken in a conical flask and more acid is added to make the solution about 4 N with respect to H_2SO_4. Granulated zinc or zinc dust is added which reacts with H_2SO_4 to produce nascent hydrogen:

$$Zn \longrightarrow Zn^{2+} + 2e$$
$$H^+ + e \longrightarrow H] \times 2$$

$$Zn + 2H^+ \longrightarrow Zn^{2+} + 2H \quad (6.8)$$

The nascent hydrogen so produced reduces Fe^{3+} to Fe^{2+} :

$$Fe^{3+} + e \longrightarrow Fe^{2+}$$
$$H \longrightarrow H^+ + e$$

$$Fe^{3+} + H \longrightarrow Fe^{2+} + H^+ \quad (6.9)$$

Complete conversion of Fe^{3+} to Fe^{2+} is tested using KSCN or NH_4SCN solution as external indicator which produces blood red colour with Fe^{3+} ion :

$$Fe^{3+} + SCN^- \longrightarrow \underset{\text{Blood red colour}}{[Fe(SCN)]^{2+}} \quad (6.10)$$

The non-appearance of red colour indicates absence of Fe^{3+}. After the reduction is complete, excess zinc is removed either by filtration or by boiling until the whole of zinc dissolves.

6.1.2 *Determination of the amount of Iron in the iron ore solution by $KMnO_4$*

Reagents Required

1. Standard potassium permanganate solution (N/10)
2. Copper sulphate solution
3. Ammonium thiocyanate solution
4. Dilute sulphuric acid
5. Granulated zinc or zinc dust.

Theory

A known volume of the iron ore solution prepared in H_2SO_4 is reduced with Zn and H_2SO_4 as described above. After the removal of excess of zinc, the solution is diluted to a definite volume. A suitable aliquot of this solution is then titrated with standard $KMnO_4$ solution :

$$Fe^{2+} \longrightarrow Fe^{3+} + e] \times 5$$
$$MnO_4^- + 8H^+ + 5e \longrightarrow Mn^{2+} + 4H_2O$$

$$5Fe^{2+} + MnO_4^- + 8H^+ \longrightarrow 5Fe^{3+} + Mn^{2+} + 4H_2O \quad (6.11)$$

$KMnO_4$ acts as self indicator and appearance of light pink colour shows the end-point.

Procedure

With a pipet, transfer 50 ml of the iron ore solution (prepared in H_2SO_4) to a conical flask. Add about 5 g of AR Granulated zinc and 2 to 3 drops of copper sulphate solution. Place a short funnel in the mouth of the flask and add about 25 ml of dilute H_2SO_4 (6–8 N). Slightly warm and allow the reaction to continue, with occasional shaking, until the solution appears pale green. Remove a drop of the reaction mixture with a glass rod and mix it with a drop of NH_4SCN solution placed on a white glazed tile. If a blood red colour appears (which indicates that reduction is incomplete), allow the reduction to proceed until a drop of the reaction mixture when mixed with a drop of NH_4SCN does not give a red colour (Fe^{2+} ions give only a faint pink tint with NH_4SCN). Now filter the solution through a plug of glass wool placed in the neck of a funnel into a 100-ml measuring flask. (Alternatively, filter under suction through a gooch crucible with an asbestos fibre bed). Rinse the conical flask 2–3 times with 5 ml portions of dilute H_2SO_4 and pass the same through the filter. Make the volume upto the mark with distilled water.

Pipet out 20 ml of this solution into a titration flask, add about 5 ml of dilute H_2SO_4 and titrate against standard $KMnO_4$ solution until a light pink colour appears in the reaction mixture. Take concordant readings.

Precautions

(1) All the zinc used for reduction should be removed before titration. If zinc dust has been used for reduction, it is better to boil the solution to completely dissolve the zinc.

(2) After the reduction is complete, the determination should be made as quickly as possible to avoid atmospheric oxidation of Fe^{2+} to Fe^{3+}.

Observations and Calculations

Weight of iron ore taken	=	W g
Volume of iron ore solution prepared	=	250 ml
Volume of ore solution taken	=	50 ml
Volume made after reduction	=	100 ml
Volume of reduced solution taken for each titration	=	20 ml
Volume of $N/10$ $KMnO_4$ used	=	C ml

100 ml reduced solution = 50 ml iron ore solution

1 ml " " $= \frac{50}{100}$ ml " "

20 ml " " $= \frac{50 \times 20}{100}$ " "

$= 10$ ml

$$\underset{\text{(Iron ore solution)}}{N_1V_1} = \underset{(KMnO_4)}{N_2V_2}$$

$$N_1 \times 10 = \frac{1}{10} \times C$$

$$\text{or} \quad N_1 = \frac{C}{10 \times 10}$$

Amount of total iron per litre of the ore solution $= \dfrac{C}{10 \times 10} \times 56$ g

% of iron in the ore $= \dfrac{C \times 56}{10 \times 10} \times \dfrac{250}{1000} \times \dfrac{100}{W}$

$$= \frac{C \times 14}{W}$$

Exercises

176. During reduction with Zn and H_2SO_4, why is a small funnel placed in the mouth of the flask?
177. What is the function of a few drops of copper sulphate solution added during reduction?
178. Can Fe^{2+} ions be determined with $KMnO_4$ in presence of HCl? If not, why?
179. An iron ore has been dissolved in HCl. What procedure should be adopted to remove the interference of HCl in the determination of total iron by $KMnO_4$?
180. Reduction with Zn and H_2SO_4 has largely been replaced with that using zinc amalgam. Why?
181. How is zinc amalgam prepared?
182. List some other reagents that are used for the reduction of Fe^{3+} to Fe^{2+} during the determination of total iron in iron ore solution.

6.2 Copper Ore and Brass

Copper occurs both in the native form and in the combined state. The common ores are :

(a) Cuprite (red) Cu_2O

(b) Malachite (green) $Cu(OH)_2 \cdot CuCO_3$

(c) Azurite (blue) $Cu(OH)_2 \cdot 2CuCO_3$

(d) Copper glance Cu_2S

(e) Copper pyrites $CuFeS_2$

In addition to iron, metals like As, Sb, Zn, Ni, etc., bitumen and silica may be present as impurities.

Preparation of Solution of the Ore

About 1 g of the finely grinded copper ore is taken in a beaker and warmed with 15–20 ml of concentrated HNO_3 until the copper passes into solution. It is then evaporated on a hot plate to a volume of about 5 ml and after adding about 10 ml of concentrated HCl and an equal volume of 1:1 H_2SO_4, it is again evaporated until dense white fumes of SO_3 appear, indicating complete removal of nitrogen oxides. About 15 ml of distilled water followed by an equal volume of saturated bromine solution is then added and the mixture is heated. This ensures complete oxidation of any As and Sb to the pentavalent state. Excess of bromine is then boiled off. (If any blackish particles are seen in the solution, it may be sand which should be filtered.) The resulting solution which, in addition to Cu^{2+} may contain Fe^{3+}, As^{5+}, Sb^{5+}, Zn^{2+}, and Ni^{2+} as impurities, is diluted to 100 ml in a measuring flask.

Brass

The alloy brass consists mainly of copper and zinc but may contain small amounts of tin, lead and iron. A known weight of the sample (as drillings or turnings) is heated with concentrated HNO_3 when copper, zinc, lead and iron pass into the solution while tin is precipitated as metastannic acid ($SnO_2.H_2O$) which is filtered, dried, ignited and weighed to get the percentage of tin in the alloy. The filtrate is evaporated with concentrated H_2SO_4 which expels nitrogen oxides. The residue is treated with distilled water and filtered. The precipitate is washed, dried, ignited and weighed as $PbSO_4$ from which percentage of lead in brass is calculated. The filtrate which may contain Fe^{3+} and Zn^{2+} in addition to Cu^{2+} is diluted to 100 ml in a measuring flask.

6.2.1 *Determination of the amount of copper in a solution of copper ore or brass*

Reagents Required

1. Standard copper sulphate solution (N/10)
2. Sodium thiosulphate solution (N/10)
3. Ammonium thiocyanate solution (10%)
4. Dilute Acetic Acid
5. 1:1 ammonia solution
6. Freshly prepared starch solution
7. Sodium carbonate solution
8. Potassium iodide
9. Ammonium bifluoride.

Theory

When an excess of KI is added to a solution containing Cu^{2+} in neutral or slightly acidic medium, quantitative liberation of iodine takes place:

$$Cu^{2+} + I^- + e \longrightarrow CuI] \times 2$$

$$2I^- \longrightarrow I_2 + 2e$$

$$2Cu^{2+} + 4I^- \longrightarrow \underset{\text{White precipitate}}{2CuI} + I_2 \qquad (6.12)$$

The liberated iodine is titrated with standard $Na_2S_2O_3$ solution using freshly prepared starch solution as indicator near the end-point:

$$I_2 + 2e \longrightarrow 2I^-$$

$$2S_2O_3^{2-} \longrightarrow S_4O_6^{2-} + 2e$$

$$I_2 + 2S_2O_3^{2-} \longrightarrow 2I^- \quad (6.13)$$

In the absence of Fe^{3+} (test with NH_4SCN), the mineral acid is neutralised by dropwise addition of a solution of Na_2CO_3 until a slight precipitate or turbidity appears:

$$2H^+ + CO_3^{2-} \longrightarrow CO_2 + H_2O \quad (6.14)$$

$$Cu^{2+} + CO_3^{2-} \longrightarrow \underset{\text{(turbidity)}}{CuCO_3} \quad (6.15)$$

The turbidity is removed by dropwise addition of dilute CH_3COOH:

$$CuCO_3 + 2CH_3COOH \longrightarrow Cu^{2+} + CO_2 + 2CH_3COO^- + H_2O \quad (6.16)$$

and this adjusts the pH value at which As (V) and Sb (V) do not interfere.

When Fe^{3+} is present, dilute NH_3 solution is added to precipitate it:

$$Fe^{3+} + 3NH_4OH \longrightarrow \underset{\text{(precipitate)}}{Fe(OH)_3} + 3NH_4^+ \quad (6.17)$$

Ammonium bifluoride is then added which complexes iron and also

$$Fe(OH)_3 + 3NH_4HF_2 \longrightarrow [FeF_6]^{3-} + 3NH_4^+ + 3H_2O \quad (6.18)$$

adjusts the pH to above 3.2 where As (V) and Sb (V) do not interfere.

Procedure

(a) *Standardisation of Sodium Thiosulphate Solution*

Pipet out 20 ml of N/10 copper sulphate solution into an iodine titration flask. Add 1 g of AR KI dissolved in 10 ml of distilled water. Stopper the flask, mix the contents and keep in dark for 1–2 minutes. Titrate against standard $Na_2S_2O_3$ solution. When the colour of the reaction mixture is straw yellow, add 2 ml of freshly prepared strach solution, mix and continue titration until the blue colour becomes faint. Add 10 ml of thiocyanate solution (10% NH_4SCN or KSCN). An intense blue colour appears. Immediately titrate further to the complete disappearance of blue colour. Repeat to get concordant readings.

(b) *Estimation of Copper*

(i) *When Fe^{3+} is absent* (Test with NH_4SCN)

Pipet out 20 ml of ore or brass solution into an iodine titration flask. Add Na_2CO_3 solution dropwise with shaking until a slight turbidity appears in the solution. Add dilute CH_3COOH dropwise until the turbidity just disappears and the solution becomes clear. Add 1 g of AR KI dissolved in 10 ml of distilled water and proceed as in (a) above.

(ii) *When Fe^{3+} is Present*

Pipet out 20 ml of the ore or alloy solution into an iodine titration flask. Add 1 : 1 ammonia solution dropwise with shaking until the solution smells faintly of ammonia. To dissolve the precipitated $Fe(OH)_3$, add 2 g of NH_4HF_2 and shake. Add 1 g of AR KI dissolved in 10 ml of distilled water and proceed as in (a) above.

Precautions

(1) When much of bitumen or sulphur is present, the ore should be roasted before treating with acid.

(2) Brass surface, if not clean, should be washed with ether and then dried in air.

(3) Large excess of ammonia or Na_2CO_3 and CH_3COOH, as the case may be, should be avoided.

(4) Direct sunlight should be avoided as it catalyses atmospheric oxidation of I^-.

(5) After addition of thiocyanate solution, the titration should not be delayed.

Observations and Calculations

(a) *Standardisation of thiosulphate solution*

Volume of N/10 $CuSO_4{\cdot}5H_2O$ solution taken = 20 ml

Let the concordant volume of $Na_2S_2O_3$ solution used = A ml

$$\underset{\text{(copper sulphate)}}{N_1V_1} = \underset{(Na_2S_2O_3)}{N_2V_2}$$

$$\frac{1}{10} \times 20 = N_2 \times A$$

Therefore, N_2 (Normality of $Na_2S_2O_3$) $= \frac{1}{10} \times 20 \times \frac{1}{A} = \frac{2}{A}$

(b) *Estimation of Copper*

Let the weight of ore or alloy taken = W g

Volume of solution prepared = 100 ml

Volume taken for each titration = 20 ml

Let the concordant volume of $Na_2S_2O_3$ used = B ml

$$\underset{\text{(copper)}}{N_1V_1} = \underset{(Na_2S_2O_3)}{N_2V_2}$$

$$N_1 \times 20 = \frac{2}{A} \times B$$

$$\text{or} \quad N_1 = \frac{2}{A} \times \frac{B}{20} = \frac{B}{10A}$$

Amount of copper per litre of the ore or alloy solution $= \dfrac{B \times 63.5}{10\,A}$ g

Therefore, % of copper in ore or alloy $= \dfrac{B \times 63.5 \times 100 \times 100}{10\,A \times 1000 \times W}$

$$= \frac{B \times 63.5}{A \times W}$$

Exercises

183. Why is it necessary to oxidise arsenic and antimony to the pentavalent state?
184. Why is it necessary to remove nitrite and nitrogen oxides?
185. Describe some methods, other than evaporating the solution with H_2SO_4, for removing nitrite and nitrogen oxides from the solution.
186. How does iron interfere in the copper estimation?
187. Name some reagent, other than ammonium bifluoride, which can complex Fe^{3+} and prevent its reaction with I^-.
188. Name the factors that catalyse atmospheric oxidation of I^- ion.
189. What is the effect of low pH on the copper estimation?
190. On adding KI to the Cu^{2+} solution, a brown precipitate is seen. What is it due to?
191. Why is thiocyanate solution added towards the end-point of the titration?
192. During the titration of I_2 with thiosulphate, it is sometimes seen that at the end-point, the blue colour appears again and again. Explain.

6.3 Silver Ore and Silver Alloy

Silver coins or other commercially available silver alloys may contain varying amounts of zinc and nickel in addition to silver and copper. The silver content of the alloy determines its value and characteristics.

Preparation of the Solution

The surface of the alloy (in the form of a wire, thin sheet or coin) is cleaned by rubbing with emery cloth and subsequently washed with dry acetone to free it of grease. The alloy is then cut into pieces and 0.4–0.5 g (accurately weighed) of it is placed in a conical flask. Through a small funnel, placed in the mouth of the flask to avoid mechanical loss, 10–15 ml of dilute HNO_3 (1:1) is added. The flask is warmed gently on a water bath or a low-temperature hot plate until the alloy is completely dissolved. The solution is now boiled for 5–10 minutes to expel lower nitrogen oxides. The solution is cooled, transferred quantitatively to a 100-ml measuring flask – washing the conical flask and funnel with distilled water into the measuring flask – and made upto the mark.

The Ores, silver glance (Ag_2S) and copper-silver glance ($Ag_2S.Cu_2S$) can be similarly brought into solution by treating with nitric acid followed by filtration.

6.3.1 *Determination of the amount of silver in a solution of silver ore or alloy*

Reagents Required

1. Standard silver nitrate solution (N/20)
2. Potassium thiocyanate solution (N/20)
3. Ferric alum indicator solution
4. Dilute HNO_3.

Theory

Silver ions in a solution containing free nitric acid can be titrated with a standard solution of KSCN or NH_4SCN using a solution of ferric alum or ferric nitrate as indicator (Volhard method). When thiocyanate solution is added, a white precipitate of silver thiocyanate results:

$$Ag^+ + SCN^- \longrightarrow \underset{\text{(White ppt., } K_{sp} = 7 \times 10^{-13})}{AgSCN} \qquad (6.19)$$

After the whole of Ag^+ ions are precipitated, a slight excess of thiocyanate reacts with Fe^{3+} ions to produce a red colour:

$$Fe^{3+} + SCN^- \longrightarrow \underset{\text{Blood-red colour}}{[Fe(SCN)]^{2+}} \qquad (6.20)$$

Procedure

(a) *Standardisation of thiocyanate solution*

Pipet out 20 ml of standard N/20 $AgNO_3$ solution into a titration flask and add 5 ml of dilute HNO_3 and 1 ml of ferric alum indicator. Titrate against KSCN solution until a permanent faint reddish-brown colour appears. Repeat to get concordant readings.

(b) *Estimation of silver in the alloy or ore solution*

Repeat the above procedure taking 20 ml of the silver alloy or ore solution and get concordant readings.

Precautions

1. All the apparatus should be washed with distilled water.
2. Nitric acid must be free of the lower oxides of nitrogen.
3. Near the end-point, the titration must be carried out very slowly and with constant shaking.

Observations and Calculations

Weight of silver alloy (or ore) dissolved = W g

Volume of the alloy (or ore) solution prepared = 100 ml

(a) *Standardisation of KSCN solution*

Volume of standard N/20 $AgNO_3$ taken = 20 ml

Volume of KSCN solution used = A ml

$$\underset{(\text{KSCN})}{N_1V_1} = \underset{(\text{N/20 AgNO}_3)}{N_2V_2}$$

$$N_1 \times A = \frac{1}{20} \times 20$$

Therefore, N_1, normality of KSCN solution $= \frac{1}{A}$

(b) *Estimation of silver in the solution*

Volume of silver alloy (or ore) solution taken = 20 ml

Volume of KSCN solution used = B ml

$$\underset{(\text{KSCN})}{N_1V_1} = \underset{(\text{Silver solution})}{N_2V_2}$$

$$\frac{1}{A} \times B = N_2 \times 20$$

Therefore, N_2, normality of silver solution $= \frac{B}{A \times 20}$

Strength of silver solution $= \frac{B}{A \times 20} \times 108$ g/l

Therefore, percentage of silver in the alloy or the ore

$$= \frac{B \times 108}{A \times 20} \times \frac{100}{1000} \times \frac{1}{W} \times 100$$

$$= \frac{B}{A} \times \frac{54}{W} = \frac{54B}{A.W}$$

Exercises

193. Why is it necessary to avoid the use of tap water?
194. What is the function of nitric acid in the titration?
195. Why is it necessary to remove the lower oxides of nitrogen from nitric acid?
196. How is nitrous acid removed from nitric acid solution?
197. How is approach of the end-point detected?
198. Why is it necessary to shake the reaction mixture vigorously near the end-point?

6.4 Pyrolusite

Manganese does not occur free in nature. Its most commonly found ore is Pyrolusite (MnO_2) which sometimes contains small amounts of Brounite (Mn_2O_3), Manganite [$Mn_2O_3 \cdot H_2O$ or MnO (OH)], Hausmanuite (Mn_3O_4) and/or some other impurities. When the ore is to be used for metallurgical purposes (i.e., as a source of Manganese metal or for the production of ferromanganese), it is graded on the basis of its manganese content which can be determined by dissolving the ore in nitric acid and oxidising manganese to permanganic acid using sodium bismuthate ($NaBiO_3$). After filtering off the excess of the bismuthate, the permanganic acid is determined with a standard ferrous ion solution.

Being an efficient oxidising agent, pyrolusite is directly put to several industrial uses such as

(i) Production of Cl_2 by the action of conc. HCl (Weldon process)

(ii) A drier in paints

(iii) A depolariser in Leclanche cell and dry cell

(iv) Bleaching agent

(v) A catalyst in the preparation of oxygen by heating potassium chlorate ($KClO_3$).

For these purposes, the ore is graded on the basis of its oxidising capacity which is reported in terms of the percentage of 'Available Oxygen' in pyrolusite.

Available oxygen of pyrolusite is that part of oxygen which is available for oxidation of a reducing agent when the ore is treated with a strong acid:

$$MnO_2 + 2H^+ \longrightarrow Mn^{2+} + H_2O + \underset{\text{(available oxygen)}}{(O)} \quad (6.21)$$

6.4.1 *Determination of available oxygen in pyrolusite iodometrically*

Reagents Required

1. Standard sodium thiosulphate solution (N/10)
2. Potassium iodide solution (5%)
3. Freshly prepared starch solution
4. Conc. HCl

Theory

A known weight of the dry ore is heated with conc. HCl. Chlorine equivalent to available oxygen is liberated as per the following equation

$$MnO_2 + 4H^+ + 2e \longrightarrow Mn^{2+} + 2H_2O$$

$$2Cl^- \longrightarrow Cl_2 + 2e$$

$$MnO_2 + 4H^+ + 2Cl^- \longrightarrow Mn^{2+} + Cl_2 + 2H_2O$$

or

$$MnO_2 + 4HCl \longrightarrow MnCl_2 + Cl_2 + 2H_2O \quad (6.22)$$

The chlorine gas thus evolved is passed into a solution of KI when an equivalent amount of iodine is liberated

$$Cl_2 + 2e \longrightarrow 2Cl^-$$
$$2I^- \longrightarrow I_2 + 2e$$
$$Cl_2 + 2I^- \longrightarrow I_2 + 2Cl^- \quad (6.23)$$

which is titrated with a standard solution of sodium thiosulphate using starch solution as indicator near the end-point.

$$I_2 + 2e \longrightarrow 2I^-$$
$$2S_2O_3^{2-} \longrightarrow S_4O_6^{2-} + 2e$$
$$I_2 + 2S_2O_3^{2-} \longrightarrow S_4O_6^{2-} + 2I^- \quad (6.24)$$

The amount of thiosulphate consumed corresponds to the oxygen available from pyrolusite.

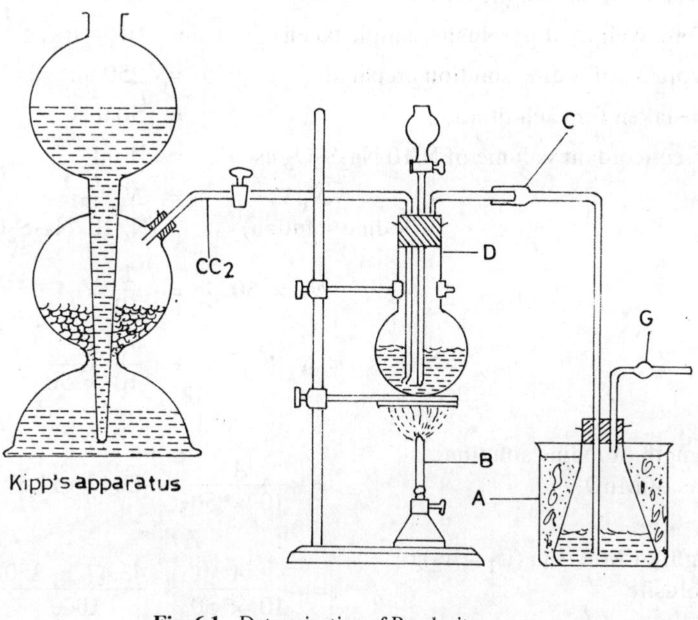

Fig. 6.1 Determination of Pyrolusite

A - Conical flask
B - Burner
C - Delivery tube
D - Distillation flask
G - Guard tube

Procedure

Set up the apparatus as shown in Figure 6.1. Place about 100 ml of a 5% solution of KI (A.R.Grade) in the conical flask (A). Weigh accurately the sample bottle containing the finely pulverised and dry pyrolusite ore. Transfer about 0.5 g of the sample to the distillation flask D and weigh the bottle again. Add about 30 ml of conc. HCl to the flask. Heat the distillation flask gently and from a Kipp's apparatus, pass a slow and steady stream of CO_2 into the contents of the distillation flask. After the ore has completely dissolved, raise the temperature and boil the contents for about 10 minutes. Now disconnect the delivery tube (C) from the distillation flask and transfer the contents of the conical flask into a 250-ml measuring flask. Rinse the conical flask, the delivery tube and the guard tube with KI solution into the measuring flask and make up the volume upto the mark. Transfer 50 ml of this solution into a conical flask and titrate against N/10 $Na_2S_2O_3$. When the colour of the solution turns straw yellow, add about 2 ml of freshly prepared starch solution and titrate further till the blue colour just disappears. Take concordant readings.

Observations and Calculations

Initial weight of sample bottle = w_1g

Final weight of the sample bottle = w_2g

Therefore, weight of pyrolusite sample taken = $(w_1 - w_2)$ g

Total volume of iodine solution prepared = 250 ml

Volume taken for each titration = 50 ml

Let the concordant volume of N/10 $Na_2S_2O_3$ used = A ml

Then

$$\underset{\text{(Iodine solution)}}{N_1 V_1} = \underset{(\text{N/10 Na}_2\text{S}_2\text{O}_3)}{N_2 V_2}$$

$$N_1 \times 50 = \frac{1}{10} \times A$$

or

$$N_1 = \frac{A}{10 \times 50}$$

(i) Strength of iodine solution in terms of MnO_2

$$= \frac{A}{10 \times 50} \times 43.47 \text{ g/l}$$

or weight of MnO_2 in $(w_1 - w_2)$g of pyrolusite

$$= \frac{A}{10 \times 50} \times \frac{43.47 \times 250}{1000} \text{ g}$$

Therefore, % of MnO_2 in pyrolusite

$$= \frac{A \times 43.47 \times 250}{10 \times 50 \times 1000} \times \frac{100}{w_1 - w_2}$$

(ii) Strength of iodine solution in terms of available oxygen

$$= \frac{A \times 8}{10 \times 50} \text{ g/l}$$

Weight of oxygen available from $(w_1 - w_2)$ g pyrolusite $= \dfrac{A \times 8}{10 \times 50} \times \dfrac{250}{1000}$ g

% Available oxygen $= \dfrac{A \times 8}{10 \times 50} \times \dfrac{250}{1000} \times \dfrac{100}{w_1 - w_2}$

$$= \frac{A \times 8}{20 \times (w_1 - w_2)}$$

(iii) Similarly, weight of Cl_2 that can be prepared from $(w_1 - w_2)$ g pyrolusite $= \dfrac{A \times 35.5}{10 \times 50} \times \dfrac{250}{1000}$ g

or Weight of Cl_2 obtainable from 100 g of pyrolusite $= \dfrac{A \times 35.5}{10 \times 50} \times \dfrac{250}{1000} \times \dfrac{100}{w_1 - w_2}$ g

$$= \frac{A \times 35.5}{20 \times (w_1 - w_2)} \text{ g}$$

Precautions

(1) To ensure complete and rapid decomposition of the ore by conc. HCl, it should be finely powdered in an agate mortar.

(2) To avoid the absorption of moisture by the powdered ore, it should be immediately transferred to a small stoppered weighing bottle.

(3) To make the apparatus airtight, all connections should be made with ground glass joints.

(4) CO_2 should be passed at a very slow rate otherwise some iodine might be lost from the solution.

(5) Before discontinuing heating at the end, disconnect the delivery tube to avoid back-suction of the absorption solution.

Exercises

199. Why is a stream of CO_2 passed through the decomposing mixture?

200. What are the contents of the guard tube ? What is its function?

201. Why is the conical flask kept immersed in ice-water?

202. What is the use of Pyrolusite in glass industry?

203. In what physical state does Pyrolusite occur in nature?

6.4.2 *Determination of available oxygen in pyrolusite by $KMnO_4$*

Theory

A known weight of the pyrolusite sample is treated with a known excess of an acidified solution of a reducing agent, such as (a) ferrous sulphate, (b) sodium oxalate, or (c) sodium arsenite. An equivalent amount of the reducing agent is oxidised as per the following reactions:

(a)

$$MnO_2 + 4H^+ + 2e \longrightarrow Mn^{2+} + 2H_2O$$

$$Fe^{2+} \longrightarrow Fe^{3+} + e] \times 2$$

$$MnO_2 + 4H^+ + 2Fe^{2+} \longrightarrow Mn^{2+} + 2Fe^{3+} + 2H_2O \quad (6.25)$$

(b)

$$MnO_2 + 4H^+ + 2e \longrightarrow Mn^{2+} + 2H_2O$$

$$C_2O_4^{2-} \longrightarrow 2CO_2 + 2e$$

$$MnO_2 + C_2O_4^{2-} + 4H^+ \longrightarrow Mn^{2+} + 2CO_2 + 2H_2O \quad (6.26)$$

(c)

$$MnO_2 + 4H^+ + 2e \longrightarrow Mn^{2+} + 2H_2O$$

$$AsO_3^{3-} + H_2O \longrightarrow AsO_4^{3-} + 2H^+ + 2e$$

$$MnO_2 + AsO_3^{3-} + 2H^+ \longrightarrow Mn^{2+} + AsO_4^{3-} + H_2O \quad (6.27)$$

The excess of the reducing agent is determined by titrating with a standard solution of $KMnO_4$, which itself acts as indicator:

(a)

$$MnO_4^- + 8H^+ + 5e \longrightarrow Mn^{2+} + 4H_2O$$

$$Fe^{2+} \longrightarrow Fe^{3+} + e] \times 5$$

$$MnO_4^- + 5Fe^{2+} + 8H^+ \longrightarrow Mn^{2+} + 5Fe^{3+} + 4H_2O \quad (6.28)$$

(b)

$$MnO_4^- + 8H^+ + 5e \longrightarrow Mn^{2+} + 4H_2O \mid \times 2$$

$$C_2O_4^{2-} \longrightarrow 2CO_2 + 2e] \times 5$$

$$2MnO_4^- + 5C_2O_4^{2-} + 16H^+ \longrightarrow 2Mn^{2+} + 10CO_2 + 8H_2O \quad (6.29)$$

(c)

$$MnO_4^- + 8H^+ + 5e \longrightarrow Mn^{2+} + 4H_2O\,] \times 2$$

$$AsO_3^{3-} + H_2O \longrightarrow AsO_4^{3-} + 2H^+ + 2e\,] \times 5$$

$$2MnO_4^- + 5AsO_3^{3-} + 6H^+ \longrightarrow 2Mn^{2+} + 5AsO_4^{3-} + 3H_2O \quad (6.30)$$

Procedure (with sodium oxalate)

Reagents Required

1. Standard sodium oxalate solution (N/10)
2. Standard potassium permanganate solution (N/10)
3. H_2SO_4 (10%)

Take two 250-ml conical flasks. Add the following reagents to each:

(i) About 0.1–0.15 g finely pulverised, dry and accurately weighed (by difference method using a weighing bottle) pyrolusite sample

(ii) 40 ml of standard N/10 sodium oxalate solution (with a pipet)

(iii) 40 ml of approximately 10% H_2SO_4 (with a measuring cylinder).

Place a small funnel in the mouth of each flask and keep them on a hot plate. Boil the contents gently until no black particles are visible in the flasks. Dilute the resultant dirty solution (usually milky) to about 100 ml and titrate slowly with N/10 $KMnO_4$ solution until a light pink colour appears. The amount of sodium oxalate consumed by pyrolusite is a measure of MnO_2 or available oxygen.

Precautions

(1) Due to presence of impurities, the solution after boiling may sometimes be coloured brown which interferes with the end-point. In such a case, the solution should be filtered before use.

(2) During boiling, the concentration of H_2SO_4 should not exceed 20% otherwise it may lead to decomposition of oxalic acid into CO and CO_2. To avoid this, some water may be added to the flask during boiling to make up for the loss.

Observations and Calculations

Flask A

Weight of sample $= w_A$ g

Volume of N/10 sodium oxalate added $=$ 40 ml

Let the volume of N/10 $KMnO_4$ consumed $= V_A$ ml

Therefore, volume of N/10 sodium oxalate used against w_A g of pyrolusite $= (40 - V_A)$ ml

and weight of MnO_2 in w_A g of sample $= \dfrac{(40 - V_A) \times 43.47}{10 \times 1000}$ g

Therefore, % MnO_2 $= \dfrac{(40 - V_A)}{10 \times 1000} \times \dfrac{43.47}{w_A} \times 100$

$= \dfrac{(40 - V_A) \times 43.47}{100 \times w_A}$

and % Available Oxygen $= \dfrac{(40 - V_A) \times 8}{100 \times w_A} = O_A$

Flask B

Weight of the sample $= w_B$ g

Volume of N/10 sodium oxalate = 40 ml

Let the volume of N/10 $KMnO_4$ used $= V_B$ ml

Therefore, % MnO_2 $= \dfrac{(40 - V_B) \times 43.47}{100 \times w_B}$

and % Available Oxygen $= \dfrac{(40 - V_B) \times 8}{100 \times w_B} = O_B$

Hence mean value of
% Available Oxygen $= \dfrac{O_A + O_B}{2}$

Exercise

204. Which of the three reducing agents suggested, namely ferrous sulphate, sodium oxalate or sodium arsenite is better and why?

6.4.3 *Determination of available oxygen in pyrolusite by KIO_3 (Andrews titration)*

Reagents Required

1. Standard potassium iodate solution (0.2N)
2. Potassium iodide solution (0.1M)
3. Conc. HCl
4. Carbon tetrachloride

Theory

In presence of a large excess of HCl, MnO_2 can oxidise iodide to iodine monochloride, via the intermediate formation of iodine :

$$MnO_2 + 4H^+ + 2e \longrightarrow Mn^{2+} + 2H_2O$$

$$2I^- \longrightarrow I_2 + 2e$$

$$MnO_2 + 4H^+ + 2I^- \longrightarrow Mn^{2+} + I_2 + 2H_2O \quad (6.31)$$

$$MnO_2 + 4H^+ + 2e \longrightarrow Mn^{2+} + 2H_2O$$

$$I_2 + 2Cl^- \longrightarrow 2ICl + 2e$$

$$MnO_2 + 4H^+ + I_2 + 2Cl^- \longrightarrow Mn^{2+} + 2ICl + 2H_2O \quad (6.32)$$

Thus the overall reaction may be written as

$$MnO_2 + I^- + 4H^+ + Cl^- \longrightarrow Mn^{2+} + ICl + 2H_2O \quad (6.33)$$

or $$MnO_2 + KI + 4HCl \longrightarrow MnCl_2 + ICl + KCl + 2H_2O \quad (6.34)$$

However, in presence of excess iodide, the reaction stops at the iodine stage (reaction 6.31). Thus, when to a known weight of the pyrolusite sample taken in a stoppered bottle is added a known excess of KI (titrated with standard KIO_3) and the mixture is shaken in presence of a large excess of HCl, a part of iodide is oxidised by MnO_2 present in the pyrolusite sample. The excess of KI is determined by titrating the resultant mixture (which may contain either I^- and I_2, or I_2 and ICl) with the standard KIO_3 in presence of chloroform or carbon tetrachloride when the iodide is first oxidised to I_2,

$$2IO_3^- + 12H^+ + 10e \longrightarrow I_2 + 6H_2O$$

$$2I^- \longrightarrow I_2 + 2e\,] \times 5$$

$$2IO_3^- + 12H^+ + 10I^- \longrightarrow 6I_2 + 6H_2O$$

or $$IO_3^- + 6H^+ + 5I^- \longrightarrow 3I_2 + 3H_2O \quad (6.35)$$

which is further converted to ICl :

$$IO_3^- + 6H^+ + Cl^- + 4e \longrightarrow ICl + 3H_2O$$

$$I_2 + 2Cl^- \longrightarrow 2ICl + 2e\,| \times 2$$

$$IO_3^- + 2I_2 + 6H^+ + 5Cl^- \longrightarrow 5ICl + 3H_2O \quad (6.36)$$

The disappearance of the colour of iodine in the organic layer indicates the end-point. The result is calculated from the amount of KI taken and the amount left after reaction with pyrolusite.

Procedure

Transfer 0.15–0.2 g of finely powdered pyrolusite sample to an iodine titration flask. Add 30 ml of about 0.1M solution of KI (AR) and 30 ml of conc. HCl. Stopper the flask and shake vigorously until the solid has dissolved. Now add about 5 ml of CCl_4 and titrate with 0.05M (0.2N) KIO_3 solution until there is no trace of violet colour of iodine in the carbon tetrachloride layer. Let the volume of KIO_3 used be V_A ml.

Blank—In another flask, take 30 ml of the same KI solution and 30 ml conc. HCl. Add 5 ml of CCl_4 and titrate with 0.05M KIO_3 solution in the same way as before. Let the volume of KIO_3 used be V_B ml.

Precautions

(1) After each addition of KIO_3 from the burette, the flask should be stoppered and shaken vigorously.

(2) The concentration of HCl should never be less than 3.5–4 N at the end-point.

Observations and Calculations

Weight of sample taken $= w$ g

Volume of 0.2N KIO_3 used in Blank $= V_B$ ml

Volume of 0.2N KIO_3 used against sample $= V_A$ ml

Therefore, MnO_2 in w g sample $\cong (V_B - V_A)$ ml of 0.2 N solution

$$= (V_B - V_A) \times 0.2 \times \frac{43.47}{1000} \text{ g}$$

% of MnO_2 in sample $$= \frac{(V_B - V_A) \times 0.2 \times 43.47 \times 100}{1000 \times w}$$

% Available oxygen $$= \frac{(V_B - V_A) \times 0.2 \times 8 \times 100}{1000 \times w}$$

$$= \frac{(V_B - V_A) \times 0.16}{w}$$

Exercises

205. What is the equivalent weight of KI and KIO_3 in the determination?

206. Why is vigorous shaking required during the titration?

207. What happens if the concentration of HCl is lower than that recommended?

6.5 Calcium Carbonate Minerals

Calcium carbonate occurs in nature in different forms — colourless, white to coloured materials and physical state being anything from a soft amorphous mass to hard rocks:

(1) *Dogtoothspar*, Icelandspar, Nailheadspar and Salinspar are varieties of calcite crystal and used as phosphors. Icelandspar is transparent and used in optical instruments.

(2) *Marble* is also crystalline and is used for building, ornamental, monumental and statuary purposes; as chips, abrasive for soaps and for neutralisation of acids.

(3) *Chalk* contains 90% calcite and may be soft, incoherent and porous to hard and crystalline. It is used for whiting, crayons, scouring and polishing preparations.

(4) *Limestone* is most widely distributed and is the most important source of $CaCO_3$. It usually contains $MgCO_3$ in amounts which may vary from traces to that present in Dolomite (containing $CaCO_3$ and $MgCO_3$ in equimolar amounts). Iron

oxide, aluminium oxide, silica (free or combined as clay or feldspar) and sulphur (as pyrites, FeS_2, or gypsum, $CaSO_4$) are other important minor constituents present in limestone and dolomite.

These minerals are used for a variety of purposes—building, whitening, manufacture of lime, portland and natural cements, fertiliser, soft glass, refractory material (dolomite), as a flux in various metallurgical processes, source of CO_2, railway ballast, macadam in lithography, cement concrete and asphalt concrete, in agriculture, as a solid diluent carrier in pesticides (dolomite).

The value and suitability of the mineral for any particular commercial purpose depends on the physical state of the mineral and to a large extent on the amount of $CaCO_3$, $MgCO_3$ and the amounts and nature of other impurities, for which a complete analysis of the mineral will be ideal. For routine work, however, the analysis of limestone and dolomite includes the following determinations:

(i) Available carbon dioxide
(ii) Loss on ignition
(iii) Impure silica or acid insoluble matter
(iv) Combined oxides ($Al_2O_3 + Fe_2O_3$)
(v) CaO, and
(vi) MgO.

(i) *Available Carbon Dioxide*

A known weight of the powdered sample is decomposed by heating with syrupy phosphoric acid. Using a current of CO_2 -free air, the CO_2 evolved from the sample is led through a water absorbent (anhydrous $CaSO_4$) into weighed bulbs containing KOH solution.

$$CaCO_3 + 2H^+ \longrightarrow Ca^{2+} + CO_2 + H_2O \tag{6.36}$$

The increase in the weight of KOH bulbs gives the weight of CO_2. The percentage available CO_2

$$= \frac{\text{Increase in the weight of KOH bulbs}}{\text{weight of sample taken}} \times 100 \tag{6.37}$$

(ii) *Loss on Ignition*

A weighed amount of the powdered air-dried sample, taken in a porcelain or platinum crucible, is dried at 110°C for 1 hour and the loss in weight, on percentage basis, is reported as superficially adsorbed moisture. The temperature of the crucible is then slowly raised to 1000–1100°C and the sample is ignited to constant weight. The percentage loss on ignition is a rough measure of the carbonate in the sample.

(iii) *Impure Silica or Acid Insoluble Matter*

It is determined by adding 1:1 HCl to about 1 g of the sample in a covered porcelain dish and heating on a water bath to dryness. The dry mass is then digested with 5–10 ml conc. HCl for about 5 minutes and diluted with distilled water. The resultant mixture is heated on a water bath for about 10 minutes and filtered using an ashless filter paper. The residue is washed with hot water, ignited along with the filter paper in a weighed silica crucible, cooled in a desiccator and weighed. The amount of the residue on percentage basis is reported as Impure Silica or Acid Insoluble Matter.

(iv) *Combined Oxides*

The combined filtrate and washings from the above determination are boiled with a few drops of conc. HNO_3 (to convert any Fe^{2+} to Fe^{3+}). About 1 g NH_4Cl is dissolved in the contents, followed by the addition of 2–3 drops of methyl red indicator. The solution is again heated to boiling and 1:1 ammonia solution added dropwise till the colour is just yellow. Boiling is continued for 2–3 minutes. Iron and aluminium are precipitated as hydroxides :

$$Fe^{3+} + 3OH^- \longrightarrow Fe(OH)_3 \quad (6.38)$$

$$Al^{3+} + 3OH^- \longrightarrow Al(OH)_3 \quad (6.39)$$

The precipitate is filtered, washed with hot water, dried and ignited $\left[2\,Fe(OH)_3 \longrightarrow Fe_2O_3 + 3H_2O,\ 2Al(OH)_3 \longrightarrow Al_2O_3 + 3H_2O\right]$ in a silica or platinum crucible to constant weight. The percentage of

$$\text{mixed oxides} = \frac{\text{Weight of residue}}{\text{Weight of sample taken}} \times 100 \quad (6.40)$$

(In addition to $Fe_2O_3 + Al_2O_3$, the residue may contain TiO_2, Mn_3O_4, $AlPO_4$ and $FePO_4$).

(v) *Calcium Oxide*

6.5.1 *Determination of the amount of calcium in limestone*

Reagents Required

1. Standard potassium permanganate solution (N/10)
2. Ammonium oxalate solution (8%)
3. 1:1 ammonia solution
4. Dilute H_2SO_4
5. Dilute HCl
6. Dilute HNO_3
7. Silver nitrate solution
8. Methyl red indicator solution

Theory

Calcium in a solution of limestone or dolomite (from which acid insoluble matter and oxides of iron and aluminium have been removed) is precipitated as calcium oxalate by adding ammonium oxalate in a medium made slightly alkaline using ammonia:

$$\left(\begin{array}{l}\text{COO}\\ |\\ \text{COO}\end{array}\right)^{2-} + \text{Ca}^{2+} \longrightarrow \underset{\text{White precipitate}}{\begin{array}{l}\text{COO}\diagdown\\ |\qquad\quad \text{Ca}\\ \text{COO}\diagup\end{array}} \qquad (6.41)$$

The precipitate is filtered, washed free of oxalate and chloride ions, and dissolved in dilute sulphuric acid :

$$\begin{array}{l}\text{COO}\diagdown\\ |\qquad\quad \text{Ca}\\ \text{COO}\diagup\end{array} + H_2SO_4 \longrightarrow \begin{array}{l}\text{COOH}\\ |\\ \text{COOH}\end{array} + Ca^{2} + SO_4^{2-} \qquad (6.42)$$

The oxalic acid liberated, which is equivalent to calcium in the original solution, is titrated with a standard solution of $KMnO_4$:

$$\begin{array}{l}\text{COOH}\\ |\\ \text{COOH}\end{array} \longrightarrow 2CO_2 + 2H^+ + 2e| \times 5$$

$$MnO_4^- + 8H^+ + 5e \longrightarrow Mn^{2+} + 4H_2O| \times 2$$

$$5\begin{array}{l}\text{COOH}\\ |\\ \text{COOH}\end{array} + 2MnO_4^- + 6H^+ \longrightarrow 10CO_2 + 2Mn^{2+} + 8H_2O \qquad (6.43)$$

Procedure

Add 2 drops of methyl red indicator to the combined filtrate and washings from the ammonia precipitation (removal of Fe_2O_3 and Al_2O_3) and acidify with dilute HCl (indicated by appearance of red colour). Reduce the volume to about 200 ml by evaporation. To the hot solution add slowly about 25 ml of a hot 8% ammonium oxalate solution with constant stirring with a glass rod (A slight precipitate appears due to the large excess of ammonium oxalate added). Add 1:1 ammonia solution dropwise until the red colour of the solution changes to yellow. Stir vigorously for 2–3 minutes and keep undisturbed for half an hour to let the precipitate settle down. Filter by decantation through a Whatman filter paper No.40 (540). Wash the precipitate repeatedly with about 5-ml portions of 0.1% ammonium oxalate solution until the filtrate is free from chloride (filtrate does not give a precipitate with $AgNO_3$ solution acidified with dilute HNO_3). Then wash the precipitate with small amounts of cold distilled water until the filtrate is free from oxalate ion (filtrate does not decolourise hot very dilute $KMnO_4$). Now place the funnel in the mouth of a 250-ml measuring flask, pierce a hole in the filter paper with a glass rod and wash down the precipitate into the flask with warm distilled water. Pour some dilute H_2SO_4 on the filter paper to dissolve any precipitate sticking to it. Now shake the flask to dissolve the precipitate (add more H_2SO_4, if necessary) and make up the volume upto the mark. Pipet out 50 ml of this solution into a titration flask, add about 10 ml of dilute H_2SO_4, heat to about 60°C and titrate with N/10 $KMnO_4$ solution until a light pink colour appears. Take concordant readings.

Alternatively dilute the combined filtrate and washings to a known volume (say 250 ml). Take 50 ml of this solution, precipitate calcium as oxalate, filter and wash free from chloride and oxalate, in a manner described above. Transfer the filter paper, along with the precipitate, to a titration flask, add 20 ml of dilute H_2SO_4 and about the same volume of distilled water. Shake to dissolve the precipitate, warm to about 60°C and titrate with N/10 $KMnO_4$ solution.

Precautions

(1) The washing liquid should always be taken in the beaker in which precipitation was carried out and then it should be transferred to the filter paper.

(2) A fresh portion of the washing liquid should be used only when the first portion has completely passed through the filter paper.

(3) For testing the filtrate for the presence of chloride or oxalate, a few drops of the filtrate should be collected directly from the funnel.

(4) Any precipitate sticking to the filter paper may be loosened by forcing on it a jet of distilled water from a wash bottle.

(5) Any precipitate sticking to the beaker after washing should be dissolved in dilute H_2SO_4 and transferred to the measuring flask.

Observations

Weight of the sample taken = w g

Volume of calcium ion/oxalate solution prepared = 250 ml

Volume taken for each titration = 50 ml

Let the concordant volume of N/10 $KMnO_4$ used = V ml

$$\underset{\text{(calcium)}}{N_1V_1} = \underset{(KMnO_4)}{N_2V_2}$$

$$N_1 \times 50 = \frac{1}{10} \times V$$

Therefore, $N_1 = \dfrac{V}{10 \times 50}$

Strength of calcium solution in terms of CaO $= \dfrac{V \times 28}{10 \times 50}$ g/l

Weight of calcium (as CaO) in w g of the sample $= \dfrac{V \times 28}{10 \times 50} \times \dfrac{250}{1000}$ g

% of calcium (as CaO) in the sample $= \dfrac{V \times 28}{10 \times 50} \times \dfrac{250}{1000} \times \dfrac{100}{w}$

$$= \frac{V \times 1.4}{w}$$

Exercises

208. Why is it necessary to wash the precipitate of calcium free of chloride and oxalate ions?

209. What improvement in the procedure is required for analysing dolomite, or limestone with higher magnesium content?

210. How is magnesium oxide determined in limestone or dolomite?

7

POLYMERS

The ability of carbon to form successive carbon-carbon bonds, i.e., catenation, has been used by man to synthesize giant molecules, macromolecules or better known as Polymers, which find extensive use (in all walks of life) as plastics, fibres, elastomers (rubbers), adhesives, etc.

The smaller and simpler substances which act as starting materials for the synthesis of polymers are known as *Monomers*. Bifunctional monomers (those having only two reactive sites) polymerise to give linear or branched molecules. Presence of polyfunctional monomers during the process of polymerisation may cause cross-linking of chains and thus result in the formation of a three-dimensional network.

Monomers containing a double bond *combine without loss* to give a product which is an exact multiple of the starting monomeric material, e.g., polymerisation of ethylene :

$$\underset{\text{(monomer)}}{nCH_2 = CH_2} \xrightarrow[\text{heat/light/high pressure}]{\text{a suitable catalyst}} \underset{\text{(polyethylene)}}{\left(CH_2 - CH_2 \right)_n}$$

The process is known as *Addition Polymerisation* and the resulting product is called a 'homopolymer'.

Simultaneous addition polymerisation of two different monomers gives rise to *Copolymers* which have properties intermediate between the homopolymers formed from individual monomers but different from the properties of their mechanical mixture.

Example :

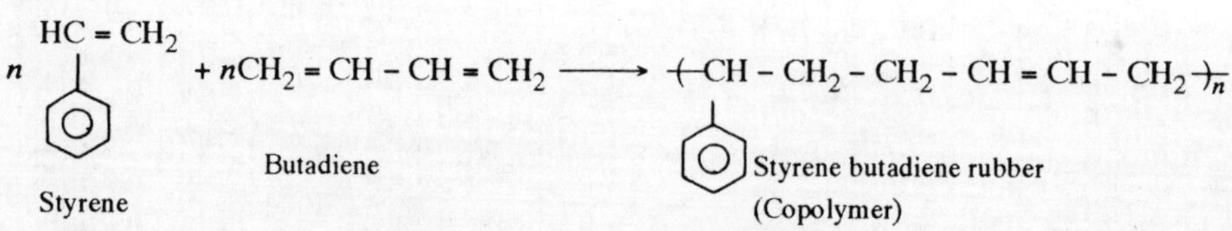

Monomers (identical or different) with suitable functional groups may interact with the elimination of small and simple molecules such as water, ammonia, HCl, H_2S, methanol, etc., to give the polymer. Such a process is called *Condensation. Polymerisation.* Notable examples are :

$$\underset{\text{Hexamethylene diamine}}{nH_2N\cdot(CH_2)_6\cdot NH_2} + \underset{\text{Adipoyl chloride}}{nClOC(CH_2)_4COCl} \longrightarrow \underset{\text{Nylon 6, 6}}{H\left[NH\cdot(CH_2)_6NH\cdot\overset{\overset{\displaystyle O}{\|}}{C}\cdot(CH_2)_4\overset{\overset{\displaystyle O}{\|}}{C}\right]_n Cl} + (2n-1)HCl$$

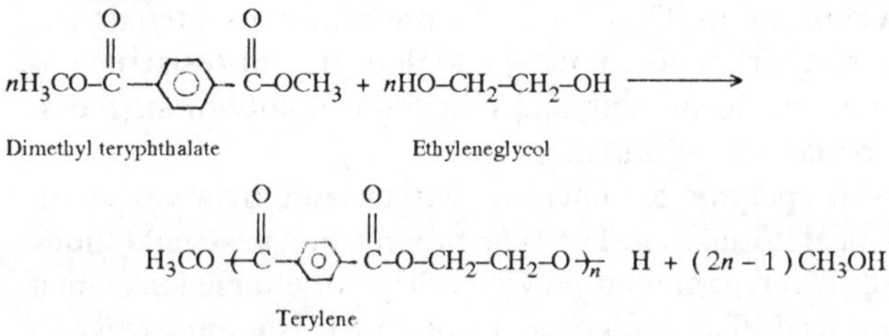

If the reactant molecules have more than two functional groups, the polymerisation process results in cross-linking between different chains which serves to turn the entire substance, in some degree, into a single three dimensional network. The most notable example of this type is the condensation between phenol and formaldehyde:

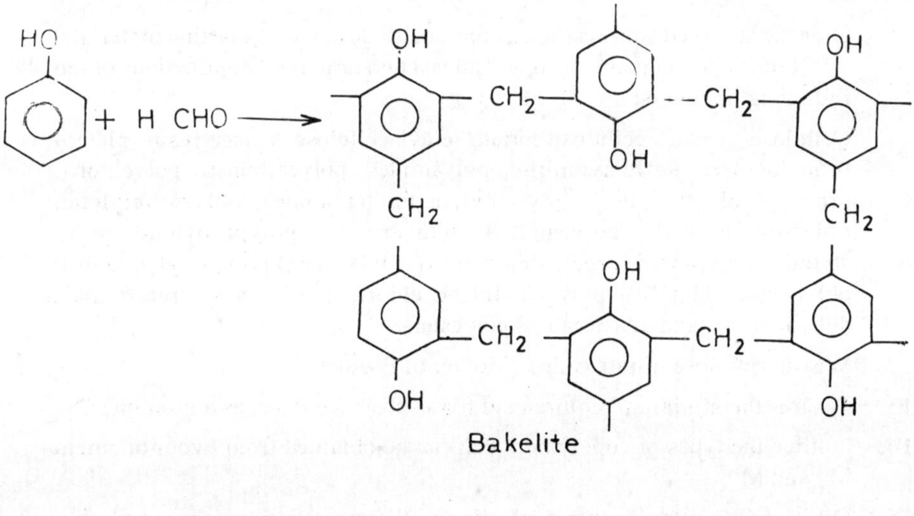

Bakelite

Classification of Polymers

Based on characteristics, intended use and performance of the finished product, polymers are broadly divided into three main categories : (i) plastics, (ii) fibres, (iii) elastomers.

Plastics are a wide variety of polymer-based composite materials which possess appreciable mechanical strength (they have stiff chains at room temperature) and are characterised by plasticity, i.e., they can be formed or moulded into useful shapes by application of heat and pressure. Materials which possess plasticity at some stage during their formation are also included in this category. Based on their thermal behaviour, plastics have been subdivided into *Thermoplastics* (that soften and flow on heating) and *Thermosetting* or thermohardening plastics (that set or harden on heating).

A *fibre* is any material whose minimum length is at least 100 times its average diameter which should be less than 0.25 mm. Fibres are natural (cotton, silk, wool) or artificially prepared (polyamides, polyesters, polyacrylics) long-chain polymers with average molecular weight of 15,000 or more. Fibre-forming materials are characterised by high softening or melting points, a high degree of resistance to chemicals and solvents, high tensile strength and very high rigidity or stiffness. They, however, undergo irreversible deformation.

Elastomers or rubbers are polymeric materials characterised by a very high degree of reversible or elastic deformation. They can be stretched to several times (7 to 8) their original length, but regain their previous shape or dimensions when the stretching force is removed. The molecular chains of elastomers exist in randomly coiled state and their elastic behaviour can be compared to that of the spring of a chest expander which uncoils and recoils on application and removal, respectively, of an elongating force. The glass-transition temperatures of elastomers are very much below their use temperature.

Exercises

211. Give the accepted abbreviations, common trade name(s), starting materials, structural/repeating unit, class and atleast two important applications of the following polymers:

     Cellulose acetate, cellulose nitrate, ethyl cellulose, epoxy resin, phenol formaldehyde, polyacryonitrile, polyamides, polycarbonate, polychloroprene, polyethylene, poly(ethylene terephthalate), polyisobutylene, polyisoprene(cis), poly(methyl methacrylate), polypropylene, polystyrene, poly(tetrafluoroethylene), poly(vinylacetate) poly(vinyl alcohol), poly(vinyl chloride), poly(vinylidene chloride), silicones, urea-formaldehyde resins and unsaturateed polyesters.

212. Why is cellulose nitrate called 'Mother of Plastics'?

213. What is the minimum requirement for a substance to act as a monomer?

214. Outline the types of copolymers that can be obtained from two monomers M_1 and M_2.

215. Differentiate between thermoplastic and thermosetting materials and give examples.

216. What are the structural requirements of a polymer to be useful as a fibre-forming material?

217. Why do the molecular chains of elastomers get recoiled when the stretching force is removed?

218. What is Glass-Transition temperature? Discuss its significance.

219. What is elastic range? How can it be widened?

220. What are the chief physical characteristics expected of an elastomer? How are these achieved in a new product through 'Molecular Engineering'?

221. How does the chain structure of rubbers compare with that of plastics?

7.1 Preparation of Polymers

Some of the important techniques employed for the production of polymers are outlined below:

(a) *Bulk Polymerization*

A free radical catalyst or initiator is dissolved in the monomer which is then heated and stirred in a suitable vessel. The polymerisation is exothermic and dissipation of heat through cooling may be required. As the reaction progresses, the system becomes viscous making stirring difficult. The method is economical and the product is of high purity. The technique is used for preparing polyvinyl chloride (PVC), polystyrene (PS) and polymethymethacrylate (PMMA).

(b) *Solution Polymerisation*

The monomer and catalyst (free radical, cationic or anionic) are dissolved in a suitable inert solvent. The resulting solution is heated and stirred. The presence of solvent helps in heat dissipation and in controlling viscosity. The solvent may interact and reduce the molecular weight of the product whose isolation from the solution is uneconomical unless it is insoluble. The technique is employed where the polymer is to be used in solution form such as in the case of adhesives and surface coatings. Polyacrylonitrile, polyisobutylene and certain block copolymers are produced by this method.

(c) *Suspension Polymerisation*

A solution of the catalyst in the monomer is dispersed as fine droplets in an inert solvent, usually water. To stabilise the suspension, water-soluble protective colloids such as polyvinyl alcohol, methyl cellulose or starch are added and the mixture is kept stirring continuously. The problems of heat dissipation and viscosity increase are absent. The method gives a fairly high molecular weight product in the form of easily separable beads that can be filtered or centrifuged and waterwashed to remove the protective colloids. The technique is employed for the production of PVC, PS and styrene divinylbenzene copolymer (used for making ion-exchange resins).

(d) *Emulsion Polymerisation*

The particle size of the monomer is reduced to colloidal dimensions by more vigorous stirring and use of synthetic surfactants (anionic, cationic or non-ionic) in place of protective colloids used in suspension polymerisation. Usually water soluble catalysts such as persulphate or hydrogen peroxide are used. Thermal dissipation and viscosity problems are absent. Both the rate of polymerisation and the molecular weight of the product formed are very high. The product, which is in the form of fine particles dispersed in water (called latex), can be used directly as adhesive or an emulsion paint, or it can be isolated by coagulating with an electrolyte. The technique is employed for the industrial production of PVC, polychloroprene (PCP), polybutadiene, polyacrylates and polymethyl methacrylate.

(e) *Melt Polycondensation*

The reactants are heated together in exact stoichiometry above the melting point of the product, at which temperature the starting materials and the product must be thermally stable. Oxygen has to be excluded from the reaction chamber to avoid oxidation at the high temperature. Increase in viscosity makes removal of the by-product extremely difficult towards the end (unless high vacuum is applied) which may prevent formation of a high molecular weight product. The molten polymer is usually sent directly for spinning, extrusion, etc. The technique is usually applied for the preparation of polyesters and polyamides.

(f) *Interfacial Polycondensation*

The reaction takes place at the interface between solutions of the reactants in immiscible solvents. Increasing the interface by thorough agitation of the two solutions substantially increases the rate of polymerisation. Exact stoichiometry is not necessary and a high molecular weight product can be easily formed. Being very simple, the method is widely used for the production of polyamides, polysesters, polyurethanes and polysulphonamides. Difference in the reactivity of materials can be utilised to prepare ordered copolymers which otherwise are very difficult to produce. The technique is however limited to reactants having highly reactive functional groups that can readily react at the ambient temperatures.

7.1.1 *Conversion of (à) methylmethacrylate, and (b) styrene into the corresponding polymers*

Reagents Required

1. Methylmethacrylate monomer
2. Styrene monomer
3. Lauroyl peroxide [dodecanoyl peroxide, $(C_{11}H_{23}COO)_2$]

Theory

Methylmethacrylate and styrene can be bulk-polymerised by heating in presence of a peroxide catalyst:

$$nCH_2 = \underset{COOCH_3}{\overset{CH_3}{\overset{|}{\underset{|}{C}}}} \xrightarrow[\text{or Benzoyl peroxide}]{\text{Lauroyl peroxide}} \left(CH_2 - \underset{COOCH_3}{\overset{CH_3}{\overset{|}{\underset{|}{C}}}} \right)_n$$

Methylmethacrylate Polymethylmethacrylate

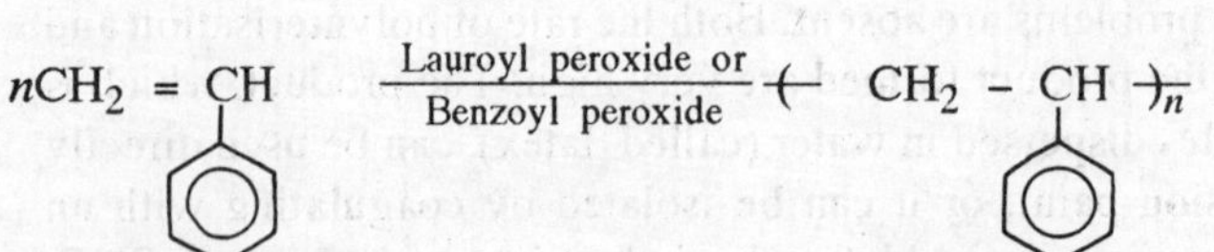

Styrene Polystyrene

Procedure

(a) *Polymerisation of methylmethacrylate*

Add a pinch (20–30 mg) of lauroyl peroxide to 3-4 ml of methylmethacrylate taken in a hard glass test tube and shake to dissolve. Using a clamp stand, secure the test tube in a water bath maintained at 60°C. As the polymerisation progresses, the liquid in the test tube slowly starts thickening and in about one hour changes to a transparent solid.

(b) *Polymerisation of styrene*

Heat a solution of lauroyl peroxide (20–30 mg) in styrene (3–4 ml) taken in a test tube at 100°C in a boiling water bath. During the course of about one hour, the clear monomer liquid gradually changes to a colourless solid.

Precautions

1. The experiment should preferably be carried out in a fume-cupboard as the monomer vapour is toxic.
2. The monomers should preferably be freshly distilled.

Exercises

222. Why should the vinyl monomers (e.g., methyl methacrylate, styrene, etc.) be distilled before polymerisation?

223. Explain why the softening temperature of polystyrene is lower than that of polyethylene although the former contains polar groups.

7.1.2 *Preparation of Nylon 6,6 and Nylon 6,10 and to draw them in the form of a thread*

Reagents Required

1. Hexamethylene diamine solution (~ 1%)
2. Adipoyl chloride solution (~ 2%)
3. Sebacoyl chloride solution (~ 3%)

Theory

When an aqueous solution of the diamine is carefully brought in contact with the solution of the diacid dichloride in an organic solvent (immiscible with water), the reactants diffuse to the interface where the polycondensation reaction takes place:

(i) $$\underset{\text{Hexamethylene diamine}}{n\mathrm{H_2N-(CH_2)_6-NH_2}} + \underset{\text{Adipoyl chloride}}{n\mathrm{Cl-\overset{\overset{\displaystyle O}{\|}}{C}-(CH_2)_4-\overset{\overset{\displaystyle O}{\|}}{C}-Cl}} \longrightarrow$$

$$\mathrm{H} \,+\!\!\!\!- \mathrm{NH \cdot (CH_2)_6 \cdot NH - \overset{\overset{\displaystyle O}{\|}}{C} - (CH_2)_4 - \overset{\overset{\displaystyle O}{\|}}{C}} -\!\!\!\!+_n \mathrm{Cl} + (2n-1)\,\mathrm{HCl}$$

Nylon 6,6 [Poly(hexa methyleneadipamide)]

(ii) $$n\mathrm{H_2N - (CH_2)_6 - NH_2} + n\,\mathrm{Cl - \overset{\overset{\displaystyle O}{\|}}{C} - (CH_2)_8 - \overset{\overset{\displaystyle O}{\|}}{C} - Cl} \longrightarrow$$

Sebacoyl chloride

$$\mathrm{H} \,+\!\!\!\!- \mathrm{NH - (CH_2)_6 - NH - \overset{\overset{\displaystyle O}{\|}}{C} - (CH_2)_8 - \overset{\overset{\displaystyle O}{\|}}{C}} -\!\!\!\!+_n \mathrm{Cl} + (2n-1)\,\mathrm{HCl}$$

Nylon 6, 10 [Poly (hexamethylenesebacamide)]

The polymer film, which is insoluble in both the solvents, is formed at the interface and, using a pair of forceps, can be drawn out in the form of a thread or a rope.

Procedure

Place about 50 ml of the solution of the appropriate diacid dichloride (in CCl_4) in a 100-ml tall-form beaker. Carefully pour about 25 ml of the aqueous solution of hexamethylene diamine along the sides of the beaker so that it forms a separate layer over the heavier CCl_4 solution layer. When the polymer film appears at the interface, hold it gently at its centre with a pair of forceps and lift out of the beaker.

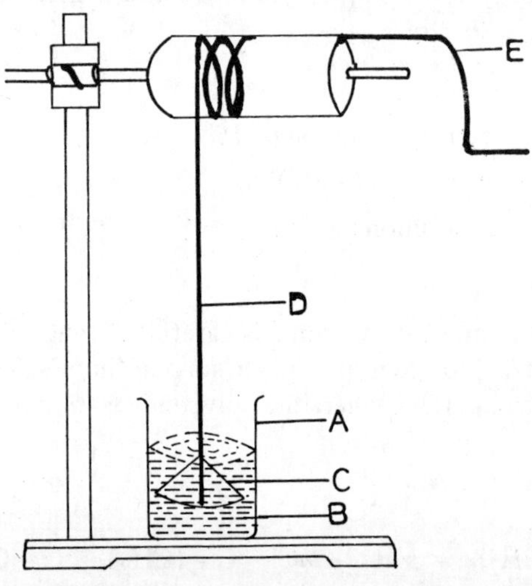

Fig. 7.1 Preparation of Nylon Thread

A - Tall-form beaker

B - Diacid dichloride solution in CCl_4

C - Hexamethylene diamine solution in water

D - Nylon thread

E - Mechanical wind-up device

As it comes out in the form of a thread or rope, wrap it around a thick glass rod or a test tube.

For continuous and automatic wrapping of the thread, a mechanical or an electrical wind-up device may be placed up the beaker.

Precautions

1. Hexamethylene diamine is corrosive and irritates the skin. So care should be taken to avoid its contact with skin.
2. The addition of the diamine solution should be very slow and along the sides of the beaker so that the lower layer of the acid chloride solution is not disturbed and there is a *clear* surface of separation between the two layers.
3. If the thread is not easily drawn, the beaker should be placed in a water bath whose temperature should be very slowly raised, otherwise the solutions may get mixed. Rise in temperature will increase the rate of polymerization resulting in a rapid increase in the length of the polymer chain.

Exercises

224. How are the Nylons named?
225. Explain why the softening point of Nylon 6, 10 is lower than that of Nylon 6,6.
226. How is molecular engineering used to prepare Nylons of very high melting points and high tensile strength?
227. What is meant by 'The Nylon Rope Trick'?

7.1.3 *Preparation of Hexamethylene diamine–Adipic acid salt*

Reagents Required

1. Adipic acid
2. Hexamethylene diamine
3. Absolute ethyl alcohol.

Theory

When an alcoholic solution of hexamethylene diamine is mixed with an alcoholic solution of adipic acid at room temperature, the salt of the diamine and diacid in the ratio of 1:1 (called the Nylon 6,6 salt) soon separate out:

$$H_2N \cdot (CH_2)_6 \cdot NH_2 + HOOC \cdot (CH_2)_4 \cdot COOH \longrightarrow [\,H_3N^+ (CH_2)_6N^+H_3\,]\,[\,^-OOC \cdot (CH_2)_4 \cdot COO^-\,]$$

Significance

In order to produce a high molecular weight nylon from the polycondensation of a diamine with a diacid, the two reactants must be taken in stoichiometric amounts. This stoichiometry can be easily achieved by preparing a balanced salt of the diamine and the diacid. Therefore, the preparation of the salt usually constitutes the first stage in the manufacture of nylon by melt condensation technique.

Procedure

Add 55–60 ml of absolute alcohol to 7.3 g of adipic acid taken in a 250-ml conical flask and dissolve by warming. Cool to room temperature and add to it a solution of 5.9–6.0 g of hexamethylene diamine dissolved in 10 ml of absolute alcohol. Mix and let stand. A white crystalline solid starts separating out. After about two hours, filter and wash the precipitate with cold absolute alcohol.

To recrystallize, dissolve the salt in the minimum amount of hot water. Cool by chilling and filter. Wash the solid with cold absolute alcohol and dry in air. Yield 12 g, m.p. 196–197°C.

Precaution

Contact of hexamethylene diamine with skin should be avoided because it is corrosive.

Exercise

228. In the preparation of the balanced salt, why the amount of diamine taken is slightly more than the exact stoichiometry.

7.2 Molecular Weights of Polymers

Polymers have high molecular weights, which may vary from 10,000 to several millions. The molecular weight of polymers is related to the chain length and the extent of cross-linking between different chains. The extent of cross-linking depends on the concentration (during polymerization) of the monomer having functionality higher than two and increases with increase in the functionality. Both the chain length and the extent of cross-linking depend on the reactivity of the monomers and increase with the reaction time in case of condensation polymerisation which proceeds via the step-growth mechanism. In case of addition polymerization, which follows a chain mechanism, the length of the chain depends on the relative rates of propagation and termination reactions. A high molecular weight results if the rate of propagation is much more than the rate of termination step. The actual length of the polymer chain in both cases depends on the random encounter between the monomer and the reactive site of the chain. Because of this randomness, some polymer chains may grow longer than the others. The product will thus not be a single chemical species but a mixture of chains of different lengths and therefore of different molecular masses. A polymer sample is thus an inhomogeneous mixture of different molecules and the range of molecular weights is fairly wide. Therefore, the experimentally determined molecular weight of a polymer sample is always some sort of an average of the molecular weights of the molecules present. The kind of average depends on the method of measurement. The two most common averages used are the number average and the weight average.

Number-Average Molecular Weight

The Number-Average molecular weight, $\overline{M}_n$, is defined by the expression:

$$\overline{M}_n = \frac{n_1M_1 + n_2M_2 + n_3M_3 + - - - -}{n_1 + n_2 + n_3 + - - - -}$$

$$= \frac{\Sigma n_i M_i}{\Sigma n_i} = \frac{\text{Total mass of the polymer sample}}{\text{Number of molecules present in the sample}}$$

where n_1, n_2, n_3, etc. are the numbers of molecular species having molecular mass M_1, M_2, M_3, etc., respectively.

A solution of known concentration is obtained by dissolving a weighted amount of a polymer sample in a suitable solvent. Each molecule of the sample, regardless of its size or weight, makes an equal contribution to the depression in freezing point (Cryoscopy), elevation in boiling point (Ebullioscopy), osmotic pressure (Osmometry) and lowering in vapour pressure. Measurement of an appropriate Colligative Property thus affords the Number Average Molecular Weight, $\overline{M}_n$. Another method by which $\overline{M}_n$ is determined is the End-group analysis method.

Weight Average Molecular Weight

The weight average molecular weight, $\overline{M}_w$, is defined by the mathematical expression

$$\overline{M}_w = \frac{W_1M_1 + W_2M_2 + W_3M_3 + ------}{W_1 + W_2 + W_3 + ------}$$

$$= \frac{n_1M_1^2 + n_2M_2^2 + n_3M_3^2 + -----}{n_1M_1 + n_2M_2 + n_3M_3 + -----}$$

$$= \frac{\Sigma n_i M_i^2}{\Sigma n_i M_i} = \frac{\Sigma n_i M_i^2}{W}$$

when $W_1 = n_1M_1$, represents the total mass of the species having molecular weight M_1;

$W_2 = n_2M_2$, the total mass of the species having molecular weight M_2, and so on, and

W = total weight of the polymer sample.

Weight-average molecular weight is determined by Light Scattering and Ultracentrifuge/Sedimentation–velocity techniques which depend mainly on the size/ weight of the polymer molecules and only to a small extent on their number.

Determination of the average molecular weight by all the above-mentioned methods (except the end-group analysis method, which is applicable only in a few cases) is the job of specialists, and expensive equipment is required. But the advantage is that the determination is not dependent on any previous measurement. These methods are thus absolute and serve as the ultimate basis for a third type of average termed as *Viscosity-Average Molecular Weight,* $\overline{M}_v$.

Exercises

229. What is meant by 'Living' and 'Dead' polymers?

230. Explain the term 'Degree of Polymerisation'.

231. A polymer sample of hypothetical composition contains 10 molecules of mol. wt. 10^3 each, 100 molecules of mol. wt. 10^4 each and 10 molecules of mol. wt. 10^5 each. Calculate the number average and weight average molecular weights. Also comment on the relative importance of low mol. wt. molecules and high mol. wt. molecules in the two types of averages.

232. What is the significance of determination of molecular weight of a polymer?

233. What is meant by 'Degree of polydispersity'?

234. Why and how is the molecular inhomogeneity of a polymer sample reduced?

7.2.1 *Determination of the molecular weight of a polystyrene sample by viscometry*

Reagents Required

Polystyrene sample; Benzene or toluene.

Theory

Addition of even a very small amount of a polymer to a solvent of low viscosity causes a sharp increase in its viscosity. The magnitude of increase in viscosity depends, in addition to other factors such as concentration and size and shape of the solute molecules, on the molecular weight of the polymer. Mathematical manipulations have reduced the problem of determination of the molecular weight of a polymer to a few simple viscometric measurements on the pure solvent and solutions of known concentration of the polymer in that solvent.

The ratio of the coefficient of viscosity of the solution (η_s) to the coefficient of viscosity of the pure solvent (η_o), at the same temperature, is known as *viscosity ratio* or *Relative Viscosity* (η_r) which, by using expression (4.11, p.87), can be written as

$$\eta_r = \frac{\eta_s}{\eta_o} = \frac{\rho_s t_s}{\rho_o t_o} \quad (7.1)$$

Where ρ_s and ρ_o are the densities of the polymer solution and the solvent respectively, and t_s and t_o are the corresponding efflux times for the flow of some specified volume of the two through the same capillary (Viscometer) of narrow bore and a long efflux time of about 200 seconds (section 4.2.1). For a dilute polymer solution (C < 0.5 g/100 ml), the density of the solution may be taken as equal to that of the solvent, i.e., $\rho_s = \rho_o$.

Then

$$\eta_r = \frac{t_s}{t_o} \quad (7.2)$$

Dividing ($\eta_s - \eta_o$), the increase in viscocity of the solvent due to the presence of the solute, by η_o, the viscosity of the pure solvent, gives *Specific Viscosity* η_{sp}, of the polymer solution:

$$\eta_{sp} = \frac{\eta_s - \eta_o}{\eta_o} = \frac{\eta_s}{\eta_o} - 1$$

$$= \eta_r - 1 = \frac{t_s}{t_o} - 1 \tag{7.3}$$

The ratio of the specific viscosity of the solution to its concentration C, expressed in grams per 100 ml (g/dl), is called *Viscosity Number, Reduced Specific Viscosity* or simply *Reduced Viscosity,* η_{red}. Therefore,

$$\eta_{red} = \frac{\eta_{sp}}{C} \tag{7.4}$$

Whereas both η_r and η_{sp} vary sharply with concentration, the variation of η_{red} with concentration is somewhat less and regular, and the plot of η_{red} versus concentration is a straight line (Fig. 7.2) given by the equation

$$\eta_{red} = \frac{\eta_{sp}}{C} = mC + \text{constant} \tag{7.5}$$

where m is the slope of the line.

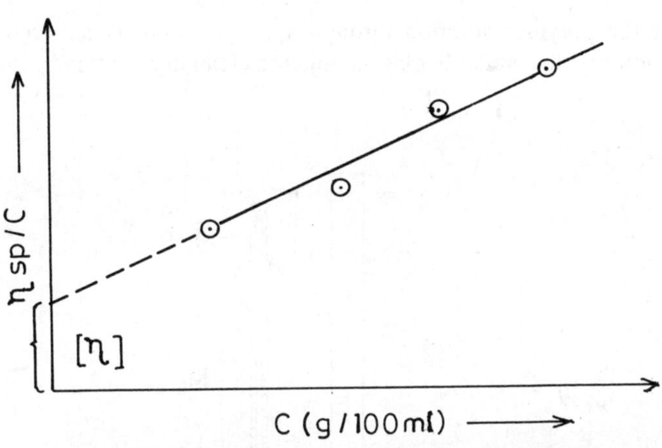

Fig. 7.2 Plot of Reduced Viscosity versus Concentration

The value of the constant is given by the intercept on the ordinate obtained by extrapolating the graph to zero concentration (infinite dilution). This constant which becomes independent of concentration is the limiting value of the reduced viscosity and is termed as *Limiting Viscosity Number* or *Intrinsic Viscosity,* [η].

Mathematically, $$[\eta] = \lim_{C \to 0} \left(\frac{\eta_{sp}}{C} \right)$$

or $$\frac{\eta_{sp}}{C} = mC + [\eta] \tag{7.6}$$

For linear polymers, the intrinsic viscosity [η] and the molecular weight M are generally found to obey the Mark-Kuhn-Houwink emperical equation

$$[\eta] = KM^a \tag{7.7}$$

where K and a are constants for a particular polymer/solvent/temperature system. K and a values are known for many systems some of which are given in Table IX. For most systems, a lies between 0.6 and 0.8 and $K \times 10^4$ lies between 0.5 and 5.

Procedure

(a) *Preparation of Polymer Solution*

Weigh accurately 500 ± 1 mg of the well-dried powdered polystyrene and transfer quantitatively to a 100-ml measuring flask. Add 90–95 ml of the solvent (benzene or toluene), stopper the flask and suspend it in the thermostat maintained at 25 ± 0.1°C. Shake occasionally to dissolve the sample. When the solution has acquired the temperature of the bath, add more solvent to fill the flask up to the graduation mark. Alternatively, take the weighed polystyrene (500 ± 1 mg) in a small bottle and add 100 ml of the solvent. Cover the mouth of the bottle with an aluminium foil and screw the cap tightly. Mount the bottle on an electric shaker and agitate until the solution is complete. [Dissolution may be facilitated by using a Shear-disk stirrer, p. 285].

Filter the polymer solution through a coarse porosity sintered glass filter without suction into a suitable glass-stoppered clean dry container.

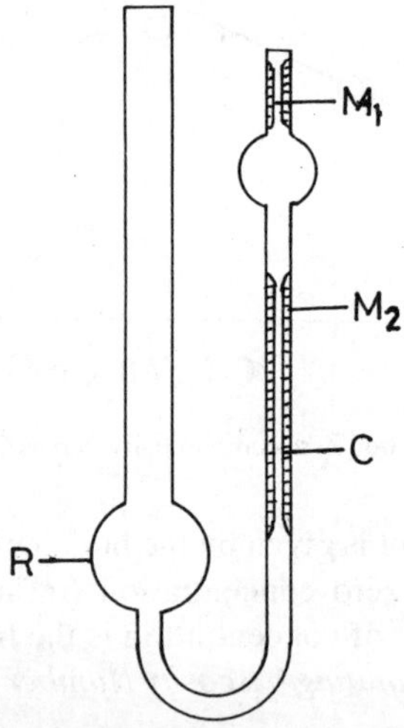

Fig. 7.3 Ostwald Viscometer

C - Capillary arm
M_1 - Upper mark
R - Reservoir
M_2 - Lower mark

Concentration of this solution is 0.5 g/100 ml (0.5%). Make solutions of lower concentrations (0.1%, 0.2%, 0.3% and 0.4%) by appropriate dilution of the above solution with well-filtered solvent.

(b) *Measurement of Flow Time with Ostwald Viscometer*

Wash the Ostawald viscometer with chromic acid/sulphuric acid mixture, tap water and then with distilled water and dry in an air oven. Pipet 20 ml of the well filtered solvent into the wide arm (reservoir, R) of the viscometer (Fig. 7.3) and attach it to the mounting support in the thermostat. Allow 5–10 min for attainment of temperature. Using a rubber ball, force air into the wide arm so as to raise the level of the liquid into the capillary arm above the higher mark M_1. Then release the pressure and with a timer, measure the time in which the liquid meniscus moves from the upper mark M_1 to the lower mark M_2. Repeat to get five values and use the average as the flow time t_o.

Now remove the viscometer from the thermostat and pour out the liquid from the wide arm, as completely as possible. Properly clean the viscometer, dry it and fill it with 20 ml of one of the solutions and determine the flow time as before. Similarly, determine the flow times for solutions of different concentrations, properly cleaning and drying the viscometer whenever the solution is to be changed.

Precautions

(1) The viscometer should be free of dust residues and other foreign matter.

(2) All measurements should be made using a constant volume of the liquid, otherwise the effective pressure head will vary from one solution to another.

(3) The viscometer should be mounted in such a way as to keep the capillary perfectly vertical throughout the experiment, otherwise the pressure head will vary even with the same volume of the liquid.

(4) A change in temperature changes the structure of the solution and hence the viscosity. Since in polymer solutions the equilibrium is reached very slowly, the solutions should be held at the predetermined temperature in a thermostat (regulated to 0.1°C) for sufficient time to establish equilibrium in solution (True viscosity values are values that do not change with time).

(5) Mouth suction should be avoided.

(6) Level of liquid on the capillary side should be raised by forcing air into the wide arm rather than pulling air out of the capillary, otherwise the liquid may get into the rubber tubing and contaminate the system.

(7) Formation of air bubbles should be avoided. Any bubble in the solution should be removed by manipulating the solution with pressure bulb (rubber bulb).

(8) Efflux time of the pure solvent should not be less than 100 sec.

(9) Measurements should be made at 4 to 5 different concentrations of the solution.

(10) The solution concentration should be restricted to a range for which η_r is 1.1 to 1.5.

Observations and Calculations

Temperature of thermostatic bath =

Solvent used =

Value of constants for the styrene/solvent/25°C system from the table k =

a =

Volume of liquid used for each measurement =

Polymer Solution Concentration C(g/ dl)	Flow time (sec.) Average of 5 measurements $\frac{t_1+t_2+t_3+t_4+t_5}{5}$	Relative viscosity $\eta_r = \frac{\eta_s}{\eta_o} = \frac{t_s}{t_o}$	Specific viscosity $\eta_{sp} = \eta_r - 1$	Reduced viscosity $\eta_{red} = \frac{\eta_{sp}}{C}$
Pure solvent	t_0			
0.1	t_{s_1}			
0.2	t_{s_2}			
0.3	t_{s_3}			
0.4	t_{s_4}			
0.5	t_{s_5}			

Determination of Intrinsic Viscosity

Plot the various values of reduced viscosity against the corresponding concentrations and extrapolate the graph to zero concentration (Fig. 7.2). Read the value of intercept on the ordinate. Record it as the value of $[\eta]$ =

Calculation of Molecular Weight

Substitute the value of $[\eta]$, K and a in equation (7.7) and calculate the value of M

$$[\eta] = KM^a$$

$$\log [\eta] = \log K + a \log M$$

$$\log M = \frac{\log [\eta] - \log K}{a}$$

$$M = \text{Antilog} \left[\frac{\log [\eta] - \log K}{a}\right]$$

Exercises

235. How are the values of K and a occurring in equation (7.7) determind ?
236. What is Ubbelohde Suspended Level Viscometer (USLV)? Describe its working and its advantages over Ostwald Viscometer.
237. Under what conditions is the intrinsic viscosity [η] independent of the molecular weight of the solute?
238. Why is a higher solution concentration (for which $\eta_r > 1.5$) not desirable?
239. Why is the plot of η_{red} versus C extrapolated to zero concentration?
240. With reference to a particular polymer, how is the power of solvation of different solvents compared?
241. What is the nature of the mol.wt. determined by Viscometry? How is it related to $\overline{M}_n$ and $\overline{M}_w$?
242. What is meant by Theta Temperature and Theta solvent?
243. With a suitable example, describe the determination of mol. wt. of a polymer by End-group Analysis.
244. Writing the polymerisation reactions for the formation of *Polyesters* from ω-hydroxy acids and formation of *carboxy terminated polybutadiene* (CTPB) from butadiene in presence of an azo initiator, predict the number of carboxyl groups per molecular chain in each case.
245. 1.232 g of a CTPB sample dissolved in toluene-ethanol mixture required 2.9 ml of 0.0965 N KOH for neutralisation to phenolphthalein end-point. Calculate the molecular weight of the sample.

7.3 Testing and Characterisation of Polymers

Testing and characterisation of polymeric materials is essential for determining their suitability for a particular application. The manufacturers and processors need it for quality control such as maintaining product uniformity and for assessing the performance of new materials in relation to the existing ones. Processors and users wanting to better understand the behaviour of the polymeric materials under various conditions are naturally interested in knowing their chemical nature.

Polymeric materials are very complex in nature. Their high molecular weights, molecular inhomogeneity and their chemical inertness often present difficulties in their identification which therefore requires special techniques and the use of advanced methods of analysis. The problem of characterisation is further complicated by the market availability of an extremely wide range of materials and the presence of compounding materials such as plasticizers, stabilizers and fillers (which change the physical properties of the product) and thus a complete identification of polymeric materials may not always be possible.

It is, however, possible to make a positive identification as to the class of polymer (such as polyolefine, polyamide, polyester, etc.) to which a given sample belongs, by carrying out some simple tests and correlating their results.

Application of these tests to the identification of a few most common polymeric materials is described in the following pages.

A. *Physical Tests*

Physical examination of the polymeric materials includes the observation of their colour, rigidity, solubility, density, etc; and the behaviour on dry heating and combustion. These tests help in preliminary sorting out and in some cases permit direct conclusions.

1. Visual Examination

Observation	*Indication*
(a) Completely transparent	PC, PMMA, PS
(b) Opaque milky-white	PTFE
(c) Light pastel shade	PF absent

2. Touch the sample with hand—A waxy feel indicates PE or PTFE.

3. Rigidity tests

(a) Scratch the sample with finger-nail

(i) Sample gets scratched or dented	A soft thermoplastic or rubber (CA, CN, CR, PE, PMMA, PS, PVAc, PVAl, PVC, SI).
(ii) Fingernail is scratched.	Hard plastic or thermoset (EP, PF, UF, UP).

(b) With a sharp knife, cut a thin silver from the edge of the sample

(i) Shows a 'soft cut' and gives a smooth and clean silver.	A thermoplastic
(ii) Hard to cut, produces powdery chips.	A thermoset

(c) Press a hot needle or metal rod against the sample — Melting indicates a thermoplastic.

(d) Try to compress and stretch the material — Flexibility indicates a rubber (PIP/NR, PCP/CR, SI).

4. Floatation Test (Density)

The test is based on Archimedes principle (p.108), from which it follows that a material will float in a liquid of the same or higher density. Suspend the sample in each of the following liquids separately and observe whether it sinks or floats in the liquid

L_1 : Water (ρ =1 g/ cm^3),

L_2 : Saturated $MgCl_2$ solution (ρ = 1.34 g/ cm^3), and

L_3 : Saturated $ZnCl_2$ solution (ρ = 2.01 g/cm^3).

(i) Sample floats in water	Materials with density <1 (PE, SI, NR/PIP).

(ii)	Sample sinks in water but floats in $MgCl_2$ solution.	Materials with density 1-1.34 (PA, PF, PMMA, PVAc, PVC, PS, PCP, CA, EP, UP).
(iii)	Sample sinks in $MgCl_2$ solution but floats in $ZnCl_2$ solution	Materials with density 1.34–2.01 (CN, CA, PVC, PF, PETP, UF, PVC, EP, UP).
(iv)	Sample sinks in $ZnCl_2$ solution.	Materials with density > 2.01 (PTFE, EP, UP).

Precautions

(1) Any air bubbles appearing on the surface of the sample should be completely removed.

(2) The sample should not swell in the liquid.

Exercises

246. How do air bubbles appearing on the surface of the sample affect floatation test?

247. Is the floatation test applicable to foams? Why?

248. Identify two other factors that affect density of a polymeric material.

249. How is the approximate density of a polymer sample determined by floatation test?

5. Solubility Test

For testing the solubility of a polymeric material in a particular solvent, take about 0.1 g of the powdered sample in a small beaker and add 5–10 ml of the solvent. Stir with a shear-disk stirrer (p.180). Appearance of streaks or 'schlieren' on stirring is an indication of solubility. Alternatively, heat the sample with the solvent in a test tube on a water bath. Decant or filter, if any insoluble particles (fillers or other additives) are left, then divide into two parts:

(i) Evaporate one part on a watch glass — any residue left is the dissolved material.

(ii) Drop the second part into excess of a non-solvent for that material — the dissolved material is reprecipitated.

Precautions

(1) The sample should preferably be in the finely powdered state.

(2) Safety glasses should be used while heating, and the mouth of the tube should point away from the face.

(3) Heating should be very slow to avoid spraying out of the liquid due to sudden boiling up.

Exercises

250. What are the factors governing the solubility of a polymeric material?

251. Why and how is a polymeric material brought into finely divided state?

252. What is dry ice? Why is it added during grinding of a polymeric material?

253. What is shear-disk stirrer? What is it used for?

6. Thermal Behaviour

On heating without direct exposure to flame, some linear polymers may become visually soft or rubbery before melting when they can flow. The softening range and the polymer-melt temperature (PMT, defined as the temperature where a polymer sample becomes molten and leaves a trail when moved across a hot metal surface with moderate pressure) provide a good clue as to the nature of the material. Changes in the physical state (swelling, colour change, decomposition, etc.) during heating and response of the escaping vapour to pH paper are also helpful in deciding the nature of the material. More indication is available from observed behaviour of the polymeric material when inserted into flame.

(a) *Softening or Melting Range* of the polymeric material is most conveniently determined with the help of a modified Dennis Bar or a Hot Stage. It consists of a metal bar along which a linear temperature gradient from 50° to about 400°C is created by means of resistance heaters. The polymeric material is directly placed on the bar in the form of

(i) a thin film,

(ii) finely ground powder spread uniformly, or

(iii) a solid plug manipulated by hand.

The temperature at which the solid leaves a trail or at the boundary between the solid and molten material is read directly from the scale attached to the bar.

(b) *Pyrolysis or pH Paper Test*: Take a small amount of powdered sample in a pyrolysis or ignition tube and place at its open end, a piece of moist pH paper (or litmus paper). Holding the tube with a clip, heat it with a Bunsen burner at low heat. Observe any changes in the physical state of the material and the change in the colour of the pH paper.

(c) *Flame or Combustion Test*: Holding with a pair of tweezers or tongs, insert a small piece of the sample into a low non-luminous flame of Bunsen burner and observe the ease of ignition, flammability of the sample in and out of the flame and the odour of any volatiles. Also observe if any droplets fall off during melting or burning.

Precautions

(1) Heating should be very slow.

(2) During heating, the open end of the ignition tube should be kept away from face.

(3) Eye shields or safety glasses should be used during pyrolysis or burning of the sample.

(4) The fumes may be toxic and so the ignition or combustion tests be better performed in a fume-cupboard.

B. *Chemical Analysis*

Under this heading are included the tests for heteroatoms and confirmatory tests for the presence of some groups and some specific polymers.

1. *Beilstein Test*

Take a stout and freshly cleaned copper wire. Turn one of its ends to get a loop. Insert the loop into a non-luminous Bunsen flame and heat until it imparts no colour to the flame. Put a small amount of the material under test on the loop (or bring the heated loop in contact with the sample, some of which adheres to the loop) and introduce it again into the flame. No appearance (or absence) of a green or bluish-green flame confirms the absence of PVC, PCP and PTFE [Halogen-containing polymers give a positive test. Certain plasticizers and flame-proofing agents contain halogens and give a positive test].

2. *Lassaigne Test*

Dry a freshly-cut pea-sized piece of sodium metal by pressing between the folds of a filter paper and transfer to a soft-glass ignition tube. Heat it gently until the sodium melts. Add carefully 50–100 mg of the finely powdered sample and reheat, first gently and then strongly, until the reaction subsides and the tube becomes red hot. Plunge the red hot tube into a china dish containing 10–15 ml distilled water. Carefully stir the contents of the china dish with a glass rod and then boil for a few minutes. Cool and filter, or separate the liquid with a pipet or by decantation. Preserve the liquid called Lassaigne's Extract (L.E.) or sodium extract for testing N, Cl, F and S.

Precautions

(1) Safety glasses should be used while heating, and the mouth of the tube should point away from the face.

(2) The sample must be completely dry. Any moisture will vigorously react with sodium.

(3) The plunging of red hot ignition tube into water in the china dish should be carried out behind a wire-gauze.

(a) *Test for Nitrogen* : Take 1–2 ml of L.E. in a test tube and add a few drops of the freshly prepared ferrous sulphate solution. If no green precipitate is observed, add 2-3 drops of sodium hydroxide solution and boil quickly. Cool and add dilute sulphuric or hydrochloric acid to dissolve the green precipitate:

(i) Appearance of a blue or bluish-green colour indicates the presence of nitrogen (PA, CN, UF indicated).

(ii) Appearance of a blood-red colour indicates the presence of both N and S (Casein indicated).

(b) *Test for Sulphur:*

(i) Add a few drops of sodium nitroprusside solution to 1–2 ml of L.E. A violet colour that slowly fades on standing indicates sulphur (Casein or vulcanised rubber).

(ii) Acidify 1–2 ml of L.E. with acetic acid and add a few drops of lead acetate solution (2M). A black precitate indicates sulphur.

(c) *Test for Chlorine*: Acidify 1–2 ml of L.E. with dilute nitric acid, boil for 1–2 minutes (filter if a precipitate is formed) and add a few drops of silver nitrate solution (2%). A flaky-white precipitate soluble in excess of ammonium hydroxide indicates chlorine (PVC or PCP present).

(d) *Test for Fluorine*

(i) Acidify 1–2 ml of L.E. with dilute hydrochloric acid and add a few drops of $CaCl_2$ solution (1M). A gel-like precipitate indicates fluorine (PTFE present).

(ii) Heat about 0.5 g of finely divided sample in a fusion tube and cool. Add 3–4 ml of conc. sulphuric acid. Non-wettability of the walls of the test tube indicates fluorine (PTFE present).

3. *Test for Silicon*

In a nickel crucible, fuse 40–50 mg sample with about 100 mg NaOH or a mixture of about 100 mg sodium carbonate and about 10 mg sodium peroxide. Cool, dissolve in a little water and boil. Acidify with dilute nitric acid and add 1–2 drops of ammonium molybdate solution. Heat nearly to boiling, cool and add a drop of benzidine solution, followed by a drop of saturated sodium acetate solution. A blue colour indicates silicon (SI).

4. *Group Tests*

(1) Molisch's test for *Cellulose:* Dissolve or suspend the sample in 1–2 ml acetone taken in a test tube. Add 2–3 drops of a 5% (w/v) solution of α-naphthol in ethanol and shake to mix. Hold the tube in an inclined position and, using a dropper pipet, carefully add about 1 ml of conc. sulphuric acid down the sides so as to form a separate layer under the acetone solution, without mixing with it. Note the colour produced at the interface:

Red to red-brown	Cellulose acetate
Green	Cellulose nitrate

(2) Lanthanum nitrate test for *Acetate*: In the pyrolysis test, insert into the mouth of the tube a cotton plug moistened with water. Remove the cotton plug after 4–5 minutes, wash it with 1–2 ml distilled water, collect the washings in a test tube and add

(i) 0.5 ml of lanthanum nitrate solution (5%, *w/v*)

(ii) 4–5 drops of iodine solution (p.3)

(iii) 4–5 drops of 1N ammonia solution.

Appearance of a deep blue colour indicates acetate (CA or PVAc).

3. Gibb's Indophenol test for *Phenol*: Immerse a filter paper in a saturated solution of 2,6-dibromoquinone-4-chlorimide in ether and then dry in air. Place the sample in a pyrolysis tube and cover its mouth with the paper prepared above. Heat the tube for 1–2 minutes. Remove the paper and moisten with 1–2 drops of ammonia solution. A blue colour indicates the presence of phenol (PF, PC, EP).

(4) Chromotropic acid test for *Formaldehyde*: Slowly heat the sample with a few crystals of chromotropic acid and 1–2 ml conc. sulphuric acid for 8–10 minutes. Appearance of violet colour indicate formaldehyde (PF or UF).

(5) Diphenylamine test for *Nitrate*: Add 1–2 drops of a freshly prepared solution (0.5% w/v) of diphenylamine in concentrated sulphuric acid to the plastic sample or its solution or suspension in acetone. Appearance of a dark blue colour indicates nitrate (CN).

Exercises

254. Explain the chemical principles on which the Beilstein's test for detection of halogens is based.
255. Giving chemical reactions, explain the chemistry of Lassaigne's test for the detection of N, S, Cl, F.
256. How can the formation of NaSCN during fusion with sodium be avoided?
257. Describe the chemistry of silicon test.
258. What are 'Virgin' polymers?
259. What are the main types of additives or compounding agents present in the processed polymers? How are these removed?

7.3.1 *Identification of individual members from a select group of PE, PMMA, PETP, PTFE and PF*

	Test	Observations				
		PE	PMMA	PETP	PTFE	PF
1.	Colour	–	Usually transparent	–	Opaque milky-white	Usually black or dark brown
2.	Touch	Waxy feel	–	–	Waxy feel	–
3.	Rigidity test	TP	TP	TP	TP	TS
4.	Solvents	*p*-Xylene Chloroform	Acetone Chloroform	Nitrobenzene Trichloroacetic acid	Insoluble	Insoluble
5.	Non-solvents	Acetone Diethyl ether	Methanol Diethyl ether	Acetone Methanol	–	–
6.	Floatation Test	Floats in L_1	Sinks in L1 Floats in L2	Sinks in L_2 Floats in L_3	Sinks in L3	Sinks in L_1 Floats or sinks in L_2
7.	Pyrolysis Test					
	(i) m.p.	105–150°C	120–150°C	250–265°C	325°C	Does not melt
	(ii) Physical change	Becomes clear	Swells, decomposes with crackling sound	Decomposes	–	Charring, decomposes with cracks

(iii) Residue	Wax-like grease	–	Dark brown	–	Black
(iv) Litmus paper	Unchanged	Unchanged	Turns red	Turns red	Unchanged or Turns blue
8. Flame Test					
(i) Ease of ignition	Ignites readily	Ignites readily	Burns in flame	Does not burn	Difficult to ignite
(ii) Out of flame	Continues burning	Continues burning	Continues burning or extinguishes	–	Extinguishes
(iii) Flame colour	Yellow, blue-centre	Bright light, blue centre	Bright, blue-edged	–	Sooty
(iv) Odour of Vapour	Burning candle	Fruity	Aromatic	Pungent, HF odour	Phenolic
(v) Dripping	Burning droplets fall off	Burning droplets fall off	Droplets fall off	–	–
9. Hetero Atom Test	–	–	–	F	–

10. *Specific Tests*

(a) *o*-Nitrobenzaldehyde test for PETP — In the pyrolysis test, cover the mouth of the tube with a filter paper moistened with a saturated solution of *o*-nitrobenzaldehyde in dilute NaOH solution (2N). A bluish-green colour indicates PETP.

(b) Millon's test for PF — Boil the sample with about 1 ml of Millon's reagent (p.6) for 2–3 minutes. Appearance of a red colour indicates PF.

(c) PF gives a positive (blue colour) indophenol test (p.188) for phenol and a positive violet colour chromotropic acid test (p.189) for formaldehyde.

7.3.2 *Indentification of individual members from a select group of CA, CN, PA, PC and UF*

	Test	Observations				
		CA	CN	PA	PC	UF
1.	Colour	–	Usually transparent	–	Usually transparent	–
2.	Rigidity test	TP	TP	TP	TP	TS
3.	Solvents	Chloroform, Acetone	Ethyl acetate, Acetone	Formic acid, Dimethyl formamide	Insoluble	Insoluble
4.	Non-solvents	Methanol Diethyl ether	Methanol Diethyl ether	Methanol Chloroform	–	–
5.	Floatation Test	Sinks in L_1 Floats or sinks in L_2	Sinks in L_2 Floats in L_3	Sinks in L_1 Floats in L_2	Sinks in L_1 Floats in L_2	Sinks in L2 Floats in L_3
6.	Pyrolysis test					
	(i) m.p.	125–175°C	80–90°C	200–260°C	220–250°C	Does not melt
	(ii) Physical change	Decomposes	Decomposes violently	Becomes clear	Colourless viscous liquid, decomposes	Charring

	(iii) Residue	Black	–	–	Brown	–
	(iv) Litmus paper	Turns red	Turns red	Turns blue	Unchanged	Turns blue
7.	Flame test					
	(i) Ease of ignition	Ignites readily	Burns vigorously	Difficult to ignite, crackling sound	Difficult to ignite	Difficult to ignite, chars with white edges
	(ii) Out of flame	Continues burning	Continues burning	Extinguishes	Extinguishes	–
	(iii) Flame colour	Yellowish green with sparks	Bright white	Blue flame with yellow tip	Bright sooty	Bright yellow
	(iv) Dripping	Burning droplets fall off	–	–	–	–
	(v) Odour of vapour	Vinegar, burning paper	Brown vapours, pungent NO_2 smell	Blue smoke, odour of burning wool or hair	–	Amonia and formaldehyde
8.	Heteroatom test	–	N	N	–	N

9. *Specific Tests*

(a) CA gives a positive (red to red brown ring) Molisch's test (p.188) for cellulose, a positive (deep blue colour) Lanthanum nitrate test (p.188) for acetate, and a red colour in the chromotropic acid test (p.189).

(b) CN gives a green ring in the Molisch's test, positive (blue colour) Diphenylamine test (p.189) for nitrate, and a red colour in the chromotropic acid test.

(c) In the *o*-nitrobenzaldehyde test (p.191), Nylon 6,6 (PA) gives a deep mauve colour.

(d) Press a cold metal rod against heated (softened) material and then withdraw. From PA, threads are easily formed.

(e) Lime water test for PC : Heat the sample with a 10% solution of NaOH for a few minutes and filter. Treat the residue with dilute sulphuric acid and pass the vapours into freshly prepared lime water. The lime water turns milky.

(f) PC gives a positive (blue colour) indophenol test (p.188).

(g) UF gives a positive (violet colour) chromotropic acid test (p.189) for formaldehyde.

7.3.3 *Identification of individual members from a select group of PS, PVAc, PVAl, PVC and EP*

	Test	Observations				
		PS	PVAc	PVAl	PVC	EP
1.	Colour	Usually transparent	–	–	–	Light coloured
2.	Rigidity test	TP	TP	TP	TP	TS
3.	Solvents	Benzene Chloroform	Methanol Acetone	Water, Dimethyl formamide	THF, Cyclo-hexanone	Insoluble
4.	Non-solvents	Methanol Aliphatic hydrocarbons	Diethyl ether Petroleum ether	Methanol Acetone	Methanol Acetone	–
5.	Floatation test	Sinks in L_1 Floats in L_2	Sinks in L_1 Floats in L_2	–	Sinks in L_1 Sinks or floats in L_2	Sinks in L_1 Sinks or floats in L_2

6.	Pyrolysis test					
	(i) m.p.	70–150°C	35–85°C	–	75–90°C	Does not melt
	(ii) Physical changes	Brittle appearance, melts	Melts	–	Melts, decomposes	–
	(iii) Residue	–	Brown	Brown	Dark brown	–
	(iv) Litmus paper	Unchanged	Unchanged	Unchanged	Turns red	Unchanged
7.	Flame test					
	(i) Ease of ignition	Ignites readily	Ignites readily	Ignites readily, material decomposes	Difficult to ignite	Burns in flame
	(ii) Out of flame	Continues burning	Continues burning	Continues burning or extinguishes	Extinguishes	Continues burning or extinguishes
	(iii) Flame colour	Bright sooty, Yellow flickers, Dense smoke	Yellow, sooty, Dense smoke	Shiny	Yellow green-edged, sooty	Yellow, black smoke

	(iv) Odour of vapour	Odour of illuminating gas, sweetish	Vinegar	Irritating	Pungent HCl smell	Pungent phenolic odour
	(v) Dripping	–	Burning droplets fall off	–	–	–
8.	Heteroatom test	–	–	–	Cl	–

9. *Specific tests*

(a) (i) Drop the sample on a hard surface — A characteristic 'metallic' ring (sound) indicates PS (Compare the sound with that from a known PS sample)

(ii) In the flames test, blow out the flame and smell the dense smoke. Characteristic odour of styrene (compare with the smell of styrene monomer).

(b) (i) PVAc gives a positive (deep blue colour) Lanthanum nitrate test (p.188) for acetate.

(ii) Pour a few drops of diluted iodine solution (p.3) over the sample. A red colour that intensifies on washing indicates PVAc.

(c) PVC — Shake the polymer sample with tetrahydrofuran and filter off the undissolved material. Add methanol to precipitate the polymer and filter again. Boil the polymer with 1–2 ml pyridine and to the hot solution add 3–4 drops of a solution (2% w/v) of NaOH in methanol. A reddish-brown colour or precipitate indicates PVC.

(d) (i) Add 1–2 ml conc. H_2SO_4 to a small amount of the sample taken in a test tube. Then add about 1 ml conc. HNO_3. Carefully add down the sides of the tube an aqueous solution (5% w/v) of NaOH so as to form a separate layer at the top of the acid solution. A red colour at the interface indicates EP.

(ii) EP gives a positive (blue colour) indophenol test (p.188).

7.3.4 *Identification of individual members from a select group of NR/PIP, PCP/CR, SI, SH and UP*

	Test	Observations				
		NR/PIP	PCP/CR	SI	SH	UP
1.	Colour	No characterstic colour due to presence of fillers			White to dark brown	–
2.	Rigidity test	Flexible rubber	Flexible rubber	Flexible rubber	TP	TS
3.	Solvents	Aliphatic and aromatic hydro-carbons	Ketones	Insoluble	Alcohol, hot soda or NH_4OH solution	Insoluble
4.	Non-solvents	–	–	–	Ketones, acid	–
5.	Floatation test	Floats in L_1	Sink in L_1 Floats in L_2	Floats in L_1	Sinks in L_2 Floats in L_3	Sinks in L_1 Floats in L_2
6.	Pyrolysis test					
	(i) m.p.	–	–	–	75–85°C	Does not melt
	(ii) Physical change	Softens, becomes sticky	Decomposes on strong heating	Decomposes on strong heating	Softens, melts	–
	(iii) Residue	–	–	White powder	–	–
	(iv) Litmus paper	Unchanged	Turns red	Unchanged	–	Turns red

No.	Test					
7.	Flame test					
	(i) Ease of ignition	Ignites readily	Burns in flame	Very difficult to ignite, glows in flame	Burns readily	Burns in flame
	(ii) Out of flame	Continues burning	Continues burning or extinguishes	Extinguishes	Continues burning	Continues burning or extinguishes
	(iii) Flame colour	Dark yellow, sooty	Yellow orange	Yellow, grey smoke	Dark yellow, sooty	Shiny, sooty
	(iv) Odour of vapour	Characterstic burnt rubber	Burnt rubber	–	Characterstic odour of burning sealing wax	Pungent
8.	Heteroatom	–	Cl	Si	–	–

9. *Specific Tests*

(a) Dissolve or suspend the sample in CCl_4 and add Wij's solution (p.7) dropwise. Decolorisation of Wij's solution indicates unsaturation — NR, PCP.

(b) Test for SH

(i) Heat the sample at 150°C for a little over half an hour – The horn-like residue is insoluble in alcohol.

(ii) Alcoholic solution of shellac turns litmus red.

(iii) Dissolve a known weight in neutral alcohol and titrate against N/10 KOH in presence of phenolphthalein. Acid value (p.101) of 60–70 indicates shellac.

8

MISCELLANEOUS

8.1 Lime

Lime is a very important industrial material which is largely used for the manufacture of bleaching powder, glass, sulphite pulp, soda lime, calcium carbide, washing soda and caustic soda. It is also used for structural purposes (cement and mortar), for white washing, in textile industry, in sugar refining, for cooking rags in paper manufacture, for water softening, for tanning of leather, for preparation of NH_3 and as a very good fertilizer for soil.

Lime is the product of calcination of lime stone:

$$CaCO_3(s) \longrightarrow CaO(s) + CO_2 \qquad (8.1)$$

There being several varieties of limestone, the limes prepared from them show corresponding variations in composition. The type of analysis required on a given sample depends to some extent on the use for which the sample is intended. Though routine analysis includes the same determinations as are carried out on limestone, there are certain specific uses which depend on the caustic value of lime (CaO content), and the impurities including $CaCO_3$ are not available for reaction. It is therefore often desirable that available CaO content of lime be determined.

8.1.1 *Determination of available lime or free CaO in industrial lime or in milk of lime or determination of caustic value of agricultural lime iodimetrically*

Reagents Required

1. Standard iodine solution (N/10)
2. Standard sodium thiosulphate solution (N/10)
3. Freshly prepared starch solution.

Theory

A known weight of the powdered sample is slaked with boiling water when CaO content reacts:

$$CaO + H_2O \longrightarrow Ca(OH)_2 \qquad (8.2)$$

A measured excess of iodine solution is then added when the base reacts with iodine:

$$I_2 + 2e \longrightarrow 2I^-$$

$$I_2 + 4OH^- \longrightarrow 2OI^- + 2H_2O + 2e$$

$$2I_2 + 4OH^- \longrightarrow 2OI^- + 2I^- + 2H_2O \quad (8.3)$$

or
$$2I_2 + 2Ca(OH)_2 \longrightarrow Ca(OI)_2 + CaI_2 + 2H_2O \quad (8.4)$$

The excess of iodine is back-titrated with a standard $Na_2S_2O_3$ solution using starch solution as indicator near the endpoint:

$$I_2 + 2e \longrightarrow 2I^-$$

$$2\ S_2O_3^{2-} \longrightarrow S_4O_6^{2-} + 2e$$

$$I_2 + 2S_2O_3^{2-} \longrightarrow 2I^- + S_4O_6^{2-} \quad (8.5)$$

Procedure

Transfer about 0.2 g of powdered and accurately weighed (by difference method) lime sample to an iodine titration flask. Add about 100 ml of boiled-out distilled water. Heat nearly to 60°C. Stopper and boil for 5–10 minutes (in the analysis of milk of lime, use 10 ml of the sample and omit boiling). Cool and add 40 ml of standard N/10 iodine solution. Stopper the flask and shake until the whole of lime has dissolved. Titrate against N/10 $Na_2S_2O_3$ using starch solution as indicator. Analyse one more sample in the same way.

Observations and Calculations

Weight of 1st sample $= w_1$ g

Volume of N/10 iodine solution added = 40 ml

Let the volume of N/10 $Na_2S_2O_3$ solution used $= V_1$ ml

Therefore, volume of N/10 iodine solution used against w_1 g of sample $= (40 - V_1)$ ml

or Weight of CaO in the sample $= (40 - V_1) \times \frac{1}{10} \times \frac{28}{1000}$ g

Therefore, % lime in the 1st sample $= \frac{(40 - V_1)}{10 \times 1000} \times \frac{28}{w_1} \times 100$

$= \frac{(40 - V_1)}{w_1} \times 0.28 = A_1$

Weight of 2nd sample $= w_2$ g

Let the volume of N/10 $Na_2S_2O_3$ used $= V_2$ ml

Therefore, % lime in the 2nd sample $= \frac{(40 - V_2)}{w_2} \times 0.28 = A_2$

Therefore, mean % available lime $= \dfrac{A_1 + A_2}{2}$

Precautions

CO_2-free distilled water should be used and excessive contact with atmosphere should be avoided (A sodalime guard tube may be used to prevent absorption of CO_2 from atmosphere).

Exercise

260. How do the impurities of $CaCO_3$ and MgO affect the estimation?

8.1.2 *Determination of available lime (after extraction with sucrose solution) with standard acid*

Reagents Required

1. Standard hydrochloric acid (N/10)
2. Sucrose solution (10%)
3. Phenolphthalein indicator solution

Theory

A known weight of the sample is slaked by boiling with water (reaction 8.2). Both calcium oxide and hydroxide are then extracted with sucrose solution and filtered. $CaCO_3$, MgO and other impurities are left behind. The base in filtrate is then titrated with standard acid using phenolphthalein indicator.

$$OH^- + H^+ \longrightarrow H_2O \qquad (8.6)$$

Procedure

Add about 25 ml of boiled-out distilled water to about 0.5 g accurately weighed (by difference method) lime sample taken in an iodine titration flask and boil for 5–10 minutes. Cool, add a few clean dry glass beads and about 100 ml of a 10% sucrose solution. Stopper the flask and shake for 1 minute after every 5 minutes for a period of half an hour. Filter, by suction, through a Buchner funnel. Wash the residue 3–4 times with 10-ml portions of a 5% sucrose solution. Collect the filtrate and washings in a 250-ml measuring flask and make up to the mark with boiled-out distilled water. Take 50 ml of this solution in a titration flask. Add 2 drops of phenolphthalein indicator and titrate with N/10 HCl solution until the pink colour disappears. Take concordant readings.

Observations and Calculations

Weight of sample taken $= w$ g

Volume of lime solution prepared = 250 ml

Volume taken for each titration = 50 ml

Let the concordant volume of N/10 HCl used $= A$ ml

$$\underset{(CaO)}{N_1V_1} = \underset{(HCl)}{N_2V_2}$$

$$N_1 \times 50 = \frac{1}{10} \times A$$

or $$N_1 = \frac{A}{10 \times 50}$$

Therefore, strength of lime solution $= \dfrac{A}{10 \times 50} \times 28$ g/l

Weight of CaO in w g of the sample $= \dfrac{A \times 28}{10 \times 50} \times \dfrac{250}{1000}$ g

% Available lime $= \dfrac{A \times 28}{10 \times 50} \times \dfrac{250}{1000} \times \dfrac{100}{w}$

$$= \frac{A \times 1.4}{w}$$

Precautions

1. Same as for Experiment 8.1.1 (p. 201).
2. While adding the sugar solution to lime, the flask should be continuously rotated to avoid granulation of lime.

Excercises

261. What is the function of suscrose in the experiment?
262. What is the purpose of adding glass beads?
263. Why does MgO not interfere in the estimation of lime by this method?
264. (a) What is meant by slaking of lime?
     (b) What are the factors that affect the rate of slaking reaction?
265. How does long exposure of a lime sample to atmosphere affect its available lime?
266. What is meant by the terms Milk of Lime and Lime Water?
267. What is meant by 'Soft' and 'Hard' limes?
268. How are limes classified?

8.2 Plant Nutrients

Of the various elements that are known to be essential for the proper growth and reproduction of plants, Nitrogen, Phosphorus and Potassium are required in relatively large amounts and so are known as major elements or macro-nutrients. A deficiency of one or more of these elements in soil may limit the growth of a particular plant or crop and may thus decrease soil fertility.

8.2.1 *Nitrogen*

It is generally present in soil as ammonium and nitrate ions, and as organic nitrogen compounds. The NH_4^+ and NO_3^- ions are not held firmly by soil and are easily leached away due to excessive irrigation or by rain water. The nitrogen content of soil therefore may get depleted either due to leaching or due to crop raising.

It is estimated that a region with moderate rainfall receives 5 to 7 pounds of nitrogen (a solution of nitrogen oxides formed by lightening) per acre per year. This amount is much less than required and has to be supplemented by

(a) Planting legume crops occasionally

(b) Applying organic manures such as animal wastes or compost.

(c) Using commercial fertilizers:

- (i) Synthetic ammonia or one of its derivatives namely NH_4Cl, $(NH_4)_2SO_4$, NH_4NO_3, $NH_4H_2PO_4$ (which provides phosphorus also).
- (ii) Other inorganic salts — $NaNO_3$,$Ca(NO_3)_2$
- (iii) Nitrogen in the organic form such as urea ($NH_2{\cdot}CO{\cdot}NH_2$)
- (iv) Calcium cyanamide which reacts with water slowly and forms urea

$$CaCN_2 + 3H_2O \longrightarrow Ca(OH)_2 + NH_2{\cdot}CO{\cdot}NH_2 \quad (8.7)$$

For a proper and economical application of manures and fertilizers, a knowledge of their nitrogen content as well as that of the soil is necessary.

8.2.2 *Determination of the total nitrogen content of soil, manure or a fertilizer*

Reagents Required

1. Standard sulphuric acid (N/10)
2. Standard sodium hydroxide solution (N/10)
3. Sodium hydroxide solution (50%)
4. Concentrated sulphuric acid
5. Salicylic acid
6. Zinc dust
7. Potassium sulphate
8. Anhydrous copper sulphate
9. Sucrose
10. Ferrous sulphate solution
11. Methyl red indicator solution
12. Red litmus paper

Theory

In the absence of nitrate, nitrogen in the sample is determined by same procedure as used for coal (p.132). Nitrate-nitrogen, however, is lost during digestion with conc.H_2SO_4:

$$4NO_3^- + 4H^+ \longrightarrow 4NO_2 + O_2 + 2H_2O \quad (8.8)$$

This loss is prevented by treating the nitrate-containing sample with a mixture of salicylic acid and conc. H_2SO_4 in cold, when the whole of nitrate is used up in nitrating salicylic acid :

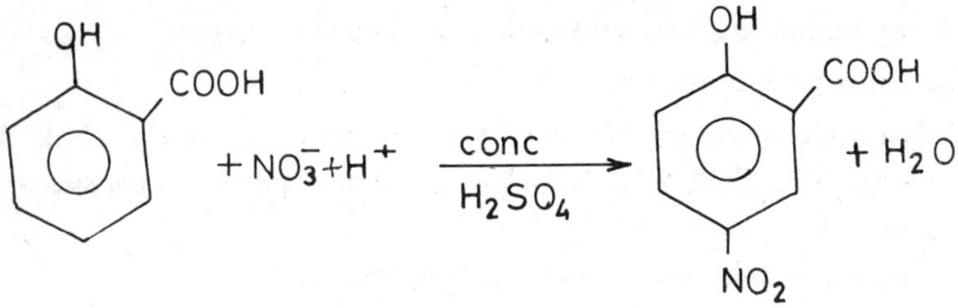

(8.9)

Zinc dust is then added which reacts with sulphuric acid to produce nascent hydrogen which reduces nitrosalicylic acid to amino salicylic acid:

$$Zn + 2H^+ \longrightarrow Zn^{2+} + 2H \quad (8.10)$$

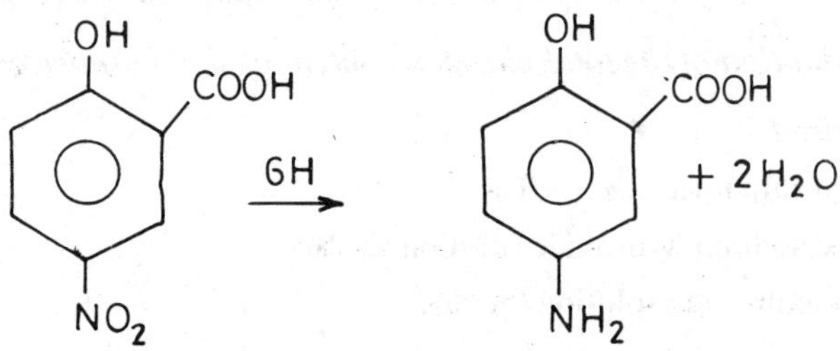

(8.11)

The latter is decomposed during digestion with conc. H_2SO_4, forming ammonium sulphate:

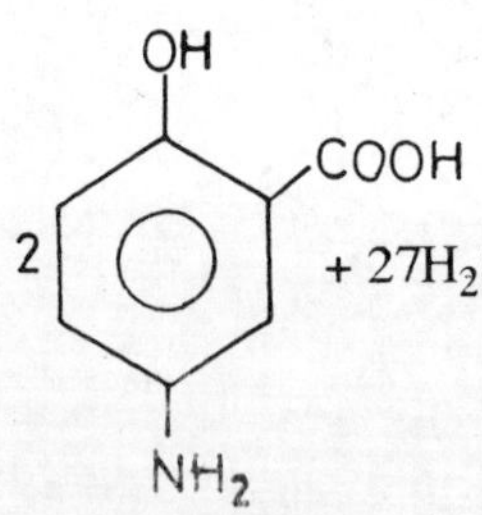

$$+ 27H_2SO_4 \longrightarrow (NH_4)_2SO_4 + 14CO_2 + 26SO_2 + 30H_2O \quad (8.12)$$

During the digestion stage, nitrogen in other forms (ammoniacal nitrogen and organic nitrogen) is also converted to ammonium sulphate. Distillation with NaOH and absorption of liberated ammonia into standard acid followed by back titration of excess of acid completes the total nitrogen determination (p.133).

Procedure

(a) *Detection of nitrate*

Shake about 5 g of finely ground representative sample with about 25 ml of hot distilled water and filter. To 1 ml of this solution in a test tube, add 1 ml of freshly prepared ferrous sulphate solution and cool. Add along the sides of the tube a few drops of conc. H_2SO_4 taking care not to disturb the solution. Formation of a brown or a reddish-coloured ring at the junction of the two liquids indicates the presence of nitrate.

(b) *Determination of nitrogen*

When nitrate is absent, proceed as for the determination of nitrogen in coal (p.133). In the presence of nitrate, proceed as under:

Transfer a known weight (0.5–5 g, depending upon the nitrogen content) of the sample into a Kjeldahl's flask. Mix 2 g of salicylic acid with about 40 ml of conc. H_2SO_4 in a beaker and add the acid mixture to the Kjeldahl's flask, in one lot so that it covers the whole of the sample at once. Immerse the flask in cold water and allow the reaction to proceed for about half an hour, occasionally shaking the reaction mixture. Take out the Kjeldahl's flask from the cold bath and add 1-2 g zinc dust in small portions with constant shaking. When the reaction subsides, heat gently until the frothing ceases. Add about 10 g of K_2SO_4 and either 0.2 g of Se powder or 0.5 g of anhydrous $CuSO_4$ or 0.3 g of cupric selenite dihydrate ($CuSeO_3 \cdot 2H_2O$), followed by about 10–15 ml concentrated H_2SO_4 and proceed as described on page 133.

Precautions

1. An appreciable rise of temperature during the digestion of the sample with sulphuric acid-salicylic acid mixture should be avoided as it may lead to loss of nitrogen.
2. See precautions 1 to 6 (page 133–4).

Observations and Calculations

Same as for nitrogen in coal (see page 134–5).

Excercises

269. How does the slow addition of acid-mixture to the nitrate sample affect the determination?
270. Name some other organic compounds that form nitroderivatives under the conditions of the experiment.
271. Name some other organic reagent that is employed in place of zinc for the reduction of nitrosalicylic acid.
272. What is meant by 'Crop Rotation'? What is its advantage?
273. What is 'Organic Farming'? What are its advantages?
274. What is compost? How is it obtained?

275. What is the effect of application of excess fertilizer?

8.2.3 *Phosphorus*

On an average, a ton of good soil holds about a pound of phosphorus. The whole of this phosphorus may not be in a form that can be taken up by the plants, yet the determination of the total phosphorus content of the soil is important because a decrease in it will point to a fall in the availability of phosphorus to plants. Luckily, the orthophosphate ion, which is an available form of phosphorus, is held tenaciously by the soil and is not easily removed by leaching as is the case with nitrate, chloride and sulphate. However, with the introduction of intensive agriculture, more and more phosphorus is taken up by the plants into their foliage and fruit and, as it is not replenished by nature, its deficiency has to be overcome by the application of a fertilizer containing calcium phosphate and/or ammonium phosphate. The usefulness and the value of a fertilizer will depend on its phosphorus content and hence the need for its determination.

Phosphate rock is the principal raw material for the manufacture of superphosphate, a water soluble mixture of $Ca(H_2PO_4)_2$ and $CaSO_4$, which is obtained by treating finely ground rock phosphate with roughly an equal amount of sulphuric acid. Determination of $Ca_3(PO_4)_2$ content of the phosphate rock is essential for fixing the price of the rock and for exploring the economic feasibility of the manufacture of superphosphate from the rock.

Preparation of Solution

(a) *For Soil and Fertilizers* — Take about 2 g of accurately weighed and finely ground representative sample in a Kjeldahl's flask and add about 25 ml of conc. HNO_3. Boil gently for about half an hour to remove all the easily oxidisable organic matter. Cool and add about 15 ml of 70% perchloric acid. Boil again until the solution becomes almost colourless and dense white fumes appear. Continue boiling for another 5 minutes and cool slightly. Add about 50 ml of distilled water and boil for a few minutes. Cool and filter into a 250-ml measuring flask. Wash the residue with hot water and make up the volume upto the mark.

When organic matter content is small, the following method with aqua regia may be adopted:

(b) *For Phosphate Rock* — Take about 2 g of accurately weighed and finely pulverised (to pass a 60-mesh sieve) representative sample of phosphate rock in a 250-ml conical flask. Add about 25 ml dilute aqua regia (1 part conc. HCl + 1 part conc. HNO_3 + 2 parts distilled water) and boil gently until the evolution of brown fumes stops. Cool and filter into a 250-ml measuring flask. Wash the residue with hot distilled water and make up the volume upto the mark.

8.2.4 *Determination of phosphate content of the given soil extract, fertilizer solution or phosphate rock solution*

Reagents Required

1. Standard hydrochloric acid (N/10)
2. Standard sodium hydroxide solution (N/10)

3. Ammonium molybdate solution (5%)
4. Conc. HNO_3
5. Ammonium nitrate
6. Potassium nitrate solution (1% or 0.1 M)
7. Phenolphthalein indicator solution
8. Methyl red indicator solution

Theory

When an excess of ammonium molybdate solution is added to the phosphate solution in the presence of nitric acid, a yellow precipitate of varying composition from $(NH_4)_3PO_4{\cdot}12MoO_3$ to $(NH_4)_3PO_4{\cdot}12MoO_3{\cdot}2H_2O$ is obtained. Upon suitably washing the precipitate with dilute KNO_3 solution the acid is completely removed and the precipitate is converted to ammonium phospho-molybdate with the composition $(NH_4)_3PO_4{\cdot}12MoO_3$.

$$PO_4^{3-} + 3NH_4^+ + 12MoO_4^{2-} + 24H^+ \longrightarrow (NH_4)_3PO_4{\cdot}12MoO_3 + 12H_2O$$

$$\text{or} \quad (NH_4)_3[PMo_{12}O_{40}] \qquad (8.13)$$

The washed precipitate is then treated with a known excess of standard NaOH which dissolves the precipitate as per the following reaction:

$$(NH_4)_3PO_4{\cdot}12MoO_3 + 23NaOH \longrightarrow$$

$$11Na_2MoO_4 + (NH_4)_2MoO_4 + NaNH_4HPO_4 + 11H_2O \qquad (8.14)$$

The excess of NaOH is titrated with standard acid using phenolphthalein indicator. The amount of phosphate in the given solution is calculated from the volume of the standard alkali consumed, using the reaction (8.14) which indicates that 1 gram-atom of phosphorus is equivalent to 23 gram-moles of NaOH.

Procedure

Pipet out 20 ml or a suitable aliquot (corresponding to 12–50 mg P_2O_5) of the given solution into a 250-ml glass-stoppered flask. Add about 8 g of NH_4NO_3 and 5 ml of concentrated HNO_3. Shake to dissolve and dilute to about 100 ml with distilled water. Take about 50 ml of a freshly prepared 5% ammonium molybdate solution into another flask. Heat both the solutions on a water bath to a temperature of 40-45°C and, while shaking continuously, add the ammonium molybdate solution slowly to the phosphate mixture. A yellow precipitate appears. Now fit the flask with the glass stopper, shake vigorously for about 10 minutes and allow to stand for half an hour on the water bath (40–45°C). Filter by decantation through a Whatman filter paper No. 5. Now transfer, as far as possible, the precipitate to the filter paper and wash the flask and the precipitate repeatedly with small amounts of 1% KNO_3 solution until the filtrate is no longer acidic. Now return the precipitate, together with the filter paper, to the flask in which precipitation was carried out. Add with a pipet, 40 ml of N/10 NaOH solution. Stopper the flask and shake to dissolve the yellow precipitate (If the whole of the precipitate is not dissolved, a further measured amount, say 10 ml, of N/10 NaOH should be added).

Dilute to about 100 ml, add 2-3 drops of phenolphthalein indicator and titrate with N/10 HC1 until the pink colour is completely discharged.

Precautions

(1) Perchloric acid should not be added to samples containing easily oxidisable organic matter, otherwise an explosive reaction might take place.

(2) If there is a turbidity in the ammonium molybdate solution, it should be filtered before use.

(3) The temperature of the solution should not be allowed to go beyond 45°C.

(4) The washing liquid should each time be taken first in the flask in which the precipitation was carried out, well shaken so that the whole of the inner surface of the flask and any precipitate sticking to it is thoroughly washed and then it should be poured over the precipitate on the filter paper.

(5) A second fraction of the washing liquid should be used only when the first fraction has completely passed through the filter paper. If the two fractions are allowed to mix, the washing will remain incomplete.

(6) The standard sodium hydroxide solution must be free from sodium carbonate.

(7) The alkali solution should be added to the flask through the funnel in which filtration took place so as to dissolve any precipitate remaining on its sides.

(8) The funnel then should be rinsed into the flask with distilled water.

Observations and Calculations

Weight of the sample taken = w g

Volume of the extract prepared = 250 ml

Volume of extract taken for determination = 20 ml

Volume of N/10 NaOH added = 40 ml

Volume of N/10 HC1 used for titrating excess of alkali = A ml

Hence volume of N/10 NaOH left unused = A ml

Or volume of N/10 NaOH consumed by ammonium phosphomolybdate obtained from 20 ml of extract = $(40 - A)$ ml

From reaction (8.14)

23 g-mole of NaOH = 1 g mole of $(NH_4)_3PO_4 \cdot 2MoO_3$

Or 23 liters of 1 N NaOH = 31 g P = 71 g P_2O_5

Or 1 liter of N/10 NaOH $= \frac{71}{23 \times 10}$ g P_2O_5

Or (40 - A) ml N/10 NaOH $= \frac{71}{23 \times 10} \times \frac{(40 - A)}{1000}$ g P_2O_5

Hence phosphate content of 20 ml of extract $= \dfrac{71}{23 \times 10} \times \dfrac{(40 - A)}{1000}$ g P_2O_5

Or strength of extract $= \dfrac{71}{23 \times 10} \times \dfrac{(40 - A)}{1000} \times \dfrac{1000}{20}$ g P_2O_5/litre

Or % of P_2O_5 in soil or fertilizer $= \dfrac{71}{23 \times 10} \times \dfrac{(40 - A)}{1000} \times \dfrac{1000}{20} \times \dfrac{250}{1000} \times \dfrac{100}{w}$

$$= \frac{71}{23 \times 10} \times \frac{(40 - A)}{w} \times 1.25$$

Results of phosphate rock are generally reported in terms of % of $Ca_3(PO_4)_2$,

$$\begin{array}{cc} P_2O_5 & \cong Ca_3(PO_4)_2 \\ (142) & (310) \\ 71 & 155 \end{array}$$

Strength of extract $= \dfrac{155}{23 \times 10} \times \dfrac{(40 - A)}{1000} \times \dfrac{1000}{20}$ g/l of $Ca_3(PO_4)_2$

% of $Ca_3(PO_4)_2$ in phosphate rock $= \dfrac{155}{23 \times 10} \times \dfrac{(40 - A)}{w} \times 1.25$

Exercises

276. Why is it necessary not to heat the solution above 45°C?
277. The filtrate from the ammonium phosphomolybdate precipitate is yellow coloured. What does it indicate?
278. What test should be performed to be sure that the filtrate contains no acid?
279. What is a fertilizer?
280. List the elements known to be essential for plant growth?
281. What is meant by N-P-K value of a fertilizer?
282. What is the general composition of a phosphate rock?
283. What does B.P.L. stand for?
284. What is meant by available or assimilable phosphorus?
285. How is citrate–insoluble phosphorus determined?
286. List the forms in which nitrogen, phosphorus and potassium are generally absorbed by plants.

8.3 Bleaching Powder

Bleaching powder, also called 'Chloride of Lime' or Bleach in trade, is the most commonly used material for bleaching cellulose, cotton yarn and textile and linen. It is a very important and cheap commercial product, used widely as a disinfectant for potable water, for making wool unshrinkable, in petroleum refining, in the manufacture of chloroform, in laboratory as a source of oxygen and chlorine, and as an oxidising agent.

It is prepared by passing chlorine gas over slaked lime at a temperature of 35–45° (considerable amounts of chlorate being formed at higher temperature). Majority of its reactions can be conveniently explained by assuming it to be a mixed calcium hypochlorite-chloride [Ca(OCl)Cl]. However, X-ray examination has shown bleaching powder to be a mixture of calcium hypochlorite [$Ca(OCl)_2 \cdot 4H_2O$] and basic calcium chloride [$CaCl_2 \cdot Ca(OH)_2.H_2O$], some free slaked lime being also present. Hypochlorite is the actual constituent which is responsible for liberation of Cl_2 when bleaching powder is acidified:

$$OCl^- + Cl^- + 2H^+ \longrightarrow Cl_2 + H_2O \qquad (8.15)$$

The bleaching, oxidizing or disinfecting value of a sample depends on the percentage of chlorine thus liberated, i.e., grams of chlorine liberated from 100 grams of the sample when treated with diluted acid. This is referred to as 'Available Chlorine' or 'Bleaching Chlorine' and forms the basis on which bleaching powder is marketed.

Continuous decomposition of bleaching powder during storage is its chief drawback. Being hygroscopic, it absorbs moisture from atmosphere and evolves chlorine:

$$OCl^- + Cl^- + Ca^{2+} + H_2O \longrightarrow Cl_2 + Ca(OH)_2 \qquad (8.16)$$

Due to this deterioration on standing, a sample of bleaching powder may always contain less amount of chlorine than what is expected and therefore, before use, every sample must be analysed for its effective or available chlorine.

8.3.1 *Determination of the percentage of 'Available Chlorine' in a given sample of bleaching powder by Bunsen's method*

Reagents Required

1. Standard sodium thiosulphate solution (N/10)
2. Potassium iodide solution (10%)
3. Dilute H_2SO_4
4. Freshly prepared starch solution

Theory

The bleaching powder solution or suspension is treated with an excess of KI solution and then acidified with dilute H_2SO_4:

$$OCl^- + 2H^+ + 2e \longrightarrow Cl^- + H_2O$$

$$2I^- \longrightarrow I_2 + 2e$$

$$OCl^- + 2I^- + 2H^+ \longrightarrow Cl^- + I_2 + H_2O \qquad (8.17)$$

The liberated iodine is titrated with a standard solution of sodium thiosulphate, using starch solution as indicator near the end-point:

$$I_2 + 2e \longrightarrow 2I^-$$

$$2S_2O_3^{2-} \longrightarrow S_4O_6^{2-} + 2e$$

$$I_2 + 2S_2O_3^{2-} \longrightarrow 2I^- + S_4O_6^{2-} \qquad (8.18)$$

Procedure

Remove and discard outside layer of the sample (which may have lost some chlorine). Mix the sample well, quickly transfer to a stoppered bottle and weigh. Transfer about 4 g of it to a porcelain mortar and weigh the bottle again. Put some distilled water on to the sample in the mortar and with a pestle rub the mixture to a smooth cream. Add more water, grind with the pestle, let settle for a while and decant the milky solution into a 500-ml measuring flask. Grind the residue with more water and pour off the liquid into the measuring flask as before. Repeat the operation until the sample is quantitatively transferred to the flask. Add distilled water to make up the volume upto the mark and mix well.

Transfer 50 ml of the above solution or suspension (with a mechanical pipet, burette or a measuring cylinder) to a 250-ml conical flask containing about 25 ml of a 10% solution of KI. Add 10 ml of dilute H_2SO_4 and run in, from the burette, standard N/10 $Na_2S_2O_3$ solution until the colour of the solution becomes very light yellow. Add 2 ml of freshly prepared starch solution, mix and complete the titration to the disappearance of blue colour. Repeat to get concordant readings.

Precautions

(1) During weighing, the sample bottle should be kept stoppered otherwise chlorine evolved due to the decomposition of the sample may corrode the material of the balance.

(2) The solution, being unstable, should be titrated immediately after its preparation.

(3) The solution should be well shaken before each aliquot is withdrawn for titration.

(4) Chlorine vapours being harmful, the solution should not be sucked into the pipet with mouth.

(5) The bleaching powder solution should not be acidified in the absence of KI otherwise some chlorine will be lost.

(6) If iron is present in the sample, phosphoric acid should be used in place of H_2SO_4 to prevent the reaction of iron with iodide.

Observations and Calculations

Initial weight of the sample bottle $= w_1$ g

Final weight of the sample bottle $= w_2$ g

Weitght of sample taken $= (w_1 - w_2)$ g

Volume of solution prepared $= 500$ ml

Volume of solution taken for each titration $= 50$ ml

Concordant volume of N/10 $Na_2S_2O_3$ used $= A$ ml

$$\underset{\text{(Bleaching Powder)}}{N_1V_1} = \underset{(Na_2S_2O_3)}{N_2V_2}$$

$$N_1 \times 50 = \frac{1}{10} \times A$$

Or
$$N_1 = \frac{A}{10 \times 50}$$

Therefore, amount of chlorine per litre of the solution
$$= \frac{A \times 35.5}{10 \times 50} \text{ g}$$

Or % Available Chlorine
$$= \frac{A \times 35.5}{10 \times 50} \times \frac{500}{1000} \times \frac{100}{(w_1 - w_2)}$$

$$= \frac{A \times 3.55}{(w_1 - w_2)}$$

Exercises

287. Write the ionic reaction representing the formation of bleaching powder.
288. Explain why starch-iodine blue colour reappears after the end-point if acetic acid is used instead of sulphuric acid for acidification.
289. In this estimation, can H_2SO_4 be replaced with HCl?
290. Outline the method that is considered to be more accurate than the iodometric method.
291. What is meant by bleach liquor? How is it prepared?
292. Describe the process of bleaching cloth.
293. List the important specifications for a good bleaching powder.
294. What is the French Scale for reporting the strength of a bleaching powder sample?
295. List some other reagents that are commonly used as bleaching agents.

8.4 Surface Tension

The molecules present in the surface of a liquid experience a resultant downward pull because the number of molecules below the surface, i.e., in the bulk of the liquid is much larger than the number of molecules above it, i.e., in the vapour phase. Consequently, the molecules in the surface tend to be drawn inside, with the result that the liquid surface tends to contract and behaves as if it were in a state of tension. This tension or force which acts along the surface of the liquid, uniformly in all directions, is known as *Surface Tension*. Numerically, it is equal to the force in dynes/cm acting along the surface at right angles to any imaginary line in the surface and is designated by γ (gamma).

As a consequence of surface tension, the drops of a liquid or bubbles of a gas are spherical in shape – rain drops and mercury drops are common examples. Smaller drops coalesce to give bigger drops because the process involves decrease in surface area (The phenomenon is utilized in collecting split Hg droplets).

The rise of a liquid into a capillary tube, which is wetted by the liquid, is directly proportional to the surface tension of the liquid. Common examples of the phenomenon are rise of oil in the wick of a lamp, rise of underground water on to the surface of earth, rise of ground water through the capillaries of plants to the twigs and leaves. Surface tension of a lubricant is important in capillary or wick lubrication methods.

High surface tension leads to superheating and bumping or spurting during boiling (Sugar and starch increase the surface tension of water and are mainly responsible for bumping in cooking pots). Decrease in surface tension of water, due to certain salts or oils in the boiler, helps in the rapid formation of steam bubbles but may lead to priming and foaming (wet steaming) which is prevented by spreading on the surface of water a layer of castor oil which increases the surface tension.

Surface tension of the liquid is utilized in transferring the liquid from the condenser to the evaporator in the heat-pipe exchange for waste heat recovery.

Surface tension measurements were used for investigating molecular structures through parachor values. The method is no longer in use.

Measurement of Surface Tension

The capillary rise method for measurement of surface tension, which gives most accurate results in absolute terms, is not in common use as it is extremely laborious, time consuming and requires a large amount of liquid.

Of the methods based on comparison with liquids of accurately known surface tension, the torsion balance or tensiometer method is the quickest and requires only a very small amount of the liquid; but if water is to be used as the standard liquid, the room must be free from organic vapours which interfere by impairing the water-air interface. This method is commonly used for industrial measurement.

The Drop Number Method, because of its practical convenience and a good degree of reliability, is commonly employed for laboratory measurement of surface tension.

8.4.1 *Determination of surface tension of a liquid, say CCl_4, benzene, alcohol etc., by drop number method*

Theory

The method is based on the principle that when a liquid is allowed to pass through a capillary tube, held vertically, at such a slow speed that the drops fall off the tip of the capillary under their own weight and are not pushed away by the kinetic force of flow or vibrations, the weight of a drop is approximately proportional to the surface tension of the liquid. Thus, if two liquids with surface tensions γ_1 and γ_2 are passed through the same capillary tube, then

$$\frac{\gamma_1}{\gamma_2} = \frac{w_1}{w_2} \tag{8.19}$$

where w_1 and w_2 are mean weights of their drops falling off the capillary end.

It is often more convenient to count the number of drops formed from a given volume of a liquid than finding the weights of single drops. Let n_1 and n_2 be the number of drops produced when equal volumes (V ml) of the two liquids are allowed to fall through the same capillary. Then.

the mean volume of a drop of one liquid $= \dfrac{V}{n_1}$

and, the mean volume of a drop of the second liquid $= \dfrac{V}{n_2}$

If ρ_1 and ρ_2 respectively be the densities of liquids 1 and 2, then the mean weights w_1 and w_2 of their drops are given by

$$w_1 = \frac{V}{n_1}\rho_1 \text{ and } w_1 = \frac{V}{n_2}\rho_2$$

Then from equation (8.19)

$$\frac{\gamma_1}{\gamma_2} = \frac{w_1}{w_2} = \frac{V}{n_1}\rho_1 \times \frac{n_2}{V\rho_2} = \frac{n_2\,\rho_1}{n_1\,\rho_2} \tag{8.20}$$

One of the liquids is usually water whose surface tension and density, at different temperatures, are accurately known. The density of the liquid under test may be determined by specific gravity bottle.

The capillary tube, through which the drops of liquid fall, forms the lower part of a pipet—as shown in Fig. 8.1 — and is then called a Drop Pipet or Stalagmometer.

Requirements

1. A wide-mouth receiving bottle with a tightly fitting rubber stopper having two holes through which pass a stalagmometer and a glass tube
2. A small rubber tubing with a screw-pinch-cock
3. A clamp stand
4. A thermometer
5. A thermostat (or a beaker full of water)
6. Distilled water, and
7. The liquid whose surface tension is to be measured.

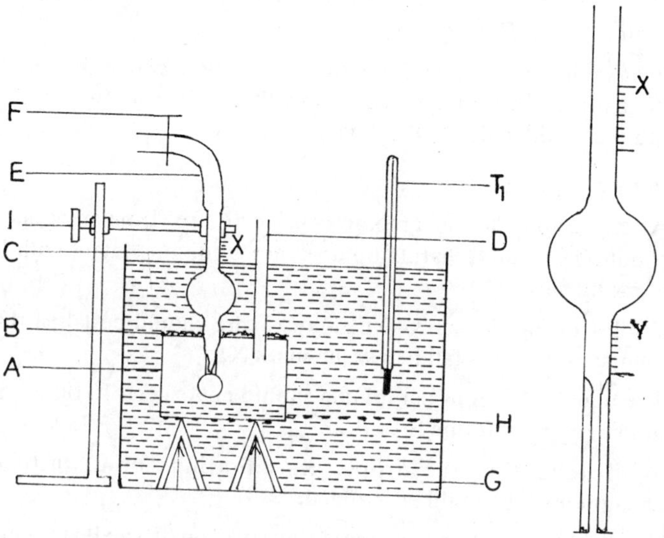

Fig. 8.1 Measurement of Surface Tension

A - Wide-mouth receiving bottle
B - Tightly fitting rubber stopper
C - Stalagmometer
D - Glass tube
E - Rubber tubing
F - Screw-pinch-cock
G - Thermostat
H - Liquid under test
I - Clamp
T_1 - Thermometer

Procedure

Thoroughly clean the stalagmometer and the receiving bottle, first with alkali and then with chromic acid mixture. Finally, rinse several times with distilled water. To the top of the stalagmometer, attach the rubber tubing with the screw-pinch-cock. Immerse the lower end of the stalagmometer into distilled water and suck through the rubber tubing until the level of water rises above mark X. Close the screw-pinch-cock, and fit the rubber stopper carrying the stalagmometer and a glass tube into the mouth of the receiving bottle. Place the whole assembly in the thermostatic bath and hold the stalagmometer vertical with the help of a clamp. Allow the stalagmometer to attain the temperature of the bath. Slightly open the screw-pinch-cock and count the number of drops as the liquid meniscus falls from the mark X to the mark Y. Repeat the process to record a number of observations.

Take out the whole assembly from the bath and dismantle the stalagmometer from the receiving bottle. Rinse the stalagmometer, the glass tube and the bottle with alcohol and dry. Now fill the stalagmometer with the liquid under test, mount it onto the receiving bottle and replace in the bath. Determine the number of drops for the fixed volume of the liquid between the two marks X and Y, as with water. Repeat to take a number of readings. Record the temperature of the bath.

For determination of Density

Properly clean the specific gravity bottle, rinse it with water, dry and weigh. Now fill it with distilled water and weigh. Empty the bottle, rinse with alcohol and dry. Fill it with the liquid under test and weigh.

Precautions

(1) After washing the stalagmometer perfectly free from any trace of grease, it should be ensured that its tip does not come in contact with hands or the working table.

(2) While closing the screw-pinch-cock, care should be taken that the liquid meniscus does not fall below the mark X.

(3) The number of drops per minute should not exceed 10 otherwise they may not be properly formed.

(4) While counting the drops, all vibrations or disturbances of the stalagmometer should be avoided.

(5) To avoid evaporation from the surface of drops, a small amount of the liquid should be taken in the receiving bottle and the drop formation should be close to the surface of the liquid.

(6) As far as possible, the temperature of the bath should remain constant.

(7) Only those observations should be taken into account for which the drop numbers do not vary by more than 0.5.

Observation and Calculations

Temperature of bath $= t\,°C$

Surface tension of water at t °C $= \gamma_2$

Weight of empty specific gravity bottle $= w_1$ g

Weight of specific gravity bottle + water $= w_2$ g

Weight of specific gravity bottle + liquid $= w_3$ g

Therefore, $\dfrac{\text{Density of liquid}}{\text{Density of water}} = \dfrac{\rho_1}{\rho_2} = \dfrac{w_3 - w_1}{w_2 - w_1}$

Surface tension of the liquid

$$\gamma_1 = \gamma_2 \cdot \frac{\rho_1}{\rho_2}\frac{n_2}{n_1} \tag{8.21}$$

$$= \gamma_2 \cdot \frac{w_3 - w_1}{w_2 - w_1} \times \frac{n_2}{n_1}$$

Liquid	No. of drops from fixed volume		Mean
	1	______	
Water	2	______	n_2
	3	______	
	4	______	
	1	______	
	2	______	
Test Liquid	3	______	n_1
	4	______	

Exercises

296. Why is the lower end or tip of the stalagmometer flattened?
297. What is the function of the narrow capillary in the stalagmometer?
298. Why is the stalagmometer tube graduated above and below the bulb?
299. What is the effect of temperature on the surface tension of a liquid?
300. What is the glass tube meant for?
301. What is Interfacial Tension?
302. Why is the value of interfacial tension between two immiscible liquids less than the value for the liquid with higher surface tension?
303. What is parachor?
304. What are surface active agents?
305. What is surface energy?

8.5 Electroplating

Electroplating or electrodeposition, usually defined as the production of metallic coatings on solid objects by the action of electric current, is the most important and widely used method for producing metal-to-metal coatings. The technique of electrodeposition has been applied.

(i) to provide decorative finish to jewelry, cutlery, table ware, light fixtures, musical instruments, etc. (Ni, Cr, Ag and Au being the common choices).

(ii) to provide protection against corrosion to a variety of industrial articles [Bumpers and most other parts of automobiles have an inner flash plate of Cu (for good adhesion), an intermediate layer of Ni (for corrosion protection), and a thin top film of Cr (primarily for appearance) or a thick Cr plating (for corrosion protection)].

(iii) for repair of worn-out articles or building up of under-size machine parts (Fe, Ni and Cr commonly used).

(iv) for electrocleansing (p.220), electrostripping (p.297) and electropolishing (p.295).

(v) for extraction and refining of metals (A1, Cu, Pb).

(vi) For *electrotyping* (manufacture of duplicates from original printing plates and blocks — thin shells of Cu, Ni or Cr backed by alloy lead) and *Electroforming* (forming metallic objects such as sculptures, busts, bells for musical instruments by electroplating a removable mandrel or matrix - Cu and Ni usually used).

(vii) for depositing rubber linings on metal sheets for use in chemical plants.

The essential requirements of an electroplating system are:

(1) A source of direct current such as a storage battery or a dry cell, though an electroplating power unit which permits control of current and voltage should be preferred.

(2) A solution containing salts of coating metal taken in glass, plastic or rubber-lined metallic tank (plating solutions can be prepared in laboratory by dissolving requisite amounts of salts in water or by mixing with water appropriate bath concentrates marketed by supply houses).

(3) An anode usually of the plating metal while the object to be plated is made the cathode.

On passing the current, metal ions from the solution migrate to the cathode (the object) where they get discharged and deposited as metal. An equivalent amount of metal dissolves from the anode and passes into the solution as ions, thus keeping the composition of the plating bath unchanged.

The potential E of an electrode for processes at reversible equilibrium ($M^{z+} + Ze \rightleftharpoons M$, a condition in which no net reaction takes place) is given by the Nernst equation:

$$E = E_0 + 2.303 \frac{RT}{2F} \log a_{M^{z+}}, \quad ..... \tag{8.22}$$

where R is the gas constant, T the absolute temperature, Z the valency of the metal ion M^{z+} being deposited, F the Faraday constant and E_0 the standard electrode potential at unit activity of the depositing metal ions ($a_{M^{z+}} = 1$). Under actual deposition conditions, almost every electrode reaction becomes irreversible to some extent when the electrode is said to be polarised. This polarisation or irreversibility causes the potential of the anode to become more noble and the cathode potential less noble. The resulting increase in the deposition potential is known as 'overvoltage'. The electrodeposition of metals from aqueous solutions is mainly limited by their decomposition potentials, their hydrogen over-voltages and several other polarisation phenomena. The magnitude of over-voltage can be reduced by increasing temperature, concentration and agitation of the plating bath.

The physical properties of the deposit such as hardness, porosity, grain size, uniformity, smoothness, ductility, etc., are a complex function of the current density, temperature, the nature of the surface to be plated, the nature of the plating metal, the bath composition, the presence of impurities and additives, and maintenance of uniform bath conditions during the plating process. The exact function of each of these factors is not fully understood.

Interdiffusion with interlocking grains (formation of an alloy but not an inter-metallic compound) of the substrate (the object to be plated) and the plating metal provides good adhesion. For achieving maximum adhesion between a given substrate and plated film, it is absolutely necessary that the surface to be plated should be perfectly free of all grease, dirt, scale, rust and other foreign matter. Otherwise, a non-adherent, less-durable deposit will result and peeling off or blistering of the deposit may occur.

The *Mechanical* means of *Surface Cleaning* include

(i) Scrapping and wiping of soil and grease with rags.

(ii) Chipping with chisels and rubbing with smooth wire-brushes.

(iii) Abrasive cleaning with sand papers, abrasive cloth and grinding wheels, and

(iv) Sand blasting.

The *Chemical cleaning* involves use of

(i) Emulsion cleaners — The surface to be plated (the 'work') is immersed in kerosene suspended in a soap solution (hot or cold) and agitated. Mineral oils are removed by this process.

(ii) Alkaline cleaners such as solutions of $NaOH$, Na_2CO_3, Na_3PO_4, Na_2SiO_3, etc. (or some proprietary mixtures) which function through saponification of grease and oil (vegetable and animal). Cleaning is done by simple immersion or making the work cathode in an electrolytic bath (Electro-cleansing) when liberation of a gas provides scrubbing action.

(iii) Organic solvents such as trichloroethylene, benzene, etc., which remove both types of oils (mineral and vegetable). The work is dipped in boiling solvent, then in cold solvent and finally withdrawn through vapour.

Pickling involves treating the work usually with an acid (or mixture of acids in presence of some inhibitor) which rapidly dissolves the surface films (oxides and scale) but attacks the base metal very slowly or not at all. The type and strength of the acid depends on the metal to be pickled and the amount of scale. The work is then rinsed in running water.

On a small scale (such as in laboratory or in a shop), the objects are shifted manually from one bath to another for cleaning, rinsing, pickling, etc. This is not possible in industry since a very large number of objects are to be handled simultaneously. The various baths are arranged in the desired order and the whole process of dipping the objects into a particular bath, raising from the bath and then shifting to the next one is automated.

8.5.1 *Electroplating of copper on a Copper plate*

Apparatus : A glass or plastic plating tank with a wooden or plastic lid, electrode suspension bar sets (bus bars), a stirrer, a rheostat (a variable resistance), an ammeter, a voltmeter, a one-way key, four copper plates (almost of the same dimensions) connecting wires, pieces of sand paper, cleaning and pickling baths.

Plating Bath Composition

Copper sulphate ($CuSO_4 \cdot 5H_2O$)	200–250 g/l
Sulphuric acid	50–75 g/l

Conditions

Temperature	20–50°C
Cathode current density	0.02–0.1 A/cm^2
Voltage	1–1.5 V at start, 2–2.5 V maximum

Theory

When electric current is passed through a solution of Cu^{2+} ions in dilute acid, using copper electrodes, the following reactions occur:

(i) Cupric ions migrate to the *cathode* where they are discharged and deposited as elemental copper

$$Cu^{2+} + 2e \rightleftharpoons Cu \text{ (metal)} \tag{8.23}$$

(ii) An equivalent amount of copper from the *anode* dissolves and passes into solution as Cu^{2+} ions

$$Cu \rightleftharpoons Cu^{2+} + 2e \tag{8.24}$$

thus maintaining electrical neutrality and keeping the bath composition unchanged. The added sulphuric acid provides good bath conductivity and prevents the formation of basic copper compounds.

Preparation of the Plating Bath

Sweep and wipe the empty plating tank to remove dust and other solid particles. If any oil or grease is seen, remove it with a rag drenched in kerosene, benzene or Na_2CO_3 solution. After rinsing with water, wash with acid pickle (10% H_2SO_4). Again rinse several times with fresh water and finally with distilled water.

Fill the tank about two-thirds with distilled water, add the requisite amount of copper sulphate and dissolve by stirring with a glass rod. Carefully add the requisite amount of concentrated sulphuric acid slowly and with constant stirring.

Preparation of the Electrodes

Clean all the copper plates by dipping them in hot 2% NaOH solution followed by rinsing in hot water. Wash in running tap water and then agitate each plate for 20-30 seconds in acid pickle (10% H_2SO_4, 2-3% HNO_3 and 1% HCl, all by volume).

Wash them well in running tap water and dry. Clean one of the plates with a piece of fine sand paper, wash it with running tap water and then with distilled water and dry. Use it as the experimental piate (the object to be electroplated). The other plates may be used as anodes or a cathode for testing the connections.

Setting up the Circuit

Draw a line diagram showing the scheme of connections (Fig. 8.2). Clean the ends of the connecting wires with sand paper. Remove the plug from the key K and connect the power supply (battery), the key, the rheostat, the electrolytic bath (alongwith suspension bars and 3 copper plates) and the ammeter in series.

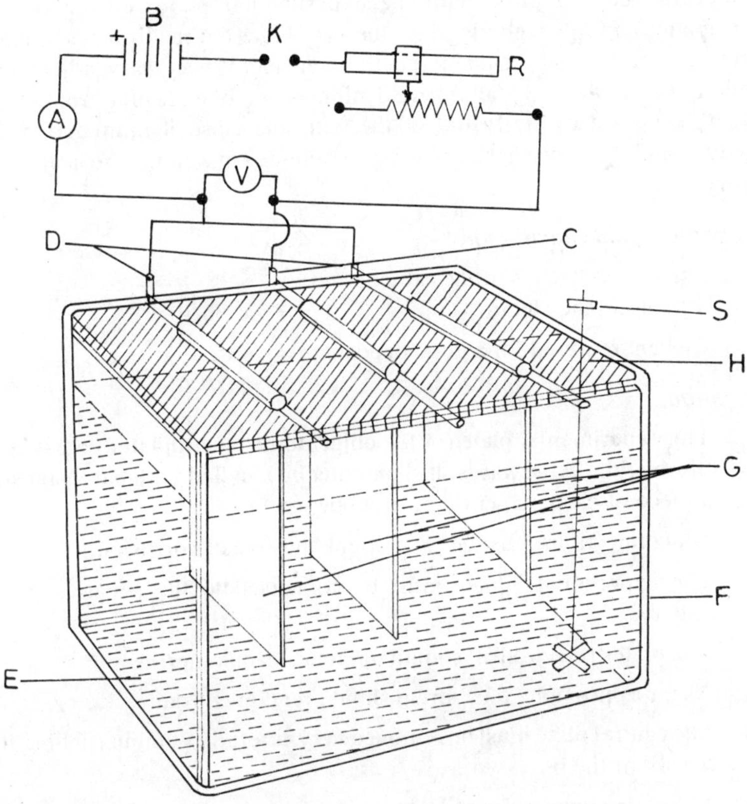

Fig. 8.2 Electroplating

A - Ammeter	B- Battery
C - Cathode bus-bar	D - Anode bus bars
E - Electrolytic bath	F - Glass or Plastic tank
G - Copper Plates	H - Lid
K - One way key	R - Rheostat
S - Stirrer	V - Voltmeter

Also connect the voltmeter in parallel. Provide the electrolytic bath with a stirrer for agitation of the plating solution.

Testing the Connections

Find the area of both the sides of the cathode plate dipping into the electrolyte and calculate the strength of the current to be passed. Put the plug in the key and adjust the value of the current approximately equal to the calculated value. See that the current remains constant for about 5 minutes. Remove the plug, take out the central plate (cothode) and inspect its surface. The connections are correct if a red copper deposit is seen.

Electroplating on the Object

Replace the central plate with the experimental plate (or the object to be electroplated) and put the plug into the key. Note the reading of the ammeter at 5-minute intervals and see that it remains constant. If necessary, adjust the current by manipulating the rheostat. After 30 minutes, remove the plug from the key. Lift the cathode plate carefully out of the bath and rinse it immediately in dilute sulphuric acid. Then wash it thoroughly in running tap water, rinse in distilled water and dry.

Observations and Calculations

Area of both sides of the cathode dipping in the electrolyte	=	x cm^2
Current required for a good deposit	=	$0.02\,x - 0.1\,x$ A

Precautions

(1) The experimental plate (or the object to be electroplated) must be cleaned thoroughly. Afterwards, it should not be handled with bare hands but with a piece of clean paper or clean rubber gloves.

(2) All connections must be clean, tight and free of corrosion.

(3) The experimental plate must be connected to the negative pole of the battery.

(4) The plates (electrodes) should not touch each other.

(5) As far as possible, the current should remain constant.

(6) The central plate must be removed very carefully without rubbing it against the slit of the lid.

(7) The plate must be rinsed in dilute H_2SO_4 as soon as it is removed from the electrolyte, otherwise the deposit will turn black.

Exercises

306. What will be the effect of using an impure copper anode for copper electroplating ?

307. What will be the effect of using an anode of a material other than the plating metal ?

308. What is the most inexpensive and easy test of cleanliness of a metal surface?
309. What is the advantage of agitating the solution during cleaning or pickling?
310. What is the most important requirement for obtaining a good finish in electroplating?
311. What is Electropolishing?
312. What is meant by Periodic Reverse Plating ? What is its effect?
313. What are brighteners? Name the most common brightener added to acid copper bath.
314. Giving suitable examples, explain the terms 'quicking' and 'striking'.
315. Outline the conditions for electrodeposition of alloys. How is brass electrodeposited?
316. What is meant by Throwing Power of an electroplating bath?
317. How is copper stripped from the surface of steel?
318. What is Electrosalvaging?
319. With the help of suitable examples, explain Electroless Plating.
320. What is the main difference between electroplaing and electroless plating?
321. Why and how are plastics electroplated?
322. Mention the chief characteristic and one formulation of the plating bath for electroplating Cu, Ag, Rh, Cd, Ni and Cr.
323. Describe in brief other methods of applying metallic coatings.
324. List some of the advantages of electroplating over other methods of applying metallic coatings.

8.5.2 *Determination of electrochemical equivalent of copper*

Apparatus : A timer in addition to the apparatus used in Experiment 8.5.1

Bath composition and conditions : Same as in Experiment 8.5.1

Theory

When electric current is passed through a solution of Cu^{2+} ions in dilute sulphuric acid, Cu^{2+} ions migrate to the cathode and are deposited as metallic copper:

$$Cu^{2+} + 2e \longrightarrow Cu \text{ (metal)}$$

The amount of copper deposited can be calculated by using Faraday's first law of electrolysis which states that the amount of a substance deposited or liberated at an electrode is directly proportional to the quantity of electricity passed through the solution.

Mathematically, $W \propto Q$

$$\text{or} \quad W = ZQ \tag{8.25}$$

where, W = weight in grams of the substance deposited,

Q = quantity of electricity in coulombs passed through the solution, and
Z = electrochemical equivalent of the substance.

When $Q = 1$ coulomb,
$W = Z$

Thus, the electrochemical equivalent of a substance is equal to the grams of the substance deposited by passage of 1 coulomb of electricity. Since the quantity of electricity Q is equal to the product of the current strength I in amperes and the time t in seconds for which the current is passed, equation (8.25) becomes

$$W = ZIt \tag{8.26}$$

The electrochemical equivalent of copper can therefore be determined by measuring the weight of copper deposited by passage of a known current I through the solution of Cu^{2+} ions for a known duration of time t and substituting the values in equation (8.26).

Procedure

Proceed as in Experiment 8.5.1. After thoroughly cleaning and drying the experimental plate, weigh it accurately. After testing the connections, replace the central (cathode) plate with the experimental plate. Put the plug into the key and simultaneously start the timer. Take several readings of the current strength during the course of the experiment. After exactly 30 minutes, take out the plug. Lift the experimental plate carefully out of the bath and rinse it immediately in dilute sulphuric acid. Then wash in running tap water, rinse in distilled water, dry and weigh accurately.

Observations and Calculations

Area of both sides of the cathode dipped in electrolyte $= x \text{ cm}^2$

Current required for good deposit $= 0.02\,x - 0.1\,x\,A$

Initial weight of the experimental plate $= w_1$ g

Final weight of the experimental plate $= w_2$ g

Weight of copper deposited, W $= (w_2 - w_1)$ g

Strength of current = (i), (ii), (iii), (iv), (v)

Mean value, I $= \dfrac{\text{(i)} + \text{(ii)} + \text{(iii)} + \text{(iv)} + \text{(v)}}{5}\ A$

Time for which current is passed $= t$ seconds

Electrochemical equivalent of copper, $Z = \dfrac{w_2 - w_1}{I \times t}$

Precautions

(1) Same as in Experiment 8.5.1.

(2) The electrolyte should be fresh and free from impurities

(3) The weighings must be accurate.

(4) Take at least 5 ammeter readings and use the mean value for calculating the result.

Exercises

325. What is meant by plating efficiency?

326. How long will it take a current of 1 ampere passing through a copper plating bath to deposit 1.5875 g of copper?

[1 Faraday = 96,500 coulombs, Eq. wt. of copper = 31.75 g].

327. A current of 0.1 A/cm^2 of the cathode surface is passed for 1 hour through a cyanide copper bath. Find the thickness of copper deposited on the cathode. [1 Faraday = 96,500 coulombs, Eq. wt. of copper 63.5, density of copper, d = 8.9 g/cm^3, Plating efficiency of the bath = 90%].

328. State Faraday's second law of electrolysis. Current from the same battery is passed through solutions of silver nitrate and copper sulphate connected in series. If 2.697 g of silver is deposited, find the weight of copper deposit. [Eq. wt. of silver 107.88, Eq. wt. of copper = 31.70].

329. Describe the use of silver coulometer for measuring the quantity of electricity passing through a circuit.

9

ANSWERS TO EXERCISES

1. A primary standard is a substance that is available in pure state, or whose purity with respect to the active component is known, and remains stable during storage, drying and weighing so that its standard solution (a solution of definite normality) can be prepared by diluting its accurately weighed amount to a definite volume.

Examples

(i) For acid-base reactions:

(a) Na_2CO_3 (Mol. wt. 106, Eq. wt. 106 for half neutralisation and 53 for complete neutralisation)

(b) Sodium tetraborate, $Na_2B_4O_7{\cdot}10H_2O$ (Mol. wt. 381.42, Eq. wt. 190.71)

(c) Potassium hydrogen phthalate, KH $C_8H_4O_4$ (Mol. wt. and Eq. wt. 204.22).

(ii) For precipitation reactions:

(a) $AgNO_3$ (Mol. wt. and Eq. wt. 169.89)

(b) KCl (Mol. wt. and Eq. wt. 74.557)

(c) KSCN (Mol. wt. and Eq. wt 97.185)

(iii) For redox reactions:

(a) $K_2Cr_2O_7$ (Mol. wt. 294.22, Eq. wt. 49.035)

(b) Sodium oxalate, $Na_2C_2O_4$

(Mol. wt. 134. 01, Eq. wt. 67)

(iv) For complexometric titrations:

(a) Disodium salt of EDTA (Mol. wt. 372.25, Eq. wt. 186.125)

(b) $CaCO_3$ (Mol. wt. 100, Eq. wt. 50)

2. A solution of slightly higher concentration than desired (say N/8 or N/9) is first prepared. It is then standardised by titrating against a standard solution of a suitable primary standard. The solution of definite normality is then prepared by appropriate dilution.

3. Strength of sodium oxalate solution = 3.35 g/l

Eq. wt. of sodium oxalate $= 67$

Normality of sodium oxalate solution, $N_1 = \frac{3.35}{67}$

$= \frac{1}{20}$

Volume of sodium oxalate solution taken, $V_1 = 20$ ml

Volume of unknown $KMnO_4$ solution used, $V_2 = 19$ ml

$$\underset{\text{(Sod. oxalate)}}{N_1V_1} = \underset{(KMnO_4)}{N_2V_2}$$

$$\frac{1}{20} \times 20 = N_2 \times 19$$

Therefore, N_2, the normality of the given solution $= \frac{1}{19}$

Normality of desired solution, $N_3 = \frac{1}{20}$

Volume of the solution desired, $V_3 = 500$ ml

Let the volume of the given $KMnO_4$ solution to be diluted $= V_2$ ml

Then $N_2V_2 = N_3V_3$

$$\frac{1}{19} \times V_2 = \frac{1}{20} \times 500$$

or $V_2 = \frac{1}{20} \times 500 \times 19 = 475$ ml

Thus, 475 ml of the given solution should be diluted to 500 ml to get N/20 $KMnO_4$.

4. Solutions prepared in boiled-out distilled water are more stable, i.e., they can be stored for much longer durations without any appreciable change in strength because boiling of water
   (i) Removes dissolved oxygen which when present may slowly react with many reducing agents.
   (ii) Removes dissolved carbon dioxide which interferes in acid-alkali reactions and may slowly decompose many reagents.
   (iii) Destroy bacteria that may slowly deteriorate many solutions due to biochemical activity.
5. 20 g of NaOH pellets are dissolved in 20 ml of boiled-out distilled water by stirring with a glass rod. The solution is transferred to a pyrex glass tube, closed with a stopper covered with tin foil and allowed to stand undisturbed. Sodium carbonate, being insoluble in concentrated solution of sodium hydroxide, settles down. To prepare approximately N/10 NaOH, about 8 ml of the supernatant liquid is taken out with a graduated pipet and diluted to 1 litre with boiled-out distilled water. [Exact normality is determined by standardisation with potassium hydrogen phthalate].

6. Because the permanganate solution is decomposed by the organic matter present in both the filter paper and the rubber tubing.
7. (a) The concentration equilibrium constant

$$K_C = \frac{[L]^1 \times [M]^m \times \cdots}{[A]^a \times [B]^b \times \cdots}$$

and the partial pressures equilibrium constant

$$K_P = \frac{[pL]^1 \times [pM]^m \times \cdots}{[pA]^a \times [pB]^b \times \cdots}$$

(b) $K_P = K_C \times (RT)^{\Delta n}$

where, Δn = No. of moles of gaseous products – No. of moles of gaseous reactants

R = Gas constant

T = Absolute temperature

8. (a) The rates of forward and backward reactions are equal.

   (b) The free energy of the reactions is equal to the free energy of the products, i.e., the free energy change of the process is zero.

9. Equilibrium is characterized by constancy of some properties but it is not the only requirement. When some of the properties of a system are constant but equilibrium does not exist, the system is said to be in a steady state. In the experimental determination of calorific value of a gaseous fuel by Boy's Gas Calorimeter, the following conditions characterise the steady state:
   (a) Air/fuel mixture under constant pressure burns at a constant rate.
   (b) The rate of flow of water through the copper tubing becomes constant.
   (c) The inlet and outlet temperature becomes constant. Though the temperature is not uniform throughout, it does not change with time.

However, it is an open system; the fuel and air are continuously entering, chemical change (combustion) is taking place and the products of combustion (CO_2, H_2O and N_2) are continuously leaving the system. Equilibrium can exist only in a closed system (a constant amount of matter) at a uniform temperature.

Another example of steady state is boiling of water in an open pan at constant temperature. The vapour pressure of water remains constant (equal to the atmospheric pressure) but the amount of water in the pan is continuously decreasing.

10. (a) A positive value of ΔH indicates that the dissociation of N_2O_4 (a colourless gas) to give NO_2 (a gas with a reddish-brown colour) is an endothermic reaction, i.e., it proceeds with absorption of heat. Heating the above system will tend to raise its temperature. In accord with Le Chatelier's principle, the equilibrium should shift in the forward direction so that a portion of heat that caused the temperature rise is absorbed. The forward shift will lead to formation of more NO_2 and therefore an increase in the intensity of the reddish-brown colour

is expected. Similarly, on cooling a reversal of the above process is expected, i.e., the intensity of the reddish-brown colour should decrease.

(b) That the predictions are correct can be shown by taking two identical glass bulbs (100-ml round-bottom flasks may be used), filling them to equal pressure of NO_2 and closing with tightly fitting stoppers. On immersing one of the bulbs (A) in ice-water and the other (B) in boiling water, it is seen that the colour in bulb A starts fading while in B it starts intensifying. If the two bulbs are now taken out and immersed in the same water bath at room temperature, it is seen that the intensity of colour in A increases and that in B decreases until the two bulbs have identical colours.

It must, however, be ensured that initial pressures (and therefore the concentrations) of NO_2 in the two bulbs are exactly equal, and sufficient time should be allowed for the equilibrium to be reached.

11. They help in predicting the effect of changing the concentration of one or more of the reactants or the products, temperature and pressure on the state of equilibrium of a given system. With this knowledge, optimum conditions for getting maximum yield of a product can be chosen and thus the process can be made most economical.

The equilibrium

$$N_2(g) + 3H_2(g) \rightleftharpoons 2NH_3(g), \; \Delta H = -22 \text{ kcal}$$

represents the well-known industrial process (Haber process) for the manufacture of ammonia.

(a) *Effect of pressure* : Increase in pressure over a system decreases the volume. The total number of moles per unit volume therefore increases. This change can be counteracted in part if some N_2 combines with H_2 to form NH_3 (as production of ammonia represents a decrease in moles from 4 to 2). Therefore, in accordance with Le Chatelier's principle, increase in pressure will increase the yield of ammonia (Increase in pressure shifts the equilibrium in a direction in which the total number of moles of gaseous substances decreases). As it is expensive to build high pressure equipment, a compromise pressure of 350 atmosphere is chosen.

(b) *Effect of temperature* : Increase in temperature of a system at equilibrium makes the reaction proceed in a direction in which heat is absorbed [Ex. 10(a)]. In accord with Le Chatelier's principle, therefore, rise in temperature will favour dissociation of ammonia and will thus lead to decrease in yield. At low temperature, however, the rate of reaction decreases. So, a compromise temperature of 500°C is chosen and a catalyst (Fe) is used to further raise the rate of reaction.

12. (i) *Formation of insoluble substances* : Use is made of the knowledge of solubility product principle, solubility product constant and the common ion effect.

Examples

(a) Removal of metal ions from industrial wastes:

$$Cu^{2+} + Ca(OH)_2 \rightleftharpoons Ca^{2+} + Cu(OH)_2 \downarrow$$

$$(K_{sp} = 8 \times 10^{-6}) \qquad (K_{sp} = 2 \times 10^{-19})$$

$$Zn^{2+} + Ca(OH)_2 \rightleftharpoons Ca^{2+} + Zn(OH)_2 \downarrow$$

$$(K_{sp}^{'} = 3 \times 10^{-17})$$

(b) Water softening by lime soda process

$$Mg^{2+} + Ca(OH)_2 \rightleftharpoons Ca^{2+} + Mg(OH)_2 \downarrow$$

$$(K_{sp} = 9 \times 10^{-12})$$

$$Ca^{2+} + Na_2CO_3 \rightleftharpoons 2Na^+ + CaCO_3 \downarrow$$

$$(K_{sp} = 5 \times 10^{-9})$$

(c) Phosphate conditioning of boiler-feed water:

$$3CaSO_4 + 2Na_3PO_4 \rightleftharpoons 6Na^+ + 3SO_4^{2-} + Ca_3(PO_4)_2 \downarrow$$

$$(K_{sp} = 2 \times 10^{-5}) \qquad (K_{sp} = 1 \times 10^{-27})$$

(ii) *Addition of a reagent that will combine with one of the ions to form a weakly ionized compound*

Examples

(a) Neutralisation of acid and alkali:

$$NaOH + HCl \rightleftharpoons Na^+ + Cl^- + H_2O$$

(b) Dissolution of precipitate of metal hydroxides

$$Fe(OH)_3 + 3H^+ \rightleftharpoons Fe^{3+} + 3H_2O$$

(iii) *Formation of complexes*

(a) Dissolution of precipitates :

$$AgCl(s) + 2NH_4OH \rightleftharpoons \left[Ag(NH_3)_2\right]^+ + Cl^- + 2H_2O$$

$$Cu(OH)_2\,(s) + 4NH_4OH \rightleftharpoons \left[Cu(NH_3)_4\right]^{2+} + 2OH^- + 4H_2O$$

$$Zn(OH)_2\,(s) + 4NH_4OH \rightleftharpoons \left[Zn(NH_3)_4\right]^{2+} + 2OH^- + 4H_2O$$

(b) Removal of cyanide from industrial wastes :

$$FeSO_4 + 6CN^- \rightleftharpoons \left[Fe(CN)_6\right]^{4-} + SO_4^{2-}$$

$$Fe^{3+} + K^+ + Fe(CN)_6^{4-} \rightleftharpoons KFe\left[Fe(CN)_6\right]$$

(iv) *Formation of a gaseous product* : The gas formed leaves the sphere of reaction and the reaction is forced to completion, e.g., dissolution of metallic sulphides such as FeS in HCl

$$FeS + 2H^+ \rightleftharpoons Fe^{2+} + H_2S\uparrow$$

(v) *Oxidation and reduction* : One or more of the ions involved are destroyed, e.g., oxidation of cyanide in industrial wastes by chlorination in alkaline medium:

$$2CN^- + 5Cl_2 + 8OH^- \rightleftharpoons 10Cl^- + 2CO_2 + 4H_2O + N_2$$

13. Ammonium hydroxide does precipitate both Cu^{2+} and Zn^{2+} as the corresponding hydroxides:

$$Cu^{2+} + 2NH_4OH \rightleftharpoons Cu(OH)_2(s) + 2NH_4^+$$

$$Zn^{2+} + 2NH_4OH \rightleftharpoons Zn(OH)_2(s) + 2NH_4^+$$

but excess NH_4OH reacts with the precipitates dissolving them by forming complexes [Ex.12 (iii) a].

14. Reactions in which the products do not combine to reproduce the reactants are termed as irreversible reactions.

Examples

(i) The decomposition of $KClO_3$ into potassium chloride and oxygen even in a closed vessel :

$$2KClO_3(s) \longrightarrow 2KCl(s) + 3O_2(g)$$

KCl and O_2 do not react to form $KClO_3$.

(ii) The decomposition of ammonium nitrite :

$$NH_4NO_2(s) \xrightarrow{\text{Heat}} N_2(g) + 2H_2O(g)$$

N_2 and $H_2O(g)$ do not react to form NH_4NO_2.

15. The volume of HIO_3 added to flask B is 10 ml more than that added to flask A and C. Therefore, addition of 10 ml less distilled water to flask B gives same volume of the total solution and hence the same bisulphite ion concentration in all the three flasks.

16. Addition of more water will decrease the concentration of both the reactants (HIO_3 and $NaHSO_3$). This is expected to reduce the rate of reaction and appearance of blue colour should be delayed.

17. Many reactions are known to occur in two or more elementary steps. In such cases, the product(s) of one step become(s) the reactant(s) for the next step. Such reactions are known as consecutive reactions or complex reactions.

    The rates of the various steps generally differ from one another. It is apparent that the rate of the overall reaction cannot be faster than the step which takes place at slowest rate. The slowest step, in the sequence of various steps, is called the 'rate-determining' step because its rate will be approximately equal to the rate of the overall reaction. This approximation is called the Principle of bottle-neck.

    In the reaction between HIO_3 and $NaHSO_3$, the rate-determining step is 2.17 :

$$3HSO_3^- + HIO_3 \longrightarrow 3HSO_4^- + HI$$

18. A clock reaction consists of a series of consecutively occurring reactions. The reaction goes on for a period of time without giving any evidence of its occurrence until a sudden colour change or formation of a precipitate indicates that the reaction has gone to completion. The reaction between HIO_3 and $NaHSO_3$ is a clock reaction whose completion is indicated, in presence of starch solution, by the sudden appearance of the blue-black colour.

19. The saw dust which is produced in considerable amount in a saw mill remains suspended in air in a finely divided state. Because of the very large surface area at which oxygen is in contact with combustible particles, the combustion, if initiated, will take place at a very fast rate and may cause an explosion.

20. (i) A faint pink colour very slowly (over a period of 5–10 minutes) appears on the surface of the particles due to the formation of HgI_2:

$$HgCl_2 + 2KI \longrightarrow \underset{\text{(red)}}{HgI_2} + 2KCl$$

    But as the surface area of contact between the reactants is very small, only a slight reaction takes place.

    (ii) On grinding, the particle size of the reactants is reduced. Fresh surface, therefore, gets exposed and more reaction takes place. The intensity of the colour will thus go on increasing as grinding is continued and finally the whole material will become red.

    (iii) On addition of water, KCl will dissolve but not HgI_2 which will appear as a red precipitate.

    (iv) On addition of more KI, the red precipitate will slowly disappear because of formation of HgI_4^{2-} complex:

$$2KI + HgI_2 \longrightarrow 2K^+ + HgI_4^{2-} \text{ (Colourless)}$$

21. (i) Atmospheric oxidation of I^- to I_2:

$$4I^- + O_2 + 4H^+ \xrightarrow{h\nu} I_2 + 2H_2O$$

(ii) Chlorination of methane:

$$CH_4 + Cl_2 \xrightarrow{h\nu} CH_3Cl + HCl$$

(iii) Decomposition of NO_2 into nitric oxide and atomic oxygen in the photochemical smog formation:

$$NO_2 \xrightarrow{h\nu} NO + (O)$$

22. According to Law of Mass Action, the rate of a reaction at any given instant depends on the concentration of the reactants at that instant. As the reaction progresses, more and more of the reactants are converted into products. Thus the concentration of the reactants falls with time and so does the rate of reaction.

23. The speed of a reaction is important as it determines whether the reaction will be of any practical use to us.

24. For determining the concentration of a reactant at a particular instant during the course of a reaction, it is necessary to completely stop the reaction or reduce its velocity to a very low value so that the concentration of the reactant does not change during the course of the titration. Addition of ice-cold water reduces the temperature which causes a marked decrease in the rate of the reaction. The process is known as Freezing (or arresting) the reaction.

25. (1) Decomposition of H_2O_2 in aqueous solution:

$$H_2O_2 \longrightarrow H_2O + \frac{1}{2} O_2$$

(2) Decomposition of ammonium nitrite in aqueous solution :

$$NH_4NO_2 \longrightarrow 2H_2O + N_2$$

(3) Inversion of cane sugar in presence of dilute acid:

$$C_{12}H_{22}O_{11} + H_2O \xrightarrow{H^+} \underset{\text{(Glucose)}}{C_6H_{12}O_6} + \underset{\text{(Fructose)}}{C_6H_{12}O_6}$$

(4) The rate of death of micro-organisms by a disinfectant is approximately proportional to the concentration of live microorganisms remaining.

(5) The rate of decomposition of organic matter in the B.O.D. test is approximately proportional to the organic matter remaining.

(6) Decay of radioactive material.

26. Molecularity of a reaction is defined as the number of molecules involved in the step leading to the chemical reaction. For simple reactions, it is equal to the order of the reaction except when one of the reactants is present in large excess. For example, in the hydrolysis of methylacetate with dilute acid,

$$\underset{}{CH_3COOCH_3} + \underset{\text{(large excess)}}{H_2O} \xrightarrow{H^+} CH_3COOH + CH_3OH$$

One molecule of methyl acetate reacts with one molecule of water and so the molecularity of the reaction is 2. However, as discussed on page 22 (Experiment 2.3.2) and verified experimentally, it is a first order reaction. Such a reaction is known as a Pseudo-unimolecular reaction.

For a complex reaction occurring in several steps (elementary reactions), each step has its own molecularity depending upon the number of molecules of the reactant or reactants taking part in that step. Molecularity has no significance for the overall reaction.

The order of a complex (multi-step) reaction is given by the order of the rate-determining step, i.e., the slowest step in the sequence of various steps involved in that reaction. The order has to be measured experimentally. Neither the order nor the molecularity of a reaction can be predicted from the stoichiometric equation of the overall reaction.

27. This is done to arrest the reaction at the instant when the concentration of residual alkali is to be determined. On adding the reaction mixture to ice-cold HCl (a known excess), the residual alkali is instantaneously neutralised and the saponification reaction altogether stops. The excess acid does catalyse the hydrolysis of ethyl acetate with water to produce acetic acid, but this reaction is extremely slow at low temperature (ice-cold) and does not cause any appreciable error.

28. (a) A reaction between two reactants becomes a first order reaction when one of the reactants is present in large excess. The concentration of this reactant remains practically unchangeed and so does not affect the rate of the reaction.

(b) The rate constant for a second order reaction between two reactants A and B to give products is given by equation (2.37):

$$k = \frac{2.303}{t\,(a - b)} \log \frac{b\,(a - x)}{a\,(b - x)}$$

where a and b are the initial concentrations of A and B, respectively. If A is taken in large excess, a will be much larger than both b and x. Then in the above equation, $(a - b)$ and $(a - x)$ can be replaced by a, and we get

$$k = \frac{2.303}{t \cdot a} \log \frac{b \cdot a}{a \cdot (b - x)}$$

or $$\frac{2.303}{t}\log\frac{b}{b-x} = ak = k'$$

which is the equation for a first order reaction in respect of B, i.e., in respect of reactant taken in smaller quantity.

29. (1) Saponification of esters:

$$\underset{\text{(Ethyl acetate)}}{CH_3COOC_2H_5} + NaOH \longrightarrow CH_3COONa + C_2H_5OH$$

(2) Decomposition of ozone to oxygen:

$$2O_3 \xrightarrow{100^{\circ}C} 3O_2$$

(3) Conversion of ammonium cyanate into urea in solution:

$$NH_4CNO \longrightarrow H_2N \cdot CO \cdot NH_2$$

30. The rate constant for a third order reaction is given by

$$k = \frac{1}{2a^2t}\cdot\frac{x(2a-x)}{(a-x)^2}$$

where a is the initial concentration of each of the reactants and x is the decrease in the concentration of each of the reactants in time t.

31. For a third or higher order reaction, three or more reacting particles (molecules or ions) must interact (collide) simultaneously. The probability for such a situation to occur is very low. Therefore, third or higher order reactions are extremely rare.

32. *Third Order Reactions*

(1) Oxidation of nitric oxide to NO_2:

$$2NO(g) + O_2(g) \longrightarrow 2NO_2(g)$$

(2) Reduction of $FeCl_3$ by $SnCl_2$:

$$2FeCl_3 + SnCl_2 \longrightarrow 2FeCl_2 + SnCl_4$$

Fourth Order Reaction :

Thermal decomposition of potassium chlorate, represented by the equation

$$4KClO_3 \xrightarrow{\text{Heat}} 3KClO_4 + KCl$$

is a fourth order reaction.

33. Reactions whose rates do not depend on the concentration of the reactants but remain constant throughout are said to be zero order reactions.

Examples

(i) Photochemical reaction between equimolar mixture of $H_2(g)$ and $Cl_2(g)$ confined in a tube dipping in water:

$$H_2(g) + Cl_2(g) \xrightarrow{h\nu} 2HCl(g)$$

(ii) In the biological waste treatment, the speed of waste utilization per unit mass of enzyme or bacteria (V/E) is found to be independent of waste or substrate concentration (s) over fairly large ranges of concentration.

34. It is the time in which the concentration of the reactant/reactants is reduced to one half of its/their initial value.

35. The half-life period of a first order reaction is given by

$$t_{1/2} = \frac{2.303}{k} \log \frac{a}{a-x} = \frac{2.303}{k} \log \frac{a}{a - \frac{a}{2}}$$

$$= \frac{2.303}{k} \log 2 = \frac{0.693}{k}$$

which does not contain any concentration term. Hence the half-life period of a first order reaction is independent of the initial concentration of the reactant. In general, the half-life period or time required for completion of any definite fraction of a reaction with equal initial concentration of all the reactants is given by

$$t_{1/2} \propto \frac{1}{a^{(n-1)}}$$

where n is the order of reaction.

36. A study of chemical kinetics helps in

(i) Regulating or making a reaction proceed at the desired rate.

(ii) Comparing the relative strengths of acids (Experiment 2.3.3),

(iii) Understanding the mechanism (description of various elementary steps) of various reactions.

37. The conversion of ammonium cyanate into urea has been

$$NH_4CNO \longrightarrow H_2N \cdot CO \cdot NH_2$$

found to be a second order reaction. Evidently, the balanced chemical equation, which shows only one reacting molecule, does not truely represent the conversion.

The reaction has been suggested to proceed via the following steps:

(i) $NH_4CNO \xrightarrow{\text{fast}} NH_4 \cdot NCO$

(Ammonium isocyanate)

(ii) $NH_4 \cdot NCO \xrightarrow{\text{fast}} H-N=C=O + NH_3$

(iii) $H-N=C=O + NH_3 \xrightarrow{\text{slow}} H_2N-CO-NH_2$

The last step, being the slowest, determines the rate of reaction. Since it involves two molecules, the order of the reaction is 2.

38. Methyl orange indicator has structure I in neutral medium. In alkaline medium, it exists as II which is orange and in acidic medium, it exists as III (red).

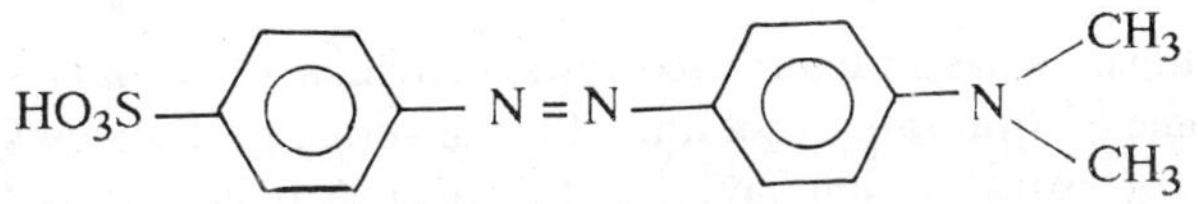

I

↓ OH^-

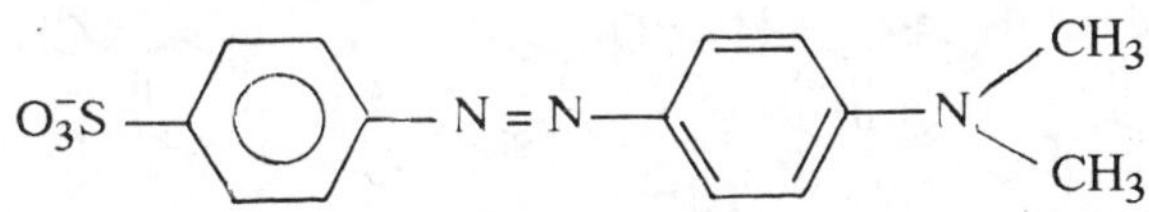

II (Orange or yellow)

OH^- ↑↓ H^+

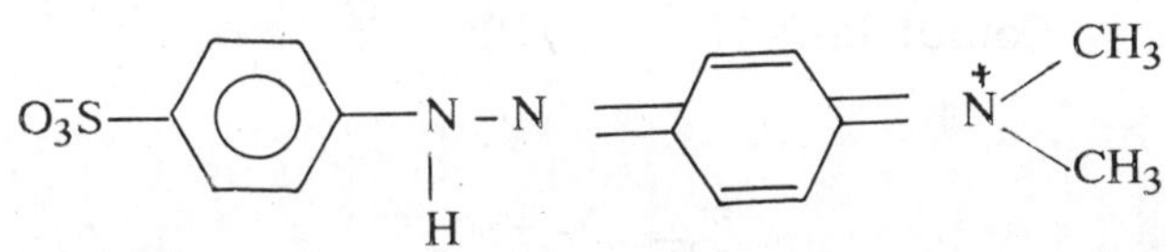

III (red)

39. Methyl red is yellow in alkali but red in acid:

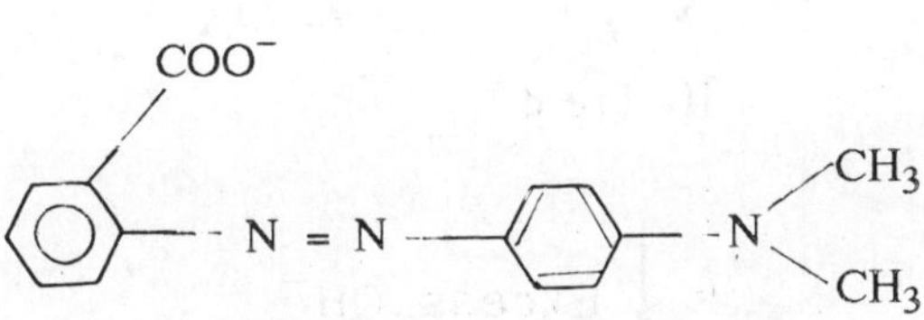

(yellow in alkali)

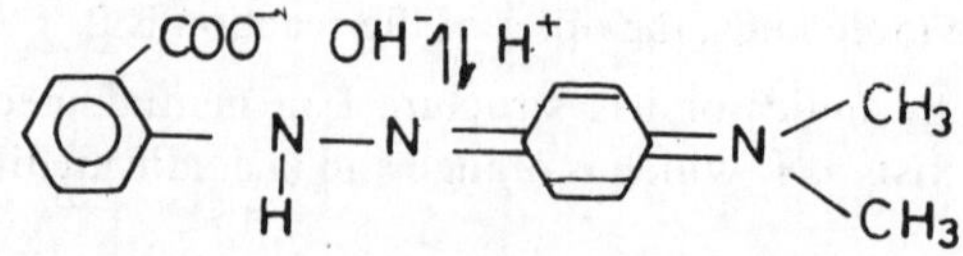

(red in acid)

40. **Phenolphthalein has structure I in neutral or acid medium and is colourless. In presence of dilute alkali solution, I is converted to II which loses water to give III (resonating ion having red colour). A large excess of a strong alkali converts III into IV which is again colourless.**

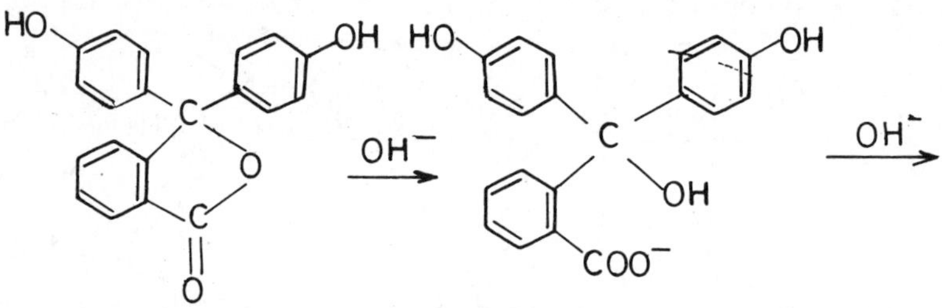

I (Colour less) II

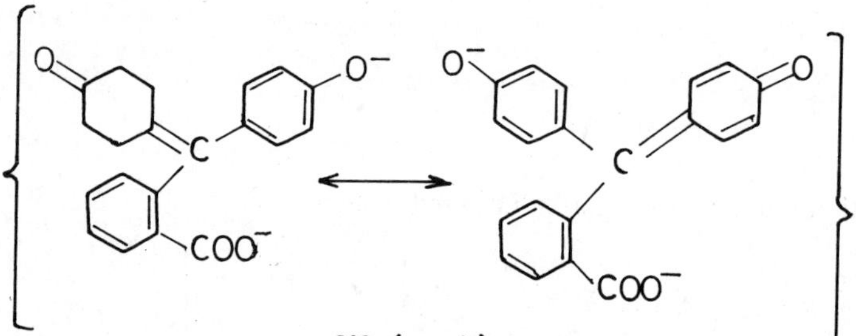

III (red)

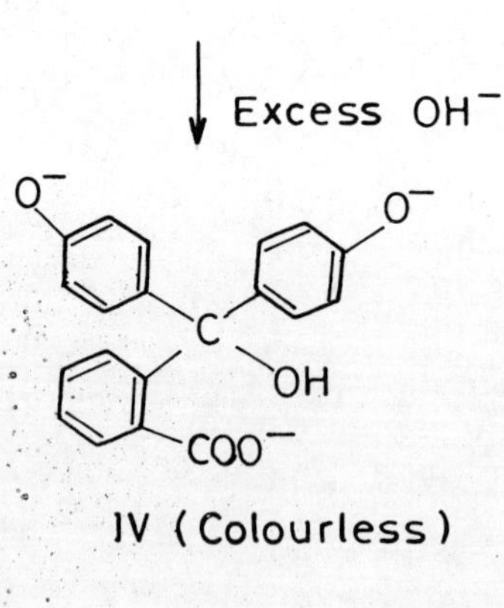

IV (Colourless)

41. Na_2CO_3 is available in market in Analytical Grade form and is quite stable. NaOH samples can absorb CO_2 from atmosphere and so often contain small amounts of Na_2CO_3 as impurity.

42. The solution should be heated to boiling to speed up the hydrolysis of metal salts so that the titration can be completed more rapidly.

43. Mineral Acidity arises from

(i) Hydrolysis of certain salts present in water (reactions 3.4 & 3.5)

(ii) Contamination with drainage from industrial area.

(iii) Waters receiving drainage from sulphide ore mines contain considerable amounts of H_2SO_4 resulting from wet oxidation of the ore:

$$\underset{\text{(iron pyrites)}}{2FeS_2\,(s)} + 7O_2 + 2H_2O \xrightarrow{\text{bacteria}} 2FeSO_4 + 4H^+ + 2SO_4^{2-}$$

(iv) Acid rains in the areas where atmosphere is highly polluted (contains a high percentage of SO_2).

44. Both acidity and alkalinity are expressed in terms of $CaCO_3$ whose equivalent weight is 50. 1 ml of N/50 solution, therefore, corresponds to 1 mg acidity or alkalinity and the calculations become simple.

45. If the sample is alkaline to phenolphalein, it should be neutralised with approximately 0.05 N H_2SO_4. If the sample is acid to methyl orange, a slight excess of pure $CaCO_3$ should be added.

46. The pH of the solution should not exceed 7.2 otherwise Ag^+ will not be precipitated because of complex formation:

$$AgCl + 2NH_4^+ + 2OH^- \longrightarrow \left[Ag(NH_3)_2\right]^+ + Cl^- + 2H_2O$$

47. The excess of the titrant that must be added before the eye can detect the colour change due to the reaction between the titrant and the indicator is called the indicator blank. This must be determined and substracted from all titrations. To an aliquot of distilled water, equal to the volume of solution at the end-point, is added the same amount of K_2CrO_4 indicator as is used in the actual determination, and the volume of $AgNO_3$ required to produce a visible amount of Ag_2CrO_4 is determined.

To avoid the indicator blank in the chloride ion determination, prepare the indicator solution as follows:

Dissolve 5 g of K_2CrO_4 in a little distilled water. Add $AgNO_3$ solution until a definite red precipitate is formed. Allow to stand for 12 hours, filter and dilute to 100 ml.

48. The solubility product of Ag_2CrO_4 increases with rising temperature and more $AgNO_3$ will have to be added to get the end-point. To avoid this, the titration should be performed at room temperature.

49. (i) To remove the colour and to clarify the sample, it is shaken thoroughly with a few drops of alumina cream [$Al(OH)_3$ suspended in water, prepared by adding NH_4OH to $Al_2(SO_4)_3$ and washed free of chloride, sulphate and ammonium ions] followed by decantation or filtration.

(ii) Interference due to H_2S may be removed by adding a crystal of $ZnSO_4$. In case it is not successful, the sample is acidified with H_2SO_4, boiled for a few minutes and neutralised with $NaHCO_3$ followed by dilution with distilled water to original volume.

50. Most accurate results are obtained by calculation from complete mineral analysis of the hard water sample (Equation 3.28) but it is a very tedious job and such type of analysis is rarely done. Of all other methods, EDTA method gives the best results, is less cumbersome and less time consuming.

51. Although Ca^{2+} (like Mg^{2+}) reacts with Eriochrome black-T indicator to form a wine red complex:

$$Ca^{2+} + HIn^{2-} \longrightarrow \underset{\text{Wine red}}{CaIn^-} + H^+,$$

the colour change from wine red to pure blue is not sharp with calcium indicator complex as with magnesium indicator complex. Therefore, Mg^{2+} ions have to be added if not present in the hard water.

52. (i) A small amount of the complexometrically neutral magnesium salt of ETDA (Na_2MgY) is added to the buffer solution.

(ii) A small amount of $MgCl_2$ may be added to the EDTA solution before it is standardised.

53. At higher pH values, $CaCO_3$ or $Mg(OH)_2$ may get precipitated and the dye (indicator) may change its colour. At lower pH values, the Mg-indicator complex becomes unstable and a sharp end-point cannot be obtained. The pH value is adjusted to about 10 by using a buffer solution of NH_4Cl/NH_4OH.

54. (a) Using equation 3.28 (p.47) :

$$\text{Total hardness} = 2.5 \times \text{ppm of } Ca^{2+} + 4.12 \times \text{ppm of } Mg^{2+}$$
$$= 2.5 \times 20 + 4.12 \times 48.6$$
$$= 50 + 200 = 250 \text{ ppm}$$

Alternatively

Impurity	ppm	epm	Hardness (ppm)	Alkalinity (ppm)
Ca^{2+}	20	20/20 = 1	1 × 50 = 50	–
K^+	39	39/39 = 1	–	–
Mg^{2+}	48.6	$\frac{48.6}{12.15} = 4$	4 × 50 = 200	–
Na^+	23	23/23 = 1	–	–
SO_4^{2-}	96	96/48 = 2	–	–
HCO_3^-	183	183/61 = 3	–	3 × 50 = 150
Cl^-	35.5	35.5/35.5 = 1	–	–

Thus, Calcium hardness = 50 ppm

Magnesium hardness = 200 ppm

Total hardness = 250 ppm

As alkalinity is less than total hardness,

Carbonate hardness (CH) = Alkalinity (Equation 3.29)

= 150 ppm

Therefore, Non-carbonate hardness (NCH) = 250 – 150

= 100 ppm

(b)

Impurity	ppm	epm	Hardness (ppm)	Alkalinity (ppm)
Mg^{2+}	24.3	$\frac{24.3}{12.15} = 2$	2 × 50 = 100	–
Ca^{2+}	40	40/20 = 2	2 × 50 = 100	–
HCO_3^-	305	305/61 = 5	–	5 × 50 = 250
			200 ppm	250 ppm

Thus, Calcium hardness = 100 ppm

Magnesium hardness = 100 ppm

Total hardness = 200 ppm

As alkalinity is more than total hardness,

Carbonate hardness (CH) = Total hardness (Equation 3.30)

= 200 ppm

Hence, there is no permanent hardness.

55. Volume of water sample taken = 100 ml
Volume of N/50 HCl used before boiling = 10.0 ml
Volume of N/50 HCl used after boiling = 0.2 ml
Mehyl orange Alkalinity before boiling (Experiment 3.1.3, p.42) $= \frac{1}{50} \times 10 \times \frac{1}{100} \times 50 \times 1000$

$= 100$ ppm

Methyl Orange Alkalinity after boiling $= \frac{1}{50} \times 0.2 \times \frac{1}{100} \times 50 \times 1000$

$= 2$ ppm

Therefore, Temporary hardness of the sample $= 100 - 2 = 98$ ppm

56. Volume of water sample taken = 100 ml
10 ml of alkali mixture ≅ 8 ml N/10 HCl
Therefore, 20 ml alkali mixture ≅ 16 ml N/10 HCl
Hence volume of alkali mixture added ≅ 16 ml N/10 HCl
≅ 16 ml N/10 alkali
Also, filtrate ≅ 13 ml N/10 HCl
≅ 13 ml N/10 alkali
Therefore, volume of N/10 alkali used in precipitating permanent hardness in 100 ml of the water sample $= 16 - 3 = 3$ ml
Therefore,
Permanent hardness of the sample $= \frac{1}{10} \times 3 \times \frac{1}{100} \times 50 \times 1000$

$= 150$ ppm

57. Volume of water sample taken = 100 ml
Volume of N/50 NaOH added = 25 ml
Also, filtrate ≅ 21ml N/50 HCl
≅ 21 ml N/50 NaOH

Therefore, volume of N/50 NaOH used for precipitating magnesium hardness of 100 ml of water sample $= 25 - 21 = 4$ ml

Therefore, Magnesium hardness of the water sample $= \frac{1}{50} \times 4 \times \frac{1}{100} \times 50 \times 1000$

$= 40$ ppm

58. (i) Lather factor = Volume of soap solution used against 70 ml of distilled water

= 0.5 ml

(ii) Strength of soap solution
1000 ml S.H.W. ≅ 0.28 g $CaCO_3$

Therefore, 70 ml S.H.W. $\cong \frac{0.28}{1000} \times 70$ g $CaCO_3$

$\cong$ 19.6 mg $CaCO_3$

But volume of soap solution used to precipitate hardness of 70 ml of S.H.W. $= 20.1 - 0.5 = 19.6$ ml

Hence,

19.6 ml soap solution $\cong$ 19.6 mg $CaCO_3$

or 1 ml of soap solution $\cong$ 1 mg $CaCO_3$

(iii) Total hardness

70 ml hard water sample $\cong 14.5 - 0.5 = 14$ ml soap solution

$\cong$ 14 mg $CaCO_3$

Therefore, Total hardness $= 14°Cl = 14 \times \frac{100}{7}$ ppm $= 200$ ppm

$= 200 \times \frac{1}{10}°F = 20°F$

(iv) Permanent hardness

70 ml boiled water $\cong 5.4 - 0.5 \cong 4.9$ ml soap solution

$\cong$ 4.9 mg $CaCO_3$

Therefore, Permanent hardness $= 4.9°Cl = 4.9 \times \frac{100}{7}$ ppm $= 70$ ppm

$= 70 \times \frac{1}{10}°F = 7°F$

(v) Temporary hardness $= 14 - 4.9 = 9.1°Cl$

$= 9.1 \times \frac{100}{7}$ ppm $= 130$ ppm

$= 130 \times \frac{1}{10}°F = 13°F$

59. (a) NO_2^- oxides I^- to free iodine in acidic medium :

$$2NO_2^- + 4H^+ + 2I^- \longrightarrow \underset{(2NO)}{N_2O_2} + 2H_2O + I_2$$

N_2O_2 is oxidised back to NO_2^- by atmospheric oxygen:

$$\underset{(4NO)}{2N_2O_2} + 2H_2O + O_2 \longrightarrow 4NO_2^- + 4H^+$$

Thus, it remains no longer possible to get a permanent end-point as the blue colour keeps on reappearing due to the formation of I_2 in a cyclic process, and higher results are obtained.

(b) Fe^{3+}, when present in amounts above 10 mg/l, also interferes by oxidising I^- to free I_2 and leads to higher results:

$$2Fe^{3+} + 2I^- \rightleftharpoons 2Fe^{2+} + I_2$$

60. Fe^{2+}, SO_3^{2-} and S^{2-} react with iodine, reducing it to iodide, and thus produce lower results:

$$2Fe^{2+} + I_2 \rightleftharpoons 2Fe^{3+} + 2I^-$$

$$SO_3^{2-} + I_2 + H_2O \longrightarrow SO_4^{2-} + 2I^- + 2H^+$$

$$S^{2-} + I_2 \longrightarrow S + 2I^-$$

$$(H_2S + I_2 \longrightarrow S + 2I^- + 2H^+)$$

61. The D.O. in a sample is utilised to oxidise Mn^{2+} to Mn^{4+} [$MnO(OH)_2$ or $MnO_2 \cdot H_2O$: reaction 3.52 or 3.52 a] in alkaline medium. This process is known as fixation of oxygen. When acidifed, $MnO(OH)_2$ gives back an equivalent amount of oxygen:

$$MnO(OH)_2 + 2H^+ \longrightarrow Mn^{2+} + 2H_2O + \frac{1}{2}O_2$$

62. When a 200 ml sample is used for titration, the number of ml of N/40 $Na_2S_2O_3$ used directly gives the D.O. in ppm.

63. At a constant temperature, the mass of a gas dissolved by a given amount of a solvent is proportional to the pressure of the gas in equilibrium with the solution. Henry's law is applicable when

(i) the temperature is not too low,

(ii) the gas pressure is not very high, and

(iii) the gas does not dissociate in or chemically react with the solvent.

64. A, volume of sample before dilution = 10 ml
B, volume of sample after dilution = 600 ml
D_1, the D.O. of diluted sample at the start of experiment = 3.9 mg/l (p.59)
D_2, the D.O. of the diluted sample after 5 days = 2.1 mg/l
B.O.D. of the sample (Eq. 3.63)

$$= \frac{3.9 - 2.1}{10} \times 600 = 108 \text{ mg/l}$$

$$= 108 \text{ ppm}$$

65. The B.O.D. test is widely used

(i) In determining the pollutional strength of domestic and industrial wastes,

(ii) In choosing and designing the treatment method for reducing the pollutional strength of such wastes,

(iii) In evaluating the purification capacity of natural water courses into which such wastes are to be discharged, and

(iv) In evaluating the efficacy of various units of the treatment process.

66. Being a first order reaction, complete BOD will be exerted only in infinite time and determination becomes impracticable. The 5-day incubation period is practicable and the BOD_5 values can be used for many considerations as they represent a fairly large percentage (70 to 80%) of the total BOD.

Also, beyond the 5-day period, significant errors may be introduced due to the growth of nitrifying bacteria (which, in raw sewage or primary effluent, are present in insufficient numbers to oxidise any appreciable amounts of reduced nitrogen in the first five days) which consume oxygen in oxidizing reduced nitrogen (reaction 3.57) which is normally present in sewage.

67. It should be prepared from distilled water by seeding with a small amount of domestic waste water to provide a mixed population of bacteria. A phosphate buffer should be added to maintain the pH at about 7. Phosphate also serves as a nutrient. Salts like $FeCl_3$ and NH_4Cl should be added to supply nutrients and small amounts of Na^+, K^+ and Ca^{2+} should also be added to serve as trace elements needed for the growth of bacteria. Finally, it should be saturated with oxygen.

68. (1) Although the dichromate reflux method is preferred because of its superior oxidising ability, applicability to a wide variety of samples, and ease of manipulation (as compared to the use of $KMnO_4$, $Ce(SO_4)_2$, KIO_3, etc.), certain compounds like

(i) straight-chain aliphatic hydrocarbons,

(ii) low molecular weight fatty acids,

(iii) pyridine, and

(iv) aromatic hydrocarbons

resist oxidation.

(2) Chloride ions, usually present in sewage in appreciable amounts, interfere because of their oxidation under the conditions of the experiment.

69. (1) Oxidation of straight-chain aliphatic hydrocarbons and low molecular weight fatty acids is effectively achieved by the use of Ag_2SO_4 catalyst. Pyridine and aromatic hydrocarbons are however not oxidised.

(2) Interference due to chloride ions is removed by complexing with $HgSO_4$:

$$Hg^{2+} + 4Cl^- \rightleftharpoons HgCl_4^{2-}$$

(very slightly dissociated)

70. The pH of the reaction mixture should be adjusted between 3 and 4 using acetic acid. At neutral pH, some forms of combined chlorine residuals do

not react with KI. However, H_2SO_4 should not be used as it increases the interference by Fe^{3+} and NO_2^-.

71. Fe^{3+}, NO_2^-, manganese in valences above 2 and organic sulphides.
72. At low pH, monochloramine also reacts in the first step and interferes in the determination of free chlorine residuals. At high pH, dissolved oxygen gives a colour and interferes.
73. Interference due to Cu^{2+} is overcome by complex formation with EDTA. By complex formation with trace metal catalysts, EDTA retards the oxidation of the DPD indicator by dissolved oxygen.
74. Chlorine Demand of a water sample is the amount of chlorine in mg that is reduced to Cl^- or converted into other less active forms by the impurities present in 1 litre of the sample.
    Chlorine Demand = Chlorine applied – Residual chlorine.
75. Chlorine-reactable materials likely to be present in water may be divided into the following categories:
    (a) Organics – Hydrocarbons, phenols, amines, humic acids, etc.
    (b) Inorganics – Ferrous, Manganous (Mn^{2+}), nitrite, sulphide (S^{2-} or H_2S), sulphite, cyanide, ammonia, etc.
    (c) Living organisms — Bacteria, algae, protozoa, etc.
76. (i) Organics and cyanide undergo substitution, addition and oxidation reactions with chlorine forming a variety of compounds (some chlorinated waters have been found to contain as many as sixty different chlorinated and other organics).
    (ii) Inorganic reducing agents like Fe^{2+}, Mn^{2+}, S^{2-}, SO_3^{2-} and NO_2^- are oxidised to Fe^{3+}, MnO_2, elemental S, SO_4^{2-} and NO_3^- respectively while chlorine is reduced to Cl^-.
    (iii) NH_3 undergoes a variety of reactions with Cl_2 giving products like NH_2Cl, $NHCl_2$, NCl_3, N_2, NO_2^-, NO_3^-, etc.
    (iv) Clorine destroys living organisms by reacting with enzymes present in their cells.
77. (i) Chlorine dosage – 0.1 to 0.2 mg/l free chlorine residuals.
    (ii) Contact period – 30 minutes.
    (iii) pH range – 6 to 7.
78. Free chlorine is readily dissipated in the distribution system and does not afford complete protection against bacteria due to short contact period. In order to prolong bactericidal action, free chlorine residuals are converted into more stable chloramines (reactions 3.70 & 3.71) by adding a little ammonia or ammonium sulphate to chlorinated water. This process is known as stabilising the chlorine.
79. Break-point chlorination is chlorination of a sample to the extent that all the ammonia present in the sample is converted into N_2 or its higher states (like

N_2O, NO_2, NO_3, NCl_3, etc.) and is marked by a minimum in the curve obtained by plotting residual chlorine versus applied dosage of chlorine (Fig. 9.1).

Explanation

When a small amount of chlorine is applied to a water sample containing reducing materials (Fe^{2+}, S^{2-} etc.), it is readily used up and so there is no residual. This stage is shown by part AB of curve II (for impure water). As the amount of chlorine applied is increased, it starts reacting with organics and also with ammonia which produces chloramines (reactions 3.70 & 3.71). Mono - and dichloramines are disinfectants and are determined as combined chlorine residuals. Part BC of the curve thus shows an increase in the amounts of these chloramines. Any further increase in chlorine dosage decomposes the chloramines possibly via the following reactions :

$$4NH_2Cl + 3Cl_2 + H_2O \longrightarrow N_2 + N_2O + 10HCl$$

$$2NHCl_2 + 3Cl_2 + 4H_2O \longrightarrow 2NO_2 + 10HCl$$

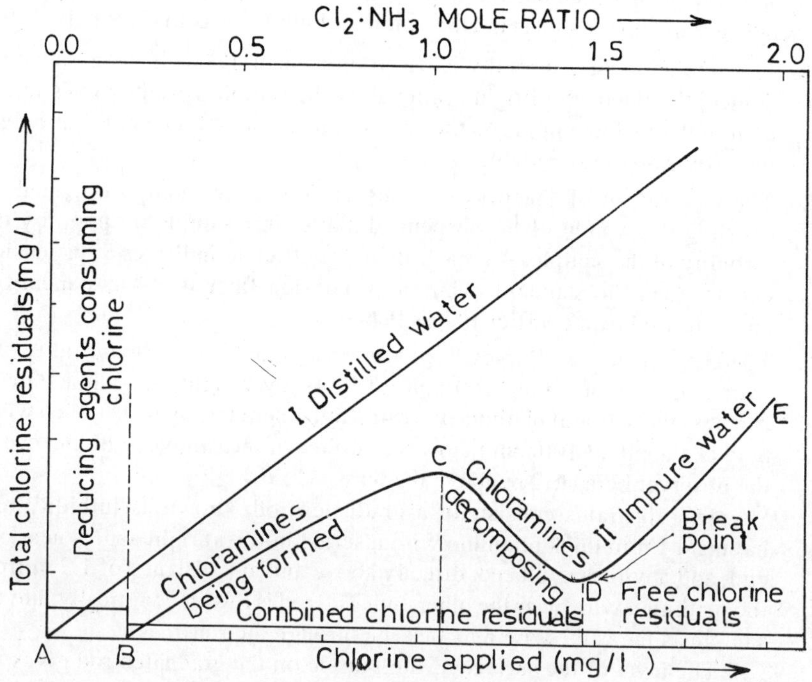

Fig. 9.1 Break-point chlorination

The amount of the combined chlorine residuals thus decreases along the curve CD. Point D indicates almost complete decomposition of chloramines. At this stage, known as Break-point Chlorination, there is marked decrease in the residual chlorine (a minimum in the curve). A higher dosage of chlorine appears as free chlorine residuals and total chlorine residuals start increasing. This is shown by the part DE of the curve.

Significance

A chlorine dosage higher than Break-point chlorination means that the chlorine demand of all the chlorine-reactable materials has been completely met with and free chlorine residuals are available for bactericidal action. It also signifies complete decomposition of NH_3, removal of colouring material and modification of taste and odour of the water sample.

80. Treatment with chlorine before filtration of the sample is known as Prechlorination. A higher chlorine dosage is required to satisfy the chlorine demand of filterable matter. This increases the cost but the quality of water obtained is superior as the chlorinated products with unpleasant tastes and odours may be adsorbed during filtration.

    Postchlorination, i.e., chlorination after filtration is cheaper than prechlorination (due to lower chlorine demand) but the treated water may have unpleasant taste and odour.

81. Jackson turbidimetry (also known as Visual turbidimetry) is based on measuring the interference caused to the passage of Light through the sample. In this method, the depth (thickness) of the sample required to obscure the image of a standard candle or bulb is measured visually.

    Nephelometric turbidity is based on measuring, with a Photoelectric device, the intensity of light scattered by the sample at right angles to the path of the incident light. As the errors of personal judgement are avoided, the results are more reliable.

82. The Coefficient of Fineness of a suspension is the number obtained by dividing the weight of the suspended matter in a sample, in ppm, by the turbidity of the sample. A greater than 1 coefficient indicates a suspension coarser than the standard, while a suspension finer than the standard is characterised by a coefficient less than one.

83. The U.S. Geological Survey have arbitrarily fixed at 100 the turbidity of a water sample containing 100 ppm of SiO_2, by weight, in such a state of fineness that a bright platinum wire of 1 mm diameter can just be seen when held at a depth of 100 mm below the surface of the sample, with the eye of the observer being 1.2 m above the wire.

    The apparatus consists of a graduated rod, known as turbidity rod, having a 1 mm thick platinum wire inserted into it at right angles near one end, and a wire ring placed, directly above and at a distance of 1.2 m from the platimum wire, near the other end. The rod is lowered vertically into the sample as far as the wire may be seen through the ring.

    The level of the surface of the sample on the graduated rod gives the turbidity of the sample. The method is used for testing the degree of fineness of the silica standard.

84. $HgCl_2$ (Mercuric chloride)

85. In addition to causing aggregation of finely divided particles dispersed in solution, the coagulants help in the removal of

    (i) True and apparent colour,

(ii) Harmful bacteria, algae and other planktons,

(iii) Taste and odour-producing substances.

86. Flocculant aids are substanaces which do not act as coagulants but when present along with a coagulant increase the rate of flocculation. They are of two types: (a) Naturally occurring materials like Bentonite (a form of clay). (b) Synthetic polymeric materials like Polyacrylamide (which acts very effectively in concentration range 0.01 – 0.1 mg/l).

87. A 50% solution of aluminium sulphate for use as a coagulant is available in the market under the trade name of 'Liquid Alum'.

88. The velocity distribution across the tube (capillary or jet) diameter does not become uniform immediately as an element of liquid enters the tube. A correction factor to minimise the error due to this drawback is known as 'Inlet Correction'. The magnitude of the correction factor decreases with increase in the length of the tube and with decrease in its diameter.

89. Though RW1 and RW2 may be used for determining relative viscosities of oils having an efflux time as low as 30 seconds, for greater accuracy a longer than 200 - seconds efflux time is recommended. This reduces the 'Kinetic Energy Correction' which depends on the volume rate flow, V/t.

90. U = 600 seconds

L = 780 seconds

H = 420 seconds

$$\text{Viscosity Index of the unknown oil} = \frac{L - U}{L - H} \times 100 = \frac{780 - 600}{780 - 420} \times 100 = \frac{180}{360} \times 100 = 50$$

91. For every degree rise in temperature, there is a decrease of roughly 2% in the coefficient of viscosity of most of liquids.

92. Lubricants with high Viscosity Index suffer only a small change in viscosity with change in temperature. Such lubricants can therefore be used over widely varying temperatures and are termed as 'all weather lubricants'.

93. (i) Silicones

(ii) Polyglycol ethers

(iii) Diesters or triesters

94. Viscosity Index of a lubricating oil is improved by adding hexanol or linear polymers like polyisobutylenes, poly-methacrylates, poly (alkyl styrenes), etc. It has been suggested that linear polymers function in the following ways:

(i) Rise in temperature increases the solubility of polymer in the oil. The increased concentration of the polymer raises the viscosity of the oil and thus compensates for the decrease in the viscosity of the oil itself.

(ii) The polymers are in the coiled form; with rise in temperature, the polymer particles tend to uncoil more and this increases the viscosity of the oil.

95. It is that state of a liquid when its viscosity does not change with rise in temperature.

96. Such a lubricant can be prepared by adding appropriate amount of a suitable linear-polymer Viscosity-Index improver.

97.

Freezing mixture	For temperature down to
Ice + water	32°F
Ice + common salt	10°F
Ice + $CaCl_2$	– 15°F
Solid CO_2 + acetone	– 70°F

98. It is the highest temperature at which an oil does not move when the standard jar containing the oil is kept in a horizontal position for 5 seconds. Pour point is taken to be 5°F higher than this.

99. For oils containing wax, pour point is the temperature at which crystallization of wax has gone to such an extent that the oil will stop flowing if cooled further. This temperature is known as 'Wax pour point'. For oils free from wax, pour point is the temperature at which the viscosity is so high that the oil will stop flowing, if cooled further, due to further increase in viscosity. This temperature is known as 'Viscosity pour point'.

100. (a) The 'Viscosity pour point' of an oil can be lowered by lowering the viscosity of the oil. This can be done either by removing some of the more viscous constituents of the oil or by adding a component of lower viscosity.

(b) 'Wax pour point' can be lowered either by dewaxing or by addidng a suitable pour point depressant.

101. Pour point depressants are materials which, when added to oils containing wax, get adsorbed on the surface of wax crystals during the initial stages of crystal formation. They thus reduce the size of the wax crystals. They also alter the crystal structure in such a way that the amount of oil held by the crystals by adsorption or by entrainment is reduced. Both these effects enhance the oil flow at lower temperature and thus reduce the 'Wax pour point' of the oil.

102. Paraflow is an important pour point depressant which, when present in concentrations of 1 to 2 per cent, may reduce the 'Wax pour point' of an oil by 50°F or more. Paraflow is a poly-alkyl naphthalene and is prepared by condensing chlorinated wax with naphthalene in the presence of anhydrous aluminium chloride which acts as a catalyst.

103. Air is a bad conductor of heat. It ensures very slow and uniform temperature change whether the oil is being heated or cooled.

104. S.I.T. is the temperature at which ignition occurs (without the introduction of a flame) when an inflammable liquid is allowed to fall in drops into a hot metal crucible.

105. (a) Presence of moisture
    (b) Vapour pressure of the oil – higher vapour pressure means lower flash and fire points.
    (c) Whether the test is made by open-cup or closed-cup method. Open-cup methods give higher flash and fire points.
    (d) Any drafts over the testing device in case of open-cup methods will tend to raise the flash and fire points.
    (e) In case of closed-cup methods, an increase in the ratio of the air space to the surface of the oil exposed increases the flash and fire points. A flash is observed when the atmosphere inside the cup contains about 2 % of oil vapour by volume.
    (f) Rate of heating affects the time available for the vapour to diffuse into air and hence affects the flash and fire points.
    (g) Frequency of application of test flame
    (h) Variations in the time of opening the shutter
    (i) Variations in the size of the test flame
    (j) Variations in the rate of stirring – temperature distribution is uneven if the stirring is too slow. Too rapid stirring produces splashing.
106. Presence of water may raise or lower the flash point. The steam formed on heating may prevent the vapour from igniting and hence raise the flash point. However, steam-distillation of low molecular weight constituents of the oil will tend to decrease the flash point. Oils containing water way split and sputter on heating and thus make the determination of the flash point extremely difficult.
107. Free water from an oil may be removed by
    (a) Settlement and decantation
    (b) Centrifugal action
    (c) Absorption by usual dehydrating agents such as anhydrous Na_2SO_4, anhydrous $CaCl_2$, etc.
    (d) Filtration through a filter paper containing Plaster of Paris.
    (e) Oils which do not lose more volatile portions on heating may be dried by passing a gentle current of air through them at 100°C.
108. 'Freaky' flash is the term used for the production of irregular flashes below the true flash point of an oil. Contamination of an oil with small amounts of volatile organic substances is responsible for this.
109. Fatty oils on heating may undergo thermal decomposition. Evolution of vapours of decomposition products and traces of free fatty acids present in the oils gives rise to freaky flashes. Flash points of fatty oils vary between 300–500°F.
110. Flash Point 'Closed' is determined by heating the oil in a closed cup; test-flame is injected into the cup through an opening produced temporarily and ignition of vapour takes place inside the cup. Thus, the vapour is not

free to diffuse to the atmosphere. Accurate and reproducible results are obtained.

Flash Point 'Open' is determined by heating the oil in a cup without cover. The surface of the oil is thus exposed to atmosphere. There is loss of vapour which is enhanced by any drafts over the testing device. The results are not reproducible and are higher, by as much as a few to 50°F, than those obtained with closed-cup methods. Cleaveland Open-Cup is the common accepted device for measuring flash points above 175°F.

111. In the Abel's apparatus, the oil cup is surrounded by an air jacket which is heated by a water bath. The temperature rises slowly at a rate of about 5°F per minute. Very precise results are obtained. The method is however slow and is recommended for oils flashing up to 120°F. In the Pensky–Marten's method, the air jacket surrounding the oil cup is heated electrically or by gas burner directly. The rate of heating is higher (10 ± 1°F/minute). It is most suitable and most commonly used method for all oils flashing above 120°F.

112. Air being a bad conductor of heat, it helps in controlling the cooling rate.

113. High-viscosity oils have less ability to penetrate rubber. So, on a time basis, there is some justification for the belief that the deteriorating action of mineral oil on rubber diminishes with increase in viscosity. However, if sufficient time is allowed, swelling is sure to occur.

114. The aniline point of an oil decreases with the increase in the percentage of its aromatic content. Aniline point is therefore inversely related to the aromatic content of the oil.

115. Paraffin-base type lubricating oils as a class have the highest aniline points. The range of their aniline points depends on the degree of refinement.

116. Aniline-point thermometers are available in the following three different ranges:
   (i) –38 to + 42°C
   (ii) 25 to 105°C
   (iii) 90 to 170°C

117. A diesel fuel with a high aniline point will have low aromatic content. This will give an easy start to the engine and will reduce knocking. The ignition quality of a diesel fuel is also reported in terms of Diesel Index which is related to the aniline point by the following expression:

$$\text{Diesel Index} = \text{Aniline point in °F} \times \frac{\text{API gravity}}{100}$$

For high-speed diesel fuel, the aniline point should be above 160°F.

118. The oil sometimes may contain materials, other than the free acids, that may be easily but slowly attacked by the alkali. Pink colour appears at the end-point due to the addition of an extra drop of alkali. In a few second, this alkali is consumed by the easily reactable materials and the pink colour fades away.

119. The esters in the oil may be hydrolysed by the moisture present and thus higher erroneous results may be obtained.
120. Appearance of a green or bluish-green colour indicates that the sample contains basic constituents; the total base value can then be determined by titrating the sample with N/100 alcoholic HCl. At the end-point, the colour changes from green to orange or yellowish-orange.
121. The presence of moisture spoils the Wij's solution by decomposing ICl:

$$ICl + H_2O \longrightarrow HCl + HIO$$

122. Sunlight catalyses the decomposition of ICl into I_2 and Cl_2:

$$2ICl \xrightarrow{h\nu} I_2 + Cl_2$$

123. To prevent the leakage of Cl_2 or I_2 vapours that may be formed by decomposition of ICl.
124. Pycnometers are vessels (5–15 ml capacity) with capillary necks (Fig. 9.2) in which a definite volume of a liquid can be weighed. The volume error in adjusting the meniscus being extremely small as compared to the total volume of the liquid being measured, the results are very precise. For determining the sp. gr. of a liquid, the pycnometer is calibrated with water.

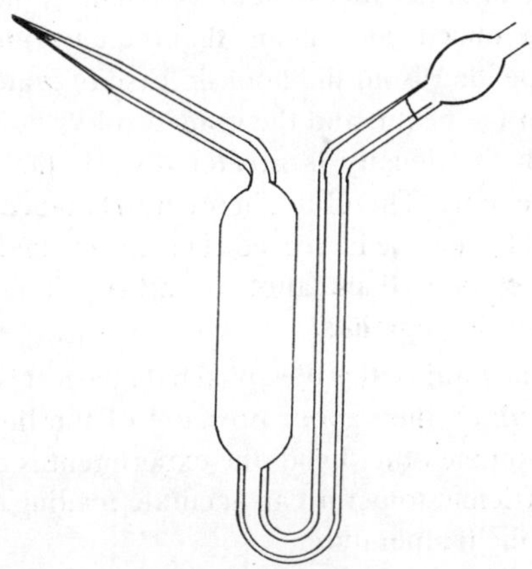

Fig. 9.2 Pycnometer

125. Yes, for this purpose, the legs of the riders are bent to form hooks. Other riders can be suspended from the hooks (Fig. 9.3).
126. It is the product of rider mass and the distance from the fulcrum to the rider, as marked on the beam.
127. The weight of the unit (large) rider is kept equal to the weight of water displaced when the plummet is just completely immersed in water at the temperature of calibration. Levelling of beam with unit rider at mark 10 indicates a sp. gr. of 1. Since the beam is divided into 10 equal parts, any

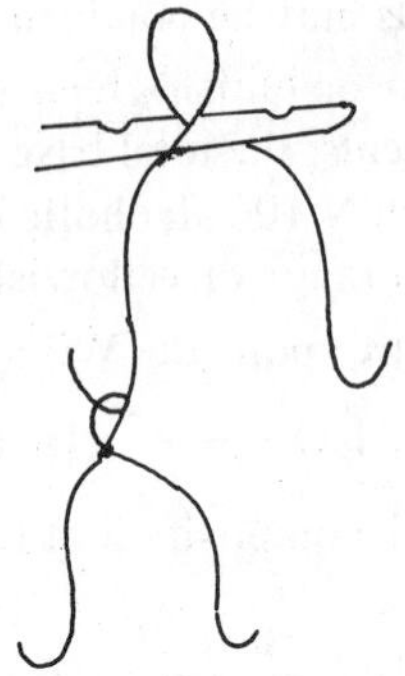

Fig. 9.3 More than one rider at the same mark

other position of unit rider on the beam gives the first digit after decimal in the value of sp. gr. of the liquid under test. When the various graded riders differ by a factor of 10, their positions on the beam directly give the 2nd, 3rd and 4th digit after decimal in the sp. gr. value. No calculations are thus required.

128. When the density is less than 1, the weight of the liquid displaced by the plummet will be less. In order to achieve balance, the weight on the beam will have to be reduced. This means that the unit rider, which is placed at mark 10 or suspended from the hook in case of water, must be moved to another mark on the beam, and the graded riders be placed at appropriate positions. When the density is greater than 1, the weight of the liquid displaced will be more. Therefore, in order to balance the beam, the weight on the beam will have to be increased. This means that one unit (large) rider will be kept at position 10 and another unit rider and/or lighter riders will be placed at appropriate points.

129. For highly volatile liquids, the Westphal balance test is conducted at a lower temperature at which the vapour pressure of the liquid is very low. For viscous liquids, on the other hand, the experiment is conducted at a higher temperature, sufficient to permit an accurate reading. A thermostat is used for maintaining the temperature.

130. Various agencies notably ASTM and American Institue of Petroleum have published conversion tables with the help of which the observed sp. gr. at any temperature can be corrected to the standard temperature of 60/60°F.

131. The average value of coefficient of expansion of oils has been accepted as 0.000348 per degree F. On this basis, the expansion capacity to be provided

$$= 1000 \text{ litres} \times 100 \text{ F} \times 0.000348/\text{F}$$

$$= 34.8 \text{ litres.}$$

132. (i) Steam Turbine lubricants

(ii) Gear lubricants

(iii) Crank Chamber oils.

133. (i) The lubricant should be refined in such a way as to eliminate the polar impurities whether initially present or formed in the early steps of the refining process.
(ii) Such additives should be used as have minimum emulsifying characteristics.
(iii) The oil should be stable towards oxidation under the conditions imposed upon it.
(iv) The lubricant should be protected againt extraneous contamination.

134. 'Soluble Oils' are oil-in-water emulsions, stablised by an emulsifying agent such as sodium salt of a carboxylic or sulphonic acid. The ratio of oil to water in the emulsion is normally of the order of 1:20 to 1:60. Since they contain a large concentration of water, they have a much higher specific heat and latent heat of vaporisation than anhydrous oils. So they are used to good advantage in machine shops during cutting, grinding, etc., of metals where cooling rather than lubricity is of paramount importance.
They are usually opaque, having milky appearance, but transparent soluble oils (cutting oils or cutting oil emulsions) have also been made which have the practical advantage that they do not mask the work from the vision of the operator.

135. Excessive foaming (forming of oil-air emulsion) of the lubricating oil is a very serious problem in diesel engines, aircraft engines and automatic transmissions. Substances which suppress foaming are called defoamants or antifoams. Silicones are the most effective antifoams (concentrations of about 10 ppm are used). Other antifoams are potassium oleate and sodium alkyl esters of sulphuric acid.

136. The cone will penetrate to a greater depth and thus the observed penetration number of the sample will be higher.

137. The cone will not fall freely and so the test will show lower penetration number.

138. The penetration number of a grease depends mainly on the structure and interaction of the gelling element, the amount of soap and nonsoap thickner present and to some extent on the viscosity of the lubricating oil present. It is also affected by the handling procedure during manufacture of grease, the temperature of filling the containers and the rate of cooling.

139. It is the separation of a liquid lubricant from a lubricating grease for any cause.

140. In a grease, the soap crystallizes in the form of threads (fibres) having lengths of the order of 20 or more times their thickness. Most soap fibres are microscopic in size and the grease appears smooth. However, when the fibre bundles are large enough to be seen with naked eye, the grease acquires a FIBROUS appearance. The fibrous structure is also noticeable when the grease is pulled apart. Greases having this fibrous structure tend to resist being thrown off gears and out of bearings. The most common example of

a FIBROUS grease is sodium-base grease. Pertolatum also has fibrous structure.

141. It is a mixture of oils and waxes stabilised by a third component and is obtained from still residues of paraffin base crudes after fractionation. It does not leave an oily stain on paper as wax is the external phase and oil the internal phase. It can be used as a grease.

142. It is the name given to a soap in which the soap crystal or fibre is formed by Co-crystallisation of a normal soap (such as metallic strearate or oleate) with a complexing agent (such as metallic salts of short chain organic acids like acetic acid or lactic acid, or the inorganic salts like carbonates or chlorides).

143. A complexing agent usually increases the dropping point of a grease.

144. (i) Weight of coal sample, w_1 = 2.5 g

Weight of empty crucible, w_2 = 19.35 g

Weight of crucible + sample, w_3 = 21.85 g

Weight of crucible + sample, after heating at 105–110°C, w_4 = 21.765 g

Loss in weight (Moisture) $= w_3 - w_4$

$= 21.85 - 21.765$

$= 0.085$ g

Therefore, % Moisture $= \dfrac{0.085}{2.5} \times 100 = 3.4$

And weight of crucible + residue, after ignition (Ash), w_5 = 19.595 g

Weight of Ash $= w_5 - w_2 = 19.595 - 19.35$

$= 0.245$ g

Therefore, % Ash $= \dfrac{0.245}{2.5} \times 100 = 9.80$

(ii) Weight of sample, w_1 = 2.5 g

Weight of empty crucible, w_2 = 19.345 g

Weight of crucible + sample, w_3 = 21.845 g

Weight of crucible + sample, after heating at 950 ± 20°C, w_4 = 20.873 g

Loss in weight, $= w_3 - w_4 = 21.845 - 20.873$

$= 0.972$ g

Therefore, % Volatile Matter (V.M.) $= \dfrac{0.972}{2.5} \times 100 - \%$ Moisture

$= 38.88 - 3.4 = 35.48$

(iii) % Fixed carbon (F.C.) = 100 – % [Moisture + Volatile Matter + Ash]

$$= 100 - [3.4 + 35.48 + 9.80]$$
$$= 51.32$$

Calorific value (Equation 5.3) $= 82 \times \text{F.C.} + \alpha \times \text{V.M.}$ (Goutel's formula)

$$= 82 \times 51.32 + 80 \times 35.48$$
$$= 7046.64 \text{ cal/g}$$

145. Presence of HCl prevents precipitation of chromates, carbonates and phosphates which are insoluble in neutral solutions. Also the precipitate formed in presence of HCl is coarse and consists of large crystals which are more readily filtrable. But excess of HCl should be avoided as solubility of $BaSO_4$ increases due to the formation of bisulphate ion:

$$SO_4^{2-} + H^+ \rightleftharpoons HSO_4^-$$

However, in presence of excess of Ba^{2+}, the solubility of $BaSO_4$ in a solution containing HCl upto 0.05 N is negligible.

146. During precipitation of an insoluble salt, some substances normally soluble in the mother liquor are carried down. This contamination of the precipitate with salts which are otherwise soluble is known as coprecipitation. Nitrate, if present, has to be decomposed by boiling the solution with a large excess of HCl before precipitation. Coprecipitation of $BaCl_2$ may however be reduced by carrying out the precipitation at (i) boiling temperature, (ii) low concentration of $BaCl_2$ and sulphate solutions, and (iii) slow mixing of the two solutions with constant stirring.

147. Due to loss of acid during digestion, the concentration of the boiling mixture with respect to ammonia may become too large and at the prevalent high temperature, loss of ammonia due to volatilisation may take place. To avoid this, sufficient sulphuric acid should be added during digestion so that 10–15 g of it remains in the free state at the end of digestion.

148. HgO and Hg have been reported to be the best catalysts but their presence leads to the formation of mercuro–ammonium compounds which are not completely decomposed by NaOH. So after digestion, 25 ml of an 8% sodium thiosulphate solution or a 4% K_2S solution is added which decomposes the mercuro-ammonium compounds and removes mercury as sulphide.

149. Zinc reacts with alkali to form hydrogen gas whose evolution helps in the regular ebullition (boiling) during distillation:

$$Zn + 2NaOH \longrightarrow Na_2ZnO_2 + H_2$$

150. (a) (i) Weight of coal sample $= 0.24$ g

Weight of CO_2 formed $= 0.792$ g

Weight of H_2O formed $= 0.0216$ g

Therefore, % carbon (C) (equation 5.9)

$$= \frac{12}{44} \times \frac{\text{weight of } CO_2}{\text{weight of sample}} \times 100$$

$$= \frac{12}{44} \times \frac{0.792}{0.24} \times 100 = 90$$

% Hydrogen (H)
(equation 5.8)

$$= \frac{2}{18} \times \frac{\text{weight of } H_2O}{\text{weight of sample}} \times 100$$

$$= \frac{2}{18} \times \frac{0.0216}{0.24} \times 100 = 1$$

(ii) Weight of coal = 1.4 g
Volume of N/10 H_2SO_4 taken = 50.0 ml
Volume of N/10 NaOH used to neutralise excess of acid = 10.0 ml
Therefore, volume of excess N/10 H_2SO_4 = 10.0 ml
Volume of N/10 H_2SO_4 neutralised by ammonia = 50 – 10 ml = 40.0 ml
Therefore, % Nitrogen(N) (equation 5.25)

$$= \frac{\text{ml of acid neutralised} \times \text{normality} \times 1.4}{\text{weight of coal}}$$

$$= \frac{40 \times \frac{1}{10} \times 1.4}{1.4} = 4$$

(iii) Weight of coal = 3.2 g
Weight of $BaSO_4$ formed = 0.233 g
Therefore, % sulphur (S)
(equation 5.19)

$$= \frac{32}{233} \times \frac{\text{Weight of Ba } SO_4}{\text{Weight of coal}} \times 100$$

$$= \frac{32}{233} \times \frac{0.233}{3.2} \times 100 = 1$$

Hence % Oxygen (O)
(equation 5.26)

$$= 100 - [C + H + N + S]$$

$$= 100 - [90 + 1 + 4 + 1]$$

$$= 4$$

Gross calorific value
(equation 5.4)

$$= \frac{1}{100}\left[8080C + 34460\left(H - \frac{O}{8}\right) + 2240\,S\right]$$

$$= \frac{1}{100}\left[8080 \times 90 + 34460\left(1 - \frac{4}{8}\right) + 2240 \times 1\right]$$

$$= \frac{1}{100}[727200 + 17230 + 2240]$$

$$= 7466.70 \text{ cal/g}$$

Net calorific value (N.C.V.)

$$\text{(equation 5.5.)} = \text{G.C.V.} - \frac{9\text{H}}{100} \times 587$$

$$= 7466.70 - \frac{9 \times 1}{100} \times 587$$

$$= 7466.70 - 52.83$$

$$= 7413.87 \text{ cal/g}$$

(b) Proximate analysis is essential to assess the suitability of a coal for a particular domestic or industrial purpose whereas ultimate analysis is essential to calculate the heating value of coal.

Proximate analysis is comparatively cheaper and simpler as against ultimate analysis which is very tedious and the cost of equipment needed is very high.

However, the two types of analysis are complimentary and, taken together, provide necessary and sufficient data to decide about the quality of coal.

151. Cuprous chloride solution being a reducing agent is oxidised to an appreciable extent and acquires a greenish black colour on standing:

$$2Cu_2Cl_2 + 4HCl + O_2 \longrightarrow 4CuCl_2 + 2H_2O$$

Addition of copper turnings, wire or gauze to the storage bottle reduces the solution back to Cu_2Cl_2 and the solution acquires a straw yellow colour:

$$CuCl_2 + Cu \longrightarrow Cu_2Cl_2$$

152. As it is not practicable to shake the apparatus, the rate of absorption of the flue gas constituents would be quite slow. In order to achieve speedy absorption, the surface area of contact between the gas and the absorbent solution is increased by placing a large number of glass tubes or beads in the pipets containing the absorbent solutions.

153. Water may preferentially dissolve some constituents of the flue gas and thus change its composition. The solubility of gases in water is considerably reduced due to the presence of NaCl.

154. They impart a colour (reddish) to the confining liquid so that its level in the water-jacketed burette may be conveniently observed.

155. The oxygen absorbent, i.e., alkaline pyrogallol solution can absorb both CO_2 and O_2. Hence the gas sample should be passed through this solution only after the CO_2 content has been removed. Also, the carbon monoxide absorbent, i.e., acid cuprous chloride can react with both O_2 and CO. So the gas sample should be led to this solution after removing O_2.

Thus, in order that only one constituent be removed by each absorbent, it is important that the flue gas should be brought into contact with various reagents in the following specific sequence or order:

(i) CO_2 absorbent – KOH solution

(ii) O_2 absorbent – Alkaline pyrogallol solution

(iii) CO absorbent – Acid cuprous chloride solution.

156. Pyrogallol, also known as pyrogallic acid, is 1, 2, 3-trihydroxybenzene and its structural formula is

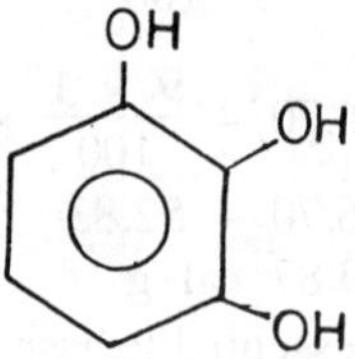

157. Acidic gases in the flue gas sample, if present, react with alkali:

$$H_2S + 2KOH \longrightarrow K_2S + 2H_2O$$

$$SO_2 + 2KOH \longrightarrow K_2SO_3 + H_2O$$

$$HCN + KOH \longrightarrow KCN + H_2O$$

They are therefore absorbed along with CO_2. Thus higher results for CO_2 are obtained.

158. These gases are not absorbed by any of the three absorbents used. Unless special modifications are made in the Orsat apparatus to measure their amounts, they will be reported as N_2. Hence the results of N_2 will be higher.

159. Temperature and pressure of the experiment affect the total volume of the gas but not the percentage composition provided they (T & P) remain constant throughout the experiment. This is because the effect of deviation of some constituents from ideal behaviour within normal temperature and pressure ranges is negligible. However, the rate of absorption may slightly be affected.

160. The water -jacket serves the purpose of a thermostat. It helps in keeping the temperature constant during the experiment.

161.

	Instrumental methods		Orsat apparatus
1.	Methods (a) Mass spectrometry (b) Gas chromatography (c) Low temperature fractional distillation	1.	Analysis by Orsat apparatus is also known as constant pressure volumetric analysis or chemical absorption analysis.
2.	They produce detailed analysis of the sample	2.	Scope is limited to the analysis of CO_2, O_2, CO and N_2 (H_2 & a few hydrocarbons with specially equipped laboratory Orsat apparatus) yet the information available is quite useful for regulating combustion.

3. Require less effort on the part of the operator and results are obtained quickly.	3. The procedure is cumbersome and time consuming.
4. Capital investment is very high.	4. The apparatus is inexpensive.
5. The apparatus is complex and demands greater maintenance.	5. The apparatus is very simple and maintenance cost is negligible.

162. A fuel gas is a fuel in the gaseous state. It contains some of the cumbustible gases like H_2, CO, CH_4 or other hydrocarbons. Small amounts of non-cumbustible gases like CO_2, N_2 and O_2 may also be present. Calorific value may vary from about 1000 k cal/m^3 to 12000 k cal/m^3.

Flue gas is the term used for the products of combustion of a fuel, coming out of the exhaust pipe of an automobile or the chimney of a furnace. Its major constituent is N_2 (usually more than 50%). CO_2, CO and O_2 are present in varying amounts depending upon the extent of combustion. Small amounts of SO_2, unburnt H_2 and/or hydrocarbons may also be present. They have high sensible heat that can be utilised with the help of regenerators. Calorific value is very low and increases with the amount of CO, e.g., Blast Furnace gas that contains up to 25% CO has a calorific value of about 800 k cal/m^3.

163. Time required to decompose the iron ore with acid decreases sharply with the decrease in the particle size of the ore. Ten minutes of proper grinding in an agate mortar often saves hours of treatment with acid.

164. (1) Any organic matter associated with the ore is oxidised to CO_2 during roasting.

(2) Roasting also decomposes any pyrites present in the ore and removes sulphur as SO_2:

$$2FeS_2 + 5O_2 \longrightarrow 2FeO + 4SO_2$$

$$4FeO + O_2 \longrightarrow 2Fe_2O_3$$

(3) Any arsenic impurity is also removed as oxide:

$$4As + 3O_2 \longrightarrow 2As_2O_3$$

165. When a large excess of $SnCl_2$ is added during reduction, a much larger amount of $HgCl_2$ will have to be added to remove the excess of $SnCl_2$. This means wastage of reagents. Also, a very thick or heavy precipitate of Hg_2Cl_2 will be formed which slowly reacts with the oxidising titrant ($K_2Cr_2O_7$ or $KMnO_4$) and also reduces ferric complex fromed during the reaction leading to inaccurate results.

166. In all the three cases, the Hg_2Cl_2, formed initially, immediately reacts with more Sn^{2+} ions present forming grey or black mercury in the finely divided state, which reacts with the oxidising titrant and also slowly reduces Fe^{3+} in presence of Cl^- ions, thus giving inaccurate results.

$$Hg_2Cl_2 + Sn^{2+} \longrightarrow 2Hg + Sn^{4+} + 2Cl^-$$

$$2Hg + 2Fe^{3+} + 2Cl^- \longrightarrow Hg_2Cl_2 + 2Fe^{2+}$$

167. (a) To avoid the atmospheric oxidation of Fe^{2+} to Fe^{3+} which is quite appreciable in hot solution.

(b) To avoid the precipitation of Hg on the addition of $HgCl_2$.

168. It indicates that stannous chloride has not been added in excess which means that reduction of Fe^{3+} to Fe^{2+} may be incomplete.

169. H_3PO_4 complexes the Fe^{3+} ions formed during the titration and thus removes the yellow colour due to $FeCl_3$. This makes detection of the end-point easier.

170. *Internal Indicator*: A substance that can be added to the reaction mixture to indicate the equivalence point of the titration.

Example: Phenolphthalein in acid-alkali titration.

External Indicator: An indicator which is not added to the reaction mixture. Rather, a drop of the reaction mixture is removed and mixed with a drop of the indicator on a glazed tile.

Example: KSCN is used as external indicator for testing complete reduction of Fe^{3+} during the reduction of iron ore solution with Zn and H_2SO_4.

Self Indicator: When one of the reactants itself acts as indicator and no external substance is required.

Example: In the titration of oxalic acid with $KMnO_4$, the titrant itself acts as indicator.

171. The external indicator used is a solution of potassium ferricyanide $K_3[Fe(CN)_6]$.

Preparation: A small crystal of $K_3[Fe(CN)_6]$ is washed repeatedly with distilled water to remove the superficial coating of the ferrocyanide. The washed crystal is then dissolved in distilled water to get an approximately 0.1 per cent solution.

172. The indicator solution is taken in the form of drops on a dry glazed tile (A glass rod or a dropper may be used). The dichromate solution is then added in small portions to the iorn ore solution and after each addition, a drop of the reaction mixture is removed with a clean glass rod and then mixed with one of the drops of the indicator. A deep blue colour appears:

$$3Fe^{2+} + 2\left[Fe(CN)_6\right]^{3-} \longrightarrow \underset{\substack{\text{Ferroferricyanide} \\ \text{(Prussiun blue)}}}{Fe_3\left[Fe(CN)_6\right]_2}$$

$$\text{or } FeSO_4 + K_3\left[Fe(CN)_6\right] \longrightarrow \underset{\text{Potassium ferroferricyanide}}{KFe\left[Fe(CN)_6\right]} + K_2SO_4$$

As the addition of dichromate solution is continued, the concentration of Fe^{2+} ions decreases, and on testing with a fresh drop of the indicator a bluish-green and then green colour appears. The titration is complete at a stage when a drop of the iron solution on being mixed with a drop of the indicator on the tile shows no trace of green – a slightly brownish tint superimposed on the yellow is observed.

173. (1) To get good drops, the glazed tile should be perfectly dry (A piece of blotting paper may be used for drying).
 (2) After each testing, the glass rod must be washed with distilled water otherwise some indicator solution sticking to the tip of the glass rod will get transferred to the reaction mixture and spoil it.
 (3) Each time the glass rod is dipped into the reaction mixture, some Fe^{2+} ions are removed and consequently lower results will be obtained. To reduce this loss, testing should be started near the end-point, i.e., after the first titration, the dichromate solution should be run into the Fe^{2+} ion solution, without testing, to within 0.5 ml of the previously determined end-point. Then the dichromate solution should be added dropwise, testing the reaction mixture after each addition.
 This helps in two ways:
 (a) Number of withdrawals for testing is reduced.
 (b) Concentration of Fe^{2+} ions near the end-point is very small and so loss of iron is not appreciable.

174. (1) Detection of the end-point is quite cumbersome.
 (2) Lower results are obtained because of loss of Fe^{2+} during testing.

175. It will immediately react with Fe^{2+} ions producing the deep blue colour of ferroferricyanide. The ferrous ions thus combined will not be available for oxidation and also the detection of the end-point will not be possible.

176. The hydrogen bubbles escaping the solution may carry along with them appreciable amounts of iron ions. When the bubbles pass through the narrow tube of the funnel, majority of them break and the loss of iron is reduced.

177. A.R. zinc reacts very slowly with dilute H_2SO_4. Copper sulphate increases the rate of dissolution of Zn and thus the production of hydrogen, by setting up a galvanic cell.

178. When HCl is present, some $KMnO_4$ may be consumed in the oxidation of Cl^- ions:

$$MnO_4^- + 8H^+ + 5e \longrightarrow Mn^{2+} + 4H_2O] \times 2$$

$$2Cl^- \longrightarrow Cl_2 + 2e] \times 5$$

$$2MnO_4^- + 16H^+ + 10Cl^- \longrightarrow 2Mn^{2+} + 5Cl_2 + 8H_2O$$

Thus $KMnO_4$ titration may give higher results.

179. Many iron ores are very difficult to decompose by H_2SO_4. So the solution has to be prepared in HCl. If it is to be titrated with $KMnO_4$, the following methods may be used:

(a) The iron ore solution is reduced with $SnCl_2/HCl$ (excess $SnCl_2$ being removed the $HgCl_2$). The solution is then titrated with standard $KMnO_4$ in presence of Zimmermann-Reinhardt reagent ($MnSO_4 \cdot 4H_2O$ + water + H_2SO_4 + H_3PO_4) which is also known as preventive solution. Presence of excess of Mn^{2+} reduces the oxidising power of $KMnO_4$. Also the reducing power of Fe^{2+} is increased by the presence of phosphoric acid which complexes Fe^{3+} ions. Thus the possibility of the oxidation of Cl^- with $KMnO_4$ is reduced.

(b) The best method is to expel HCl by evaporating the iron ore solution with H_2SO_4:

$$2H(FeCl_4) + 3H_2SO_4 \longrightarrow Fe_2(SO_4)_3 + 8HCl$$

and dissolving the residue in dilute H_2SO_4 followed by reduction with zinc and H_2SO_4.

180. Reduction with amalgamated zinc is much more rapid than with zinc and H_2SO_4:

$$2Fe^{3+} + Zn \longrightarrow 2Fe^{2+} + Zn^{2+}$$

As no hydrogen is evolved, there is no loss of iron.

181. AR granulated zinc is taken in a beaker and covered with a 2% solution of $HgCl_2$. The mixture is stirred with a glass rod for about 10 minutes. The solution is then decanted and the residual zinc amalgam formed is washed repeatedly with distilled water by decantation.

182. 1. Sulphurous acid or SO_2 (Na_2SO_3/H_2SO_4)
2. H_2S
3. Titanous chloride ($TiCl_3$)
4. Mg/H_2SO_4

183. Both As(III) and Sb(III) react with I_2 as per the following reactions:

$$H_3AsO_3 + I_2 + H_2O \longrightarrow H_3AsO_4 + 2H^+ + 2I^-$$

$$H_3SbO_3 + I_2 + H_2O \longrightarrow H_3SbO_4 + 2H^+ + 2I^-$$

Thus, if arsenic and antimony are present in the trivalent state, a part of iodine liberated by copper will be used up and the results will be lower. Though the reactions are reversible, the backward reaction, namely, the oxidation of I^- by As(V) and Sb(V) above a pH of 3.2 (maintained by buffer) is not appreciable and so no interference is caused.

184. Even when present in trace amounts, NO_2^- and nitrogen oxides cause very marked interference by oxidising I^- to I_2:

$$NO_2^- + 2H^+ + e \longrightarrow NO + H_2O\,] \times 2$$

$$2I^- \longrightarrow I_2 + 2e$$

$$2NO_2^- + 4H^+ + 2I^- \longrightarrow 2NO + I_2 + 2H_2O$$

Nitric oxide is readily oxidised by air to NO_2

$$2NO + O_2 \longrightarrow 2NO_2$$

which in turn reacts anew with iodide to form more iodine and NO and the cycle is then repeated. For this reason, when the solution is titrated with thiosulphate in presence of starch, the blue colour returns again and again and no permanent end-point can be obtained.

$$NO_2 + 2H^+ + 2I^- \longrightarrow NO + I_2 + H_2O$$

185. Both nitrate and nitrogen oxides are eliminated by heating with urea in acid medium

$$2NO_2^- + NH_2\cdot CO\cdot NH_2 + 2H^+ \longrightarrow 2N_2 + CO_2 + 3H_2O$$

$$6NO_2 + 4NH_2\cdot CO\cdot NH_2 \longrightarrow 7N_2 + 4CO_2 + 8H_2O$$

$$NO_2 + NO + NH_2\cdot CO\cdot NH_2 \longrightarrow 2N_2 + CO_2 + 2H_2O$$

Nitrous acid (nitrite in acid medium) can also be eliminated by treating the solution with sulphamic acid:

$$HNO_2 + NH_2\cdot SO_3H \longrightarrow N_2 + H_2SO_4 + H_2O$$

186. Any iron present in the ore or alloy gets oxidised to Fe^{3+} during preparation of solution. Fe^{3+} oxidises I^- to I_2:

$$Fe^{3+} + e \rightleftharpoons Fe^{2+}\,] \times 2$$

$$2I^- \rightleftharpoons I_2 + 2e$$

$$2Fe^{3+} + 2I^- \rightleftharpoons 2Fe^{2+} + I_2$$

Thus, results will be high.

187. Ethylenediaminetetra-acetic acid (EDTA) and pyrophosphate form stronger complexes with Fe^{3+} than with Fe^{2+} and thus avoid interference by Fe^{3+}.

188. Atmospheric oxidation of I^- is catalysed by low pH,

$$2I^- \longrightarrow I_2 + 2e\,] \times 2$$

$$O_2 + 4H^+ + 4e \longrightarrow 2H_2O$$

$$4I^- + O_2 + 4H^+ \longrightarrow 2I_2 + 2H_2O$$

Cu^+ ion, nitrite ion, nitrogen oxides and sunlight.

189. (a) At low pH values As(V) and Sb(V) will interfere by oxidising I^-.

(b) The reaction between Cu^{2+} and I^- goes to completion only under conditions where CuI remains insoluble. If the pH is very low, CuI dissolves, the reaction becomes reversible:

$$2Cu^{2+} + 2I^- \rightleftharpoons 2Cu^+ + I_2,$$

and results are inaccurate,

(c) Low pH catalyses atmospheric oxidations of I^-.

190. On adding KI to Cu^{2+} solution, a white precipitate of CuI is formed but the white colour is not visible because of brown colour of I_3^- produced by the reaction of liberated iodine with excess I^-:

$$I_2 + I^- \longrightarrow I_3^-$$

191. A small amount of I_2 gets adsorbed on the surface of CuI. On the addition of thiocyanate, the surface layers of CuI are converted into less soluble cuprous thiocyanate:

$$CuI + CNS^- \longrightarrow Cu(CNS) + I^-$$

In the process, the adsorbed iodine gets desorbed (which intensifies the blue colour) and is immediately titrated to the disappearance of the blue colour.

192. Reappearance of blue colour at the end-point indicates

(a) the occurrence of one or more of the factors that catalyse atmospheric oxidation of I^- (Ex.188)

(b) reaction between the oxidising agent and I^- is slow and sufficient time has not been allowed.

193. Tap water contains considerable amounts of Cl^- ions which interfere with the estimation.

194. (i) It prevents the hydrolysis of Fe^{3+} ion

(ii) It discharges the brown colour of ferric alum indicator.

195. Nitrous acid, if present, will react with thiocyanic acid and produce a red colour which may be mistaken for the end-point.

196. A.R. nitric acid is diluted with an equal volume of distilled water and boiled until it is perfectly colourless. Boiling expels any lower oxides of nitrogen.

197. As a drop of thiocyanate solution is added to Ag^+ ion solution containing Fe^{3+} ion, a reddish-brown cloud is first formed which quickly disappears on shaking because of its reaction with Ag^+ ion:

$$Fe^{3+} + SCN^- \longrightarrow \underset{\text{reddish-brown}}{[Fe(SCN)]^{2+}}$$

$$Ag^+ + Fe(SCN)^{2+} \longrightarrow AgSCN + Fe^{3+}$$

As the concentration of Ag^+ ions decreases, the above reaction becomes slow, i.e., the reddish-brown colour disappears only slowly when the end-point is approaching.

198. Some Ag^+ ions get adsorbed on the surface of the flocculent precipitate of AgSCN and escape reaction unless vigorously shaken. Thus a premature end-point may be obtained and the results will be low.

199. (i) The stream of CO_2 drives the Cl_2 produced to the conical flask containing KI solution.
 (ii) It dilutes Cl_2 and thus helps in more efficient absorption.
 (iii) It also inhibits the back-suction of KI solution into the distillation flask.

200. The guard tube contains some glass beads moistened with KI solution. Any trace of Cl_2 or I_2 vapour that escapes absorption in the conical flask is retained by the KI solution in the guard tube.

201. Cl_2 gas coming from the distillation flask will raise the temperature of the KI solution which may lead to some loss of iodine.

202. It is used for bleaching the yellow colour due to the presence of Fe_2O_3. MnO_2 imparts a purplish colour which by combination with yellow colour makes the glass appear colourless.

203. It occurs as a black-coloured rock.

204. Both sodium oxalate and sodium arsenite (prepared from A.R. As_2O_3) are more convenient because they are stable in air and are primary standards. Ferrous sulphate is not a primary standard and is oxidised by atmospheric oxygen. So the operation has to be carried out in the absence of air. Sodium arsenite though poisonous, costlier and preparation of its solution being cumbersome, gives more reliable results than sodium oxalate because oxalic acid is decomposed at higher temperatures into CO and CO_2. The extent of decomposition at a sulphuric acid concentration below 20% is however very small.

205. (a) The half cell reaction for the oxidation of iodide to iodine monochloride is given by

$$I^- + Cl^- \longrightarrow ICl + 2e,$$

which shows that there is 2-electron change per iodide ion.

$$\text{Hence the equivalent weight of KI} = \frac{\text{Mol. wt. of KI}}{2}$$

$$= \frac{166}{2} = 83$$

(b) The half cell reaction for iodate is given

$$IO_3^- + 6H^+ + Cl^- + 4e \longrightarrow ICl + 3H_2O$$

which shows 4-electron change per iodate ion.

Hence the equivalent weight of $KIO_3 = \frac{\text{Mol. wt.}}{4} = \frac{214}{4} = 53.5$

206. Iodine is only slightly soluble in water but highly soluble in the organic layer which is immiscible with water. A drop of KIO_3 added from the burette remains in the aqueous layer and the reaction between I_2 and KIO_3 will not take place unless they are brought into intimate contact by vigorous shaking.

207. At lower HCl concentrations, the rate of reaction near the end-point becomes very slow and may lead to an over consumption of KIO_3.

208. Any chloride and/or extra oxalate ions contaminated with the precipitate will consume extra $KMnO_4$ solution and thus the results will be higher.

209. When magnesium content of limestone is high, magnesium oxalate gets coprecipitated with calcium oxalate. In such a case, redissolve the precipitate of calcium oxalate on the filter paper and in the beaker in 20–25 ml of hot dilute HCl. Dilute to about 100 ml and precipitate with about 5 ml of hot 8% ammonium oxalate and proceed as per the normal procedure.

210. The combined filtrate and washings from the oxalate precipitation (calcium estimation) are diluted to a known volume (say, 250 ml). 50 ml of this solution is titrated with a standard solution of EDTA at a pH around 10, using Eriochrome Black-T as indicator (3.3.2). The results are reported as Magnesium oxide.

211. Please see Table X.

212. Cellulose nitrate (celluloid) was the first plastic of industrial significance discovered about the middle of the nineteenth century by Hyatt of New York.

213. The minimum requirement for a substance to act as a monomer is to have two bonding or reactive sites, i.e., it must be at least bifunctional.

214. The following types of copolymers are possible from two monomers M_1 and M_2 :

(a) Random copolymers:

$$-M_1 - M_2 - M_1 - M_1 - M_2 - M_1 - M_2 - M_2 - M_2 - M_2 - M_1$$

(b) Alternating copolymers:

$$-M_1 - M_2 - M_1 - M_2 - M_1 - M_2 - M_1 - M_2 -$$

(c) Block copolymers:

$$-M_1 - M_1 - M_1 - M_1 - M_2 - M_2 - M_2 - M_2 - M_1 - M_1 - M_1 - M_1 - M_1$$

(d) Graft copolymers:

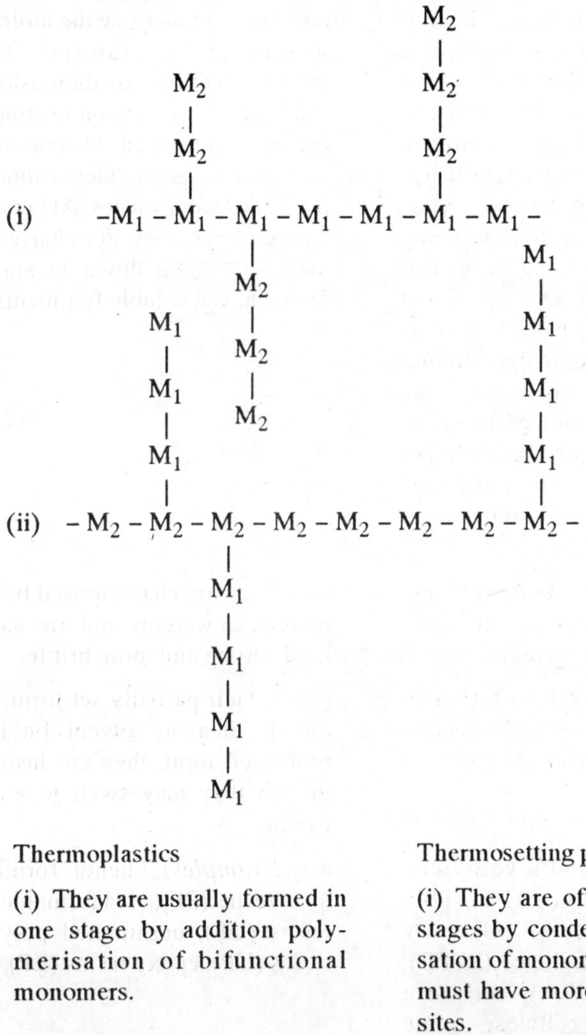

215.

Thermoplastics	Thermosetting plastics
(i) They are usually formed in one stage by addition polymerisation of bifunctional monomers.	(i) They are often prepared in two stages by condensation polymerisation of monomers, some of which must have more than two bonding sites.
(ii) They consist of long straight-chain or branched-chain molecules with negligible cross-links.	(ii) They have linear chains before curing. Because of the presence of functional groups, heating during the second stage (curing process in presence of additional monomer and usually a catalyst) causes cross-links between adjacent chains and leads to the formation of three-dimensional space networks.

(iii) Individual chains are held together by weak secondary forces (cohesive forces) which break easily by heat. Thermoplastics, therefore, soften on heating, when they can be given the desired shape, but harden on cooling. If heating is restricted to below decomposition temperature, the process of softening, reshaping and hardening can be repeated any number of times without any significant change in mechanical properties, as the heating or moulding process does not change the chemical structure or the molecular weight of the material. Thermoplastics are thus reversible and can be reclaimed from wastes.	(iii) Cross-links formed between adjacent chains during curing or processing change the chemical structure and increase the molecular weight of the material. These plastics therefore, are dimensionally stable, do not soften on heating and can not be reshaped. Their setting is irreversible. Also, they cannot be reclaimed from wastes. At very high temperatures, they get charred and may be broken down to smaller, meltable and soluble fragments.
(iv) They are often soft, less brittle, weak and usually have low molecular weights.	(iv) They are characterised by high molecular weights and are usually hard, strong and more brittle.
(v) They swell and finally dissolve in many liquids such as organic solvents.	(v) In their partially set form, they dissolve in many solvents but in the processed form, they are insoluble though they may swell to a large extent.
(vi) *Examples:* Ployethylene (PE), polystyrene (PS), poly (vinylchloride) (PVC), poly (tetrafluoroethylene) (PTFE), polyamides (PA), cellulose acetate (CA), cellulose nitrate (CN), poly (vinylacetate) (PVAc).	(vi) *Examples:* Phenol formaldehyde resins (PF), urea formaldehyde resins (UF), unsaturated polyester resins (UP), epoxy resins (EP).

216. (i) High molecular weight and long molecular chains.

(ii) Absence of bulky sidechains that prevent alignment and close packing.

(iii) Structural regularity and stiffening units such as alicyclic or aromatic rings in the chains to provide a high degree of crystallinity, and

(iv) Polar groups in the repeating units to impart high inter-chain cohesive forces.

217. Under a stretching force, the randomly coiled molecular chains of elastomers straighten out and achieve a certain degree of alignment or crystallinity (ordered arrangement) causing a decrease in entropy. The chains spontaneously return to the original random state of higher entropy.

218. Cooling of an amorphous polymer melt increases its viscosity to such an extent that alignment of molecules to produce a crystal lattice is not possible. Over a certain temperature range (of about 10–20°C interval) during the process of cooling, the coefficient of viscosity of the liquid approaches that of a solid (~ 10^{13} poise) and the polymer gradually acquires the characteristics of glass (or thermoplastic) such as hardness, stiffness and brittleness. The average of this temperature range is taken as the Glass-Transition Temperature, T_g.

When an amorphous linear polymer is heated, at a temperature around T_g, a freer vibratory motion of long segments of molecules begins and the polymer gradually changes from the glassy to the highly elastic rubbery state.

These changes being reversible, the glass-transition temperature determines whether a particular polymeric material will behave as a glass (hard and stiff plastic) or elastomer at a given temperature. In Table A are listed the approximate T_g values for some of the polymers from which it is clear that rubbers have their T_g considerably below room temperature, whereas T_g for plastics is much above room temperature. T_g for fibres is also substantially higher than room temperature.

Table A: T_g values for some polymers

Rubbers	T_g (°C)	Plastics & Fibres	T_g (°C)
Polybutadiene	– 85	Polystyerene	> 100
Natural rubber	– 70	PMMA	100
Butyl rubber	– 70	PVC	80
Polychloroprene	– 40	Nylon 6, 6	50
Silicone rubber	– 120	PETP	80

219. On cooling to a temperature around its T_g, an elastomer becomes hard and brittle. On heating, on the other hand, it becomes sticky and then changes to a highly viscous fluid state at a temperature reported as flow temperature T_f (or PMT). The temperature range (T_g – T_f) within which the polymer exhibits its elestic properties is called its Elastic Range or 'Use temperature range'. The range can be widened by lowering T_g and elevating T_f:

(i) T_g can be lowered by (a) copolymerising with small quantities of a suitable comonomer, (b) compounding with a plasticiser (a mutually compatible high boiling liquid), and (c) Milling or oxidative mustication (which reduce the chain length).

(ii) T_f can be increased by cross-linking (e.g., through vulcanisation).

220. Molecular engineering of polymers means isolation of a new product with desired properties through manipulation of the length of molecular chains, location and degree of cross-links, the number and size of branches and the nature of the repeating unit.

An optimum combination of softness, stretchability, resilience and toughness can be achieved in a new elastomer by providing for chain flexibility, free segmental mobility and resistance against slipping of chains (permanent deformation) under prolonged stretching.

While the chain flexibility is achieved by avoiding stiffening units such as alicyclic or aromatic rings in the molecular chains, free segmental mobility can be imparted by

(a) selecting repeating units containing C–C and C–O–C linkages around which free rotation is possible,

(b) reducing interchain attractive forces by avoiding polar groups, and

(c) increasing interchain free volume by inserting small sidechains which prevent close packing.

A suitable number of crosslinks is provided (through vulcanisation) to guard against permanent deformation through slipping of molecular chains past one another. Degree of rigidity can be controlled by varying the degree of cross-linking.

221. The structure of rubbers is intermediate between that of thermoplastics and thermosets and can best be described as a loosely-linked network where the interchain linkages are comparatively low. It allows considerable extension to occur, without breaking the chemical bonds, when subjected to a tensional stress.

222. Vinyl monomers usually contain hydroquinone as inhibitor to prevent polymerisation during handling or storage. Therefore, the inhibitor should be removed before polymerisation as otherwise it may not be possible to initiate the polymerisation process. The inhibitor can be removed either by distillation under reduced pressure (to avoid polymerisation due to excess heating) or by washing with 95% KOH solution (using a separating funnel) followed by washing with water and drying over anhydrous $CaCl_2$ or Na_2SO_4.

223. In polystyrene, the bulky phenyl groups attached to the chains prevent a closer approach of the polymer chains and hence lessen the attractive forces between them.

224. The term Nylon is the generic name given to long chain synthetic polyamides. Individual members of the group are designated or distinguished by two numbers — the first indicating the number of C-atoms in the diamine monomer while the second indicating that number in the diacid monomer. Thus, the nylon made from tetramethylenediamine (containing 4 C-atoms) and sebacic acid (containing 10 C-atoms) is known as Nylon 4,10. Nylons made from *w*-amino acids or from lactams are designated by just one number which is the number of C-atoms in the monomer:

$$nH_2N \cdot (CH_2)_4 \cdot NH_2 + nHOOC \cdot (CH_2)_8 \cdot COOH \longrightarrow$$

tetramethylene diamine — Sebacic acid

$$H \left[NH \cdot (CH_2)_4 \cdot NH - OC(CH_2)_8 - \overset{\overset{\displaystyle O}{\|}}{C} \right]_n OH + (2n - 1)\, H_2O$$

Nylon 4, 10

$$n\, H_2N(CH_2)_5\text{—}COOH \longrightarrow H \left[NH\text{—}(CH_2)_5CO \right]_n OH + (2n - 1)\, H_2O$$

ε Aminocaproic acid — Nylon 6

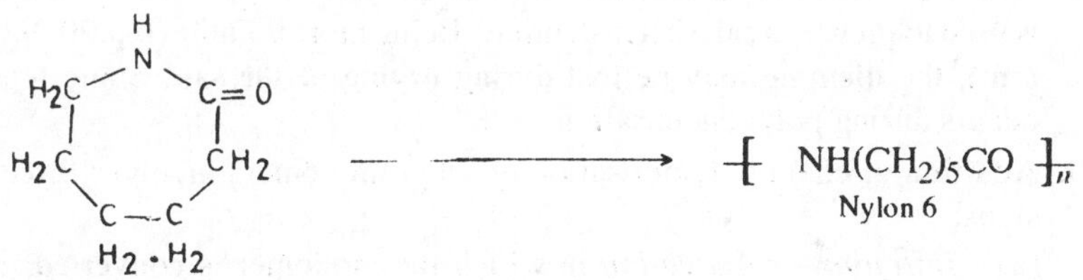

Caprolactam

225. Nylons are crystalline and have high melting points. This is explained on the basis of extensive hydrogen bonding between the adjacent chains through > NH and C=O groups. Lengthening of the hydrocarbon part in the chain 'dilutes' the hydrogen bonding resulting in lowering of melting or softening point:

Nylon type	Nylon 4,6	Nylon 6,6	Nylon 6,10
m.p. (°C)	308	265	220

226. Melting point and tensile strength of Nylons can be increased by (a) increasing the extent of hydrogen bonding by shortening the hydrocarbon chain in the diamine and the diacid (as disscussed above), and (b) introducing stiffening groups in the chains, e.g.,

(i) Poly(ethyleneterephthalamide) melts above 400°C:

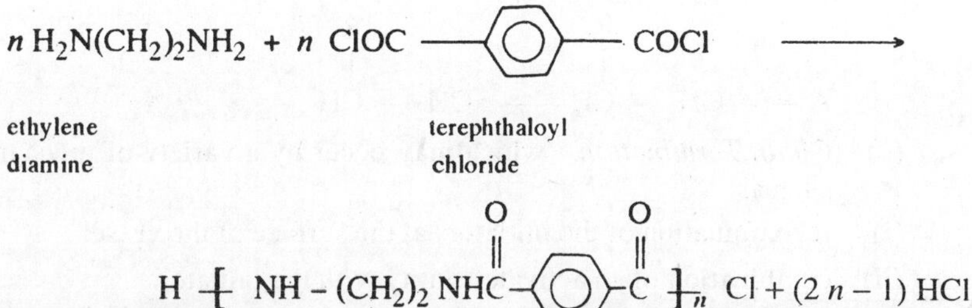

(ii) Nylons prepared from aromatic diamines and aromatic diacids (called aramides) melt above 500°C and have very high tensile strength:

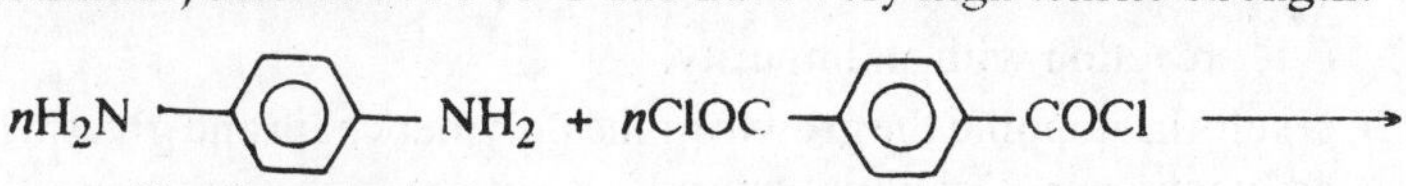

Phenylene diamine

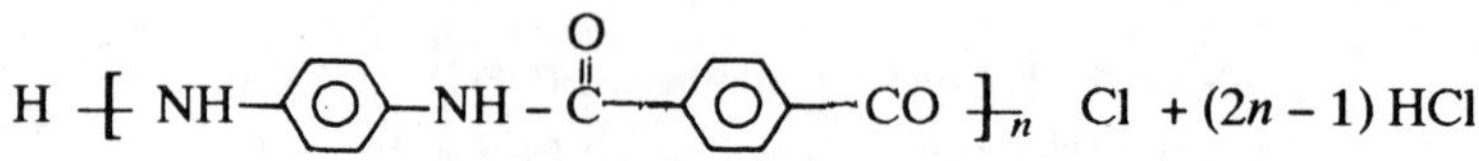

Poly (paraphenyleneterephtalamide)
(Kelvar)

227. In the beaker method of Morgan and Kvolek (p.174–5) for the formation of Nylons from a diamine and a diacid-dichloride, an almost inexhaustible thread or rope can be drawn from the transparent solution. As the rope seems to come out of nothing, it is called 'The Nylon Rope Trick'.

228. About 20% excess (than needed for stoichiometry) hexamethylene diamine is used to prepare a salt rich in diamine. Being more volatile (b.p.90–92°C/14 mm), the diamine may be lost during drying of the salt. Some loss also occurs during polycondensation.

229. Addition or chain polymerisation of vinyl monomers involves three main steps:

(a) *Initiation or Activation* in which the monomer is converted into an activated entity by the action of heat, light, or by the addition of an initiator (a catalyst — free radical or ionic):

$$\underset{\text{Initiator}}{A^*} + \underset{\text{monomer}}{CH_2 = \overset{\overset{R}{|}}{CH}} \longrightarrow \underset{\text{activated entity}}{A - CH_2 - \overset{\overset{R}{|}}{CH^*}}$$

(b) *Propagation or Growth* in which the activated entity adds on to a new monomer unit to produce a new activated entity, a product capable of further interaction with the initial monomers:

$$A - CH_2 - \overset{\overset{R}{|}}{CH^*} + CH_2 = \overset{\overset{R}{|}}{CH} \longrightarrow A - CH_2 - \overset{\overset{R}{|}}{CH_2} - CH_2 - \overset{\overset{R}{|}}{CH^*}$$

This step may be repeated a very large number of times to give

$$A \left(CH_2 - \overset{\overset{R}{|}}{CH} \right)_n CH_2 - \overset{\overset{R}{|}}{CH^*}$$

(c) *Chain Termination,* which may occur by a variety of mechanisms such as:

(i) recombination of the initiators at the surface of the vessel.

(ii) combination of an activated entity with the initiator.

(iii) combination of two activated entities.

(iv) disproportionation.

(v) elimination.

(vi) reaction with an impurity,

results in stopping the growth of the polymer chain and gives what is known as Inactive or 'Dead' polymer:

$$A\{CH_2-\underset{R}{CH}\}_n\,CH_2-\underset{R}{CH^*}+A^* \longrightarrow A\{CH_2-\underset{R}{CH}\}_n\,CH_2-\underset{R}{CH}-A$$

'Dead' polymer

Chain growth may temporarily stop when all the monomer units are used up. This happens particularly in case of polymerisation of carefully purified reactants (say styrene) by anionic initiators represented as A^-H^+ [NH_2 Li (Lithium amide), C_4H_9Li (*n* - butyl lithium), $C_{10}H_8Na$ (Sodium naphthalene)] or by Ziegler–Natta catalyst ($Et_3Al/TiCl_3/TiCl_4$ – coordination catalyst). The chains do not 'die' when the monomer units disappear. The product is known as Active or 'Living' polymer.

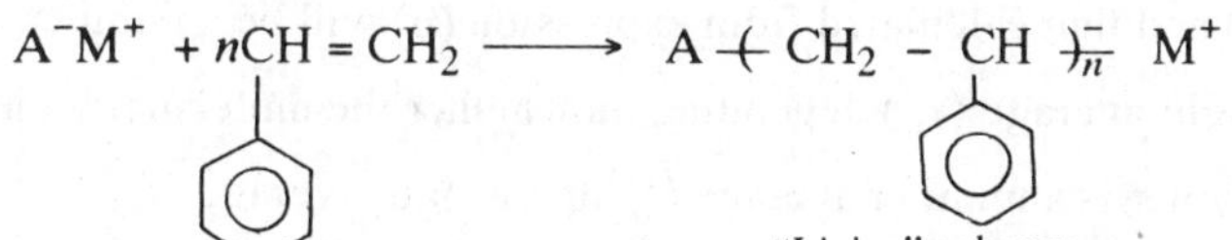

"Living" polymer

Addition of more monomer permits the chain to grow:

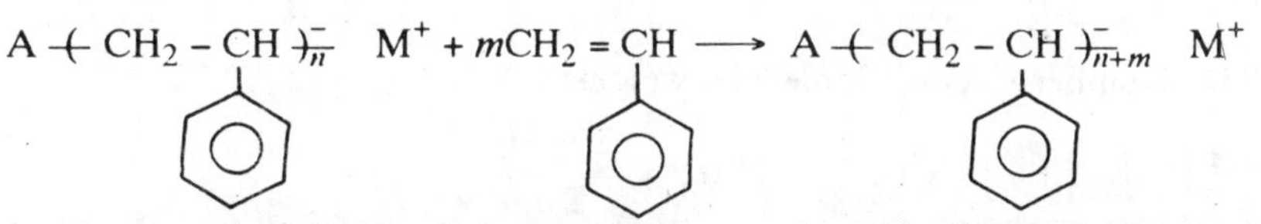

At the end of the reaction, the polymer may be 'killed' by the addition of a terminating agent, e.g., water:

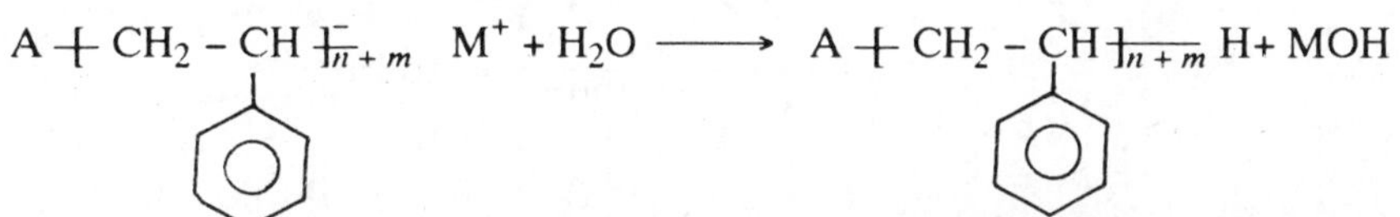

'Dead' polymer

The method is used for the preparation of Block Copolymers or Stereospecific polymers.

In case of condensation or step-growth polymerisation (p.169), the growing polymer chain continues to have one active group at each end and the polymer does not 'die'. Chain lengthening, however, may stop due to decreased activity as a result of increase in the size and molecular weight of the polymer and due to cyclization.

230. The degree of polymerisation (DP) represents the number of repeat structural units in a given polymer molécule. It thus specifies the length of the polymer chain. The degree of polymerisation, x, may be calculated from the expressions :

$$x = \frac{N_0}{N} \quad (i)$$

or $$= \frac{M}{M_0} \quad (ii)$$

where N_0 = number of monomer molecules before polymerisation,
N = number of molecules present in the polymer product,
M = molecular weitht of the polymer, and
M_0 = molecular weight of the monomer units.

In view of the inherent molecular heterogeneity of a polymer product, it will be more appropriate to call x as the Average Degree of Polymerisation. Evidently, the value calculated using expression (i) is a number average ($\bar{x}_n$) and that calculated from expression (ii) will be a number average or a weight average ($\bar{x}_w$) depending on whether the molecular weight M of the polymer is a number average $\overline{M}_n$ or a weight average $\overline{M}_w$.

Polymers are termed as *High Polymers* or *Oligopolymers* accordingly as the average degree of polymerisation is high or low.

231. Number Average Molecular weight,

$$\overline{M}_n = \frac{\Sigma\, n_i\, M_i}{\Sigma\, n_i}$$

$$= \frac{10 \times 10^3 + 100 \times 10^4 + 10 \times 10^5}{10 + 100 + 10}$$

$$= \frac{201 \times 10^4}{120} = 16750$$

Contribution of low mol. wt. (10^3) molecules

$$= \text{Number fraction} \times \frac{\text{Mol. wt.}}{\overline{M}_n} \times 100$$

$$= \frac{10}{120} \times \frac{10^3}{16750} \times 100 = 0.4975\%$$

Contribution of high mol. wt. (10^5) molecules

$$= \frac{10}{120} \times \frac{10^5}{16750} \times 100 = 49.75\%$$

Weight Average Molecular Weight,

$$\overline{M}_w = \frac{\Sigma n_i M_i^2}{\Sigma N_i M_i}$$

$$= \frac{10 \times (10^3)^2 + 100 \times (10^4)^2 + 10 \times (10^5)^2}{10 \times 10^3 + 100 \times 10^4 + 10 \times 10^5}$$

$$= 54731$$

Contribution of low mol.wt. (10^3) molecules

$$= \text{Wt. fraction} \times \frac{\text{Mol.wt.}}{\overline{M}_w} \times 100$$

$$= \frac{10 \times 10^3}{201 \times 10^4} \times \frac{10^3}{54731} \times 100 = 0.0091\%$$

Contribution of high mol. wt. (10^5) molecules

$$= \frac{10 \times 10^5}{201 \times 10^4} \times \frac{10^5}{54731} \times 100 = 90.90\%$$

As is evident from the results, the contribution of low mol. wt. molecules towards $\overline{M}_n$ is much larger (0.4975%) than towards $\overline{M}_w$ (0.0091%) while that of high mol. wt. molecules is much larger towards $\overline{M}_w$ (90.90%) than towards $\overline{M}_n$ (49.75%).

232. Many of the important properties of a polymer such as mechanical strength, softening temperature, melt viscosity, solubility in a particular solvent, etc., directly depend on the molecular weight (or degree of polymerisation, DP) of the molecules composing the material. Every polymer has its own Threshold Value (TV) of DP (or mol. wt.) below which it does not possess any strength. Similarly, when a fairly high molecular weight (different for different polymers) is reached, the mechanical strength becomes more or less constant. For most of the polymers, the useful range of mol. wt. varies from 20,000 to 200,000. During the preparation of the polymer, therefore, the polymerisation process is continued until the mol. wt. of the product has reached the useful range. Thus, the determination of the mol. wt. is an important step in the preparation of the polymer.

233. As the mol. wt. determined by any one of the different methods (p.176) is always some sort of an average, it does not convey anything about the spread of the molecular weight or molecular inhomogeneity or polydispersity. Polymer samples with the same mol. wt. average (whether $\overline{M}_n$, $\overline{M}_w$ or $\overline{M}_v$) may contain molecules with much lower or much higher mol. wt. than the average, and thus may differ in some of the properties. Therefore, to properly know a polymer, both its average mol. wt. and the spread of mol. wt. must be known.

If all the molecules in a polymer sample are of the same size, $\overline{M}_n = \overline{M}_w$. However, when there is a size distribution, $\overline{M}_w > \overline{M}_n$ (as shown in Ans. 231 above) and the ratio $\overline{M}_w / \overline{M}_n$, termed as the degree of polydispersity, gives a rough but quick indication of the molecular weight distribution. $\overline{M}_w / \overline{M}_n$ values for typical polymers range from 1.5–2 to 20-50.

234. In order to have some sort of certainty about the behaviour of polymer sample under the use conditions, the mol.wts. of the individual molecules of the sample must be quite close to the average mol. wt., or the degree of polydisperisity must be low ($\overline{M}_w/\overline{M}_n$ < 1.2). Since this value is very high in as prepared polymers (Ans. 233 above), the molecular inhomogeneity of the sample is reduced by *Fractionation.* Fractionation is also needed for determining the values of K and a (Ans. 235).

It is the process of obtaining fairly homogeneous mol. wt. samples, with progressively increasing or decreasing average mol.wts., from a highly polydisperse system. A non-solvent is slowly added to a dilute polymer solution until a turbidity appears due to the precipitation of the highest mol. wt. molecules. This is allowed to settle and then separated by decantation or centrifugation. To the remaining solution, more non-solvent is added and the whole process repeated. The fractions so obtained are usually redissolved and refractionated.

235. Changing equation (7.7) to the logrithmic form, we get

$$\log |\eta| = \log K + a \log M$$

Thus, a plot of log $|\eta|$ versus log M will be a straight line whose slope and the intercept on the ordinate represent the values of a and log K respectively (Fig. 9.4).

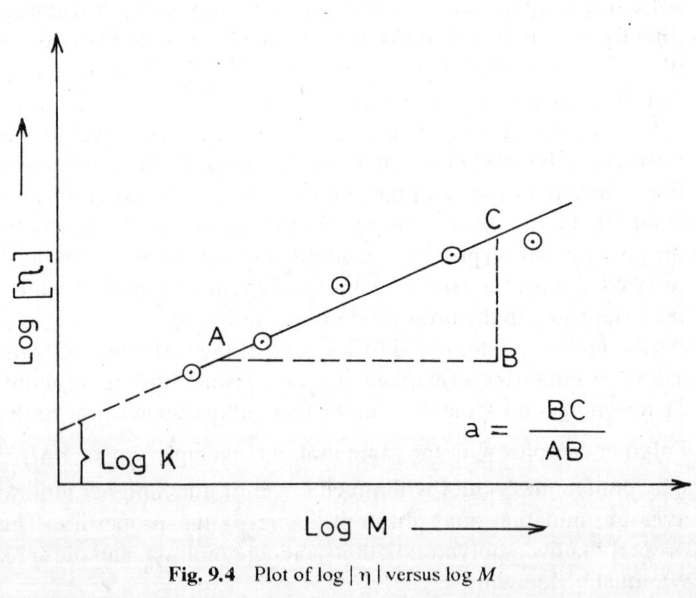

Fig. 9.4 Plot of log | η | versus log M

Since a polymer sample is never available in a truely monodisperse state, it is fractionated to obtain a number of fractions that have a narrower molecular weight distribution than the original sample. The average molecular weights M_1, M_2, M_3, etc., of these fractions are then determined by one of the absolute methods such as Light Scattering or Ultracentrifuge, and the corresponding intrinsic viscosity $|\eta|_1, |\eta|_2, |\eta|_3$, etc., are also measured and plotted to get the graph.

236. Also called Ubbelohde Dilution Viscometer, it is an improvement over the Ostwald Viscometer. In it, the bottom end of the capillary is above the level of the liquid (Fig. 9.5) and so the pressure head is independent of the volume of liquid in the reservoir.

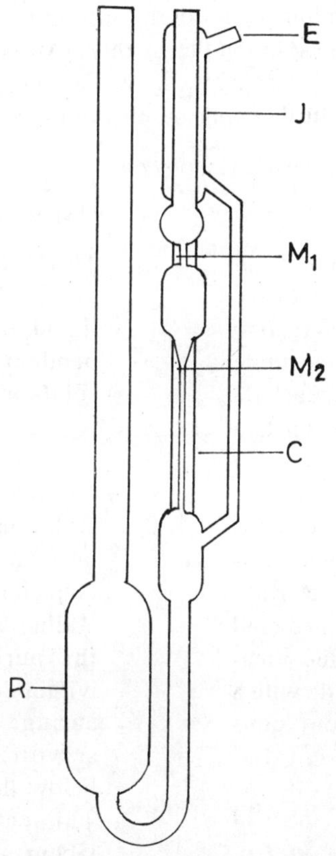

Fig. 9.5 Ubbelohde Suspended Level Viscometer

C - Capillary arm
J - Outer Jacket
M_1 - Upper mark
E - Exit hole
R - Reservoir
M_2 - Lower mark

Working : Properly clean and dry the viscometer and pipet 20 ml of the well-filtered solvent into the dilution chamber or reservoir R. Suspend the viscometer in the thermostatic bath. When the temperature equilibrium has been reached, gently force the liquid from the reservoir into the other side. When the outer jacket J is almost filled, close the exit hole E with your thumb and increase the pressure on the liquid to raise its level above the higher mark M_1 on the capillary. Release the pressure and open the exit hole and, using a timer, measure the time in which the liquid meniscus moves from the upper mark M_1 to the lower mark M_2. Make 5 measurements and take the average as the flow-time t_o for the solvent. Remove the viscometer from the thermostat and pour out the solvent. Properly clean the viscometer, dry it and pipet 20 ml of the concentrated polymer solution (0.5 g/dl) into the reservoir. After attainment of temperature equilibrium, determine the flow time t_s as before.

Now add with a pipet 5 ml of the appropriate solvent into the dilution chamber and mix the solution thoroughly by repeatedly forcing it into the outer jacket and then releasing it. Again, measure the flow time. Repeat the measurements after further appropriate dilutions of the solution.

Advantages of USLV over Ostwald viscometer

	Ostwald Viscometer		USLV
1.	The effective pressure head varies with the volume of the liquid in the viscometer; so to get reproducible and accurate results, all measurements must be made with a constant volume of the liquid.	1.	The bottom end of the capillary being above the level of the liquid, the pressure head is independent of the volume of the solution in the reservoir.
2.	The experimental procedure is tedious and more time consuming as the viscometer has to be emptied, cleaned and refilled everytime the measurement is to be made with a solution of different concentration.	2.	Only one standard solution is needed in the beginning. New concentrations are achieved by adding appropriate volumes of the pure solvent and mixing within the viscomerter itself (ensuring, however, that the reservoir level always remains below the capillary bottom end) This saves a lot of time and labour.

237. If the molecules of the solute in a solution behave as hard spheres, intrinsic viscosity is independent of the molecular weight of the solute. The value of a, the exponent in the Mark-Houwink equation (7.7), then becomes equal to zero, i.e., $[\eta] = KM^0 = K$. Examples:

(i) All globular proteins
(ii) Highly branched glycogen which is nearly sphere-like.

238. For two liquids with different drainage (efflux) times, the average pressure heads are different and the relationship (7.2) becomes approximate. For good results, therefore, the reference liquid should have approximately the same drainage time as the liquid under test. Hence the difference in the efflux time for solution (t_s) and for pure solvent (t_0) should not be very high or $\eta_r = t_s/t_0$ should not be higher than 1.5.

239. Extrapolation of the graph of η_{red} versus C to zero concentration eliminates the effect of concentration on the viscosity of the solution.

240. Under otherwise identical conditions (at a given mol.wt., degree of polydispersity, extent of branching and temperature), the value of the exponent a in equation (7.7), and hence the value of $[\eta]$, is a measure of the quality of the solvent. Large values of a and $[\eta]$ indicate a good solvent whereas lower values of a and $[\eta]$ indicate a poor solvent. Examples :

Polymer	Solvent	Value of a at 30°C	Quality of solvent
Poly (vinyl chloride)	chlorobenzene	0.59	poor
	tetrahydrofuran	0.83	good
Poly (vinyl acetate)	methonol	0.60	poor
	acetone	0.72	good

241. The Viscosity Average Molecular Weight, $\overline{M}_v$, as is the mol. wt. calculated from Mark-Houwink equation (7.7) called, is influenced by the shape of the polymer molecules in solution which depends on the solvent-solute interactions. $\overline{M}_v$ is a complicated average given by the empirical equation

$$\overline{M}_v = \left(\frac{\Sigma n_i M_i^{1+a}}{\Sigma n_i M_i}\right)^{1/a}$$

where a is the exponent in equation (7.7). The value of $\overline{M}_v$ lies somewhere between $\overline{M}_n$ and $\overline{M}_w$ ($\overline{M}_n < \overline{M}_v < \overline{M}_w$) but is nearer to $\overline{M}_w$ which is therefore preferred for calibration, i.e., for the determination of constants K and a (Ans.235). For a = 1, $\overline{M}_v = \overline{M}_w$.

242. For a given polymer/solvent system, the temperature below which the solvent behaves as a non-solvent and above which it behaves as a poor solvent is called Flory Theta (θ) Temperature. At this temperature, the association forces (cohesive forces between the polymer segments) are equal to the solvation forces (solvent dispersion forces) and the polymer

molecules are just at the point of incipient precipitation : under such conditions, the exponent a in equation (7.7) acquires a value of 0.5.

Examples:

Polymer/solvent system	Theta temperature
Polystyrene/cyclohexane	34°C
Polyisobutylene/benzene	24°C
Polyacrylic acid/Dioxane	30°C
Polymethylmethacrylate/acetone	25°C

The solvent at theta temperature is called Theta Solvent for that polymer. Solvent and non-solvent mixtures in suitable proportions have been found to behave like theta solvents for some polymers.

243. It is a chemical method used for the determination of mol. wt. of polyesters, polyamides and other linear polymers which contain a reactive group at one or both ends of the polymer chain. The method gives number-average mol. wt. and is particularly useful for determination of $\overline{M}_n$ values less than 25,000 at which range other methods become less reliable. The limitation is that the mechanism of polymer formation must be known with certainty so that the location of reactive groups can be predicted. Losses of end-groups by side-reactions, production of additional end-groups by branching and solvent interference are some of the sources of error.

The experimental procedure used depends on the nature of the end-group present. The mol.wt. of polymers containing – COOH groups is determined by titrating a known weight of the polymer dissolved in an inert solvent against a standard solution of KOH using phenolphthalein indicator.

Observations and Calculations

Let the wt. of polymer dissolved $= w$ g

Volume of KOH consumed $= V$ ml

Normality of KOH solution $= N$

Number of moles of KOH that have reacted with w g of polymer $= \frac{VN}{1000}$

Hence the number of moles of the polymer weighing w g $= \frac{VN}{1000}/n$

$$= \frac{VN}{1000 \times n},$$

where n is the number of – COOH groups per chain.

244. (a) *Formation of polyester :*

$$n\text{HO} - (\text{CH}_2)_a - \text{COOH} \longrightarrow$$
$$\omega - \text{hydroxy acid}$$

$$HO(CH_2)_a \cdot COO - [(CH_2)_a\, COO]_{n-2} - (CH_2)_a - COOH + (n-1)\, H_2O$$

Polyester

$\therefore$ No. of – COOH groups per chain = 1.

(b) *Formation of polybutadiene*

$$HOOC - R - N = N - R - COOH \longrightarrow N_2 + 2 \cdot R - COOH$$

Azo initiator

$$2 \cdot R - COOH + nCH_2 = CH - CH = CH_2$$

Butadiene

$$\longrightarrow HOOC{-}R \leftarrow\!\!\!\!\!\!\!\!\left(CH_2 - CH = CH - CH_2\right)_n R - COOH$$

CTPB

$\therefore$ No. of — COOH groups per chain = 2

245 Wt. of sample, $W = 1.232$ g

Vol. of KOH used, $V = 2.9$ ml

Normality of KOH solution, $N = 0.0965$

No. of carboxyl groups per molecule, $n = 2$

$$\overline{M}_n = \frac{W \times 1000 \times n}{V \times N} = \frac{1.232 \times 1000 \times 2}{2.9 \times 0.0965} = 8805$$

246. Air bubbles on the surface of the material tend to decrease its density.
247. Because of gas bubbles entrapped in their body, foams have lower apparent density. So, floatation test is not applicable.
248. (i) The absence or presence of branching and its extent (depending on the method of preparation of the polymer) greatly affect the density of the polymeric material.

(ii) The other important factor is the nature and amount of fillers (paper, fabric, carbon black, glass fibres, silica, etc.) compounded in the polymeric material.

249. Two miscible liquids of different densities are mixed in various proportions to give a continuous range of density. The sample is then lowered in mixtures of gradually increasing density. The mixture in which the sample just floats gives the density of the sample. For polymers lighter than water, methanol-water mixtures in different proportions may be used.
250. The solubility of a polymeric material depends very much on its chemical nature and the nature of the solvent. The solubility in a particular solvent largely decreases due to an increase in

(i) the degree of cross-linking, and

(ii) the molecular weight of the polymer.

Increase in the degree of crystallinity and hydrogen bonding also tend to decrease the solubility.

251. Use of polymer in the finely divided state usually facilitates its dissolution and offers better and quick responses to most other tests. The polymer is brought into finely divided state by grinding in presence of dry ice.

252. Dry ice is the name given to solid CO_2 as the solid passes into the gaseous state without liquification (sublimation). During the grinding process, a polymeric material may get over-heated and may even soften. Chilling by dry ice prevents softening and makes tough and elastic materials become brittle so that they can be easily grinded.
253. A shear-disk stirrer is a stainless steel shaft terminated by a steel disk (0.5" – 1" dia, 1/8" – 3/16" thick). It can be attached to a motor for rotation at high speed. Its use substantially increases the rate of polymer dissolution.

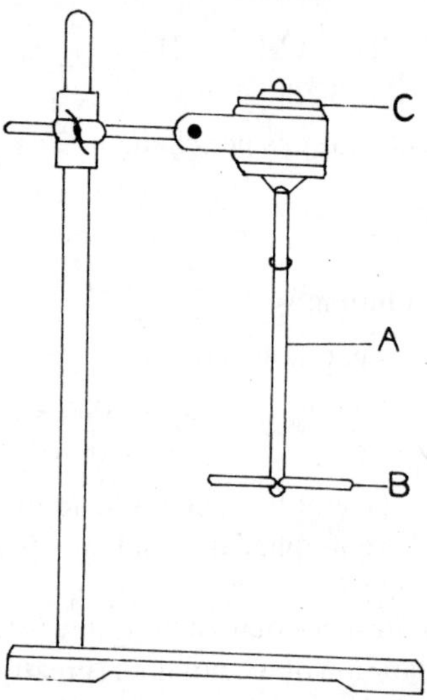

Fig. 9.6 Shear-disk Stirrer

A – Steel shaft B- Steel disk
C – Variable speed motor

254. When the copper wire is heated alone in the bunsen flame, the volatile impurities are lost and a film of pure copper oxide is formed on the wire. On subsequent heating in contact with the polymer, the halogen present in the polymer reacts with the copper oxide producing the corresponding copper halide.

$$2Cu + O_2 \xrightarrow{\text{Heat}} 2CuO$$

$$CuO + 2H + 2X \xrightarrow{\text{Heat}} \underset{\text{(Volatile)}}{CuX_2} + H_2O$$

Copper halides are volatile and impart an intense green or bluish-green colour to the flame.

255. On fusion with sodium, N, S and halogens present in the polymer are converted into corresponding ionisable inorganic compounds to which the ionic tests of inorganic qualitative analysis can be applied:

$$\underset{\text{(Present in the polymer)}}{C, H, O, N, S, Cl, F,} + Na \xrightarrow{\text{Heat}} NaCN + NaS + NaSCN + NaCl + NaF + NaOH$$

(a) *Test for nitrogen*

On heating the alkaline Lassaigne's extract with ferrous sulphate, cyanide is converted into ferrocyanide:

$$FeSO_4 + 2OH^- \longrightarrow Fe(OH)_2 + SO_4^{2-}$$

$$Fe(OH)_2 + 6CN^- \longrightarrow [Fe(CN)_6]^{4-} + 2OH^-$$

A part of $Fe(OH)_2$ is oxidised to $Fe(OH)_3$:

$$2Fe(OH)_2 + \frac{1}{2}O_2 + H_2O \longrightarrow 2Fe(OH)_3$$

Both ferrous hydroxide and ferric hydroxide get dissolved on addition of dilute acid (HCl or H_2SO_4) and the ferric ions released react with ferrocyanide forming ferriferrocyanide resulting in blue or bluish-green colour:

$$3[Fe(CN)_6]^{4-} + 4Fe^{3+} \longrightarrow Fe_4[Fe(CN)_6]_3$$

When both N and S are present, the SCN^- formed react with ferric ions to produce blood-red colour :

$$Fe^{3+} + SCN^- \longrightarrow [Fe(SCN)]^{2+}$$

(b) *Test for sulphur*

(i) The violet colour appears due to the reaction between sulphide and nitroprusside :

$$S^{2-} + [Fe(CN)_5NO]^{2-} \longrightarrow \underset{\text{(Violet)}}{[Fe(CN)_5NOS]^{4-}}$$

(ii) Sulphide reacts with lead acetate to produce a black precipitate of lead sulphide:

$$S^{2-} + Pb^{2+} \longrightarrow \underset{\text{Black ppt.}}{PbS}$$

(c) *Test for chlorine*

Interference due to $CN^-/S^{2-}/SCN^-$ is eliminated by boiling L.E. with dilute nitric acid:

$$CN^- + HNO_3 \longrightarrow HCN + NO_3^-$$

$$S^{2-} + 2HNO_3 \longrightarrow H_2S + 2NO_3^-$$

or

$$3S^{2-} + 8HNO_3 \longrightarrow 3S + 2NO + 6NO_3^- + 4H_2O$$

$$SCN^- + 2HNO_3 \longrightarrow HSO_4^- + 2NO + HCN$$

Cl^- reacts with $AgNO_3$ to precipitate AgCl which dissolves in NH_4OH:

$$Cl^- + AgNO_3 \longrightarrow \underset{\text{(white ppt.)}}{AgCl} + NO_3^-$$

$$AgCl + 2NH_4OH \longrightarrow [Ag(NH_3)_2]^+ Cl^- + 2H_2O$$

(d) *Test for fluorine*

Fluoride in the L.E. reacts with $CaCl_2$ to give a gel-like precipitate of CaF_2:

$$2F^- + CaCl_2 \longrightarrow CaF_2 + 2Cl^-$$

256. When both S and N are present in the polymer, an excess of Na should be used during fusion. This will decompose thiocyanate if formed:

$$NaSCN + 2Na \longrightarrow Na_2S + NaCN.$$

257. Upon fusion, Si in a sample is converted into Na_2SiO_3. Dissolution followed by acidification changes it to silicic acid:

$$Na_2SiO_3 + H_2O + 2HNO_3 \longrightarrow H_4SiO_4 + 2NaNO_3$$

Ammonium molybdate converts silicic acid into silicomolybdic acid:

$$H_4SiO_4 + 12MoO_4^{2-} + 24H^+ \longrightarrow H_4[SiMo_{12}O_{40}] + 12H_2O$$

The silicomolybdic acid reacts with benzidine in acetic acid solution to produce 'molybdenum blue' and a blue-coloured oxidation product of benzidine.

258. Polymers are said to be 'virgin' when they are in their pure form and, after their isolation and purification, no extraneous material has been added to modify their characteristics.

259. The main compounding agents are:

(a) *Plasticisers*: These are usually low mol. wt. high-boiling liquids or non-volatile materials added to 'virgin' polymers. Through partial neutralisation of intermolecular forces of attraction between different chains, they improve flexibility and workability of the polymeric material. However, chemical resistance and tensile strength of the material is somewhat lowered.

Examples: Non-drying vegetable oils, tricresyl phosphate, dibutyl phthalate, dioctyl sebacate, etc.

(b) *Fillers*: They are materials usually added to reduce shrinkage on setting and cost of the finished product, and to improve the opacity, hardness and tensile strength.

Examples: Saw-dust, paper pulp, cotton fibre or rags, carbon black, ZnO, Al powder, etc. Some special characteristics can also be imparted through the addition of suitable fillers, e.g.,

(i) Heat and corrosion resistance — Asbestos

(ii) Extra hardness — Mica, carborundum, silica

(iii) U.V. deactivation — Carbon black

(iv) Stoppage of X-rays — Barium salts.

(c) *Stabilisers*: These are the substances added to protect the finished product from thermal, oxidative and photodegradation.
Examples: Salts of calcium, barium and lead, organotin compounds, amines, etc.

(d) *Colouring agents*: These are suitable organic dyes or opaque inorganic pigments added to impart attractive colours to the finished products.

Removal of Compounding Agents

(a) Organic agents such as plasticisers, dyestuffs and some of the stabilisers can be removed by extraction with ether or some other solvent in which the polymer is not soluble. The sample in the powdered state is heated with the solvent under reflux. A Soxhlet extractor, if available, should be preferred.

(b) For removing fillers, inorganic stabilizers and opaque inorganic pigments, the sample is dissolved in a suitable solvent, dissolved components are removed by filtration, and the polymer is recovered from the solution by addition of a non-solvent.

260. $CaCO_3$ does not react with iodine and so does not interfere. MgO reacts very slowly and so slightly higher results may be obtained.

261. Both calcium oxide and calcium hydroxide are more soluble in sucrose solution than in water. $CaCO_3$ is relatively insoluble and is removed by filtration.

262. Glass beads help in grinding of lime and thus facilitate its extraction with sucrose solution.

263. Firstly, MgO does not react with water to any great extent, except on long standing. Secondly, solubility of $Mg(OH)_2$ is very very low and any small amount formed is retained by the filter paper.

264. (a) Slaking is hydration of quick-lime to give calcium hydroxide (reaction 8.2). When lumps of quick-lime are added to water,

(i) about 2 parts by weight of quick-lime react with 1 part of water.

(ii) the lumps disintegrate and crumble to give a fine powder called 'Slaked' or 'Hydrated' lime.

(iii) about 2.75 Kcal of heat are evolved per kg of lime slaked.

(iv) the breaking of lumps and conversion of water into steam due to the high heat of reaction produce crackling and hissing sound.

(v) the volume of slaked lime is about 2 to 3 times the volume of quick-lime used.

(b) The rate of slaking reaction decreases with

(i) increase in the percentage of MgO

(ii) increase in the amount of other impurities

(iii) increase in lump size

(iv) decrease in porosity

(v) decrease or increase in the temperature of calcination of lime-stone during manufacture of lime.

265. Quick lime slowly absorbs moisture and CO_2 from atmosphere:

$$CaO + H_2O \longrightarrow Ca(OH)_2$$

$$CaO + CO_2 \longrightarrow CaCO_3$$

Its available lime therefore goes on decreasing on continued exposure to air.

266. A suspension of slaked lime in water is termed as milk of lime. Its filtrate, which is the clear aqueous alkaline solution of calcium hydroxide is known as lime water.

267. When lime-stone is calcined at low temperature, the decomposition of $CaCO_3$ is incomplete. The product, which is porous and whose volume is not much less than that of the original lime-stone, is known as 'soft' lime. Because of lower CaO content, soft limes slake slowly.
Calcination of lime-stone at high temperature produces a more compact and dense product which is known as 'Hard' or 'Over-burnt' lime. Some lime reacts with SiO_2 and Al_2O_3 producing silicates and aluminates. Chemical reactivity or rate of slaking reaction is lower because of decreased surface area (due to higher density) and lower CaO content.

268. On the basis of amount of CaO present and the nature of the associated materials such as MgO, SiO_2 and Al_2O_3, limes are classified as:

(a) *Fat or High-calcium Limes*

(i) contain about 95% CaO and small amounts of impurities like MgO, SiO_2, Al_2O_3, Fe_2O_3, etc.,

(ii) slake rapidly with evolution of large amounts of heat and a large increase in volume,

(iii) have good sand-carrying capacity and mortar sets to a strong mass.

(b) *Lean or Low-calcium Limes*

(i) contain 70–80% CaO, the remaining being MgO, SiO_2, Fe_2O_3 and Al_2O_3,

(ii) slake slowly with smaller increase in volume, require less water and evolve less heat,

(iii) have low sand-carrying capacity and mortar develops lower strength.

(c) *Magnesium Limes*

(i) contain 10–25% MgO; the percentage of impurities like SiO_2, Al_2O_3 and Fe_2O_3 being small,

(ii) slake slowly with less evolution of heat and less expansion as compared to fat limes,

(iii) have lower sand-carrying capacity; mortar is plastic, can be easily worked, sets slowly to a harder but more smooth surface; so used as finishing coat in lime plastering.

(d) *Dolomitic or High-magnesium Limes*

(i) contain 25–40% MgO,

(ii) slake very slowly with very little expansion and evolution of heat.

(iii) are too costly to be used in masonary work; mainly used as flux in metallurgical operations and for production and repair of basic refractories.

(e) *Hydraulic Limes*

(i) contain 70–90% CaO, 10–30% clayey matter ($SiO_2 + Al_2O_3 + Fe_2O_3$) and less than 2% MgO,

(ii) slake very slowly with very small expansion and evolution of heat,

(iii) set under water without shrinkage (do not crack) and are used as inferior grade natural cements,

(iv) are further graded into:

Feebly hydraulic	–	10 –15% clayey matter
Moderately hydraulic or Semihydraulic	–	15–20% " "
Eminently hydraulic	–	20–30% " "

269. There may be some loss of nitrogen on addition of the acid-mixture if the whole of the sample does not come in immediate contact with salicylic acid.

270. Phenol or benzoic acid may be added to fix the nitrate nitrogen in the form of the corresponding nitroderivative but they are less satisfactory than salicylic acid.

271. After the reaction of the sample with acid mixture, anydrous sodium thisoulphate is added which reacts with H_2SO_4 to produce SO_2:

$$S_2O_3^{2-} + 2H^+ \longrightarrow S + SO_2 + H_2O$$

The SO_2 so produced reduces nitrosalicylic acid to aminosalicylic acid:

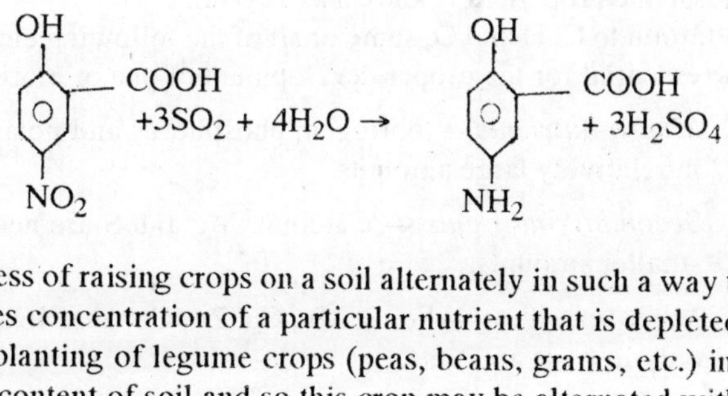

272. It is the process of raising crops on a soil alternately in such a way that one crop increases concentration of a particular nutrient that is depleted by the other. Thus planting of legume crops (peas, beans, grams, etc.) increases the nitrogen content of soil and so this crop may be alternated with wheat or other crops needing large amounts of nitrogen for their growth.

273. Organic farming includes the use of crop rotation and organic manures such as animal wastes for increasing soil fertility. For organic farming to be feasible, farmers must have plenty of livestock. An organic manure of recent discovery is obtained by drying algae, fungi and mosses and converting them into powder. Large scale organic farming may directly help in energy conservation and also reduce environmental pollution.
274. Compost is a cheap organic manure containing appreciable amounts of nitrogen and phosphorus. It is easily prepared by the farmers on the farm by burying waste plants, leaves and animal dung in pits and covering them with soil, where anaerobic decay takes place.
275. With a hope to increase the crop yield, farmers tend to overfertilize their fields. Use of fertilizers more than the optimum amounts

    (a) leads to economic loss

    (b) may become toxic to certain plants

    (c) may increase weeds, and

    (d) may become the cause of environmental pollution – nutrients drained from fields to lakes cause eutrophication (excessive algal growth).
276. Heating the solutions above 45°C may hydrolyse molybdate to molybdic acid which will be filtered along with ammonium phosphomolybdate and will consume additional alkali. Hence, the results will be higher.

$$MoO_4^{2-} + 2H_2O \rightleftharpoons \underset{(MoO_3 \cdot H_2O,\ \text{white ppt.})}{H_2MoO_4} + 2OH^-,$$

277. The yellow colour indicates that the precipitate of ammonium phosphomolybdate is in the finely divided state and is not being retained by the filter paper. The precipitate should be made afresh.
278. (a) The filtrate should not decolourise 1 ml of distilled water containing 1 drop of N/10 NaOH and 1 drop of phenolphthalein.

    (b) The colours produced by 5 drops of filtrate and 5 drops of washing liquid (1% KNO_3) with one drop of methyl red should match.
279. Any organic or inorganic material of natural or synthetic origin which is added to the soil to supply elements essential to plant growth and thereby increase the crop yield is known as a fertilizer.
280. In addition to C, H and O, some or all of the following elements are known to be essential for the proper development of one or more plants.

    (a) *Macro-nutrients* — Nitrogen, phosphorus and potassium are needed in relatively large amounts.

    (b) *Secondary nutrients* — Calcium, Mg and S are needed in relatively smaller amounts.

    (c) *Micro-nutrients* — Fe, B, Mn, Cu, Zn, Co, Se, Mo and Cl are needed only in trace amounts.

281. The N.P.K. value of a fertilizer represents the percentage of N, P_2O_5 and K_2O equivalent to nitrogen, phosphorus and potassium respectively, present in the fertilizer:

Thus, the N-P-K value of urea	= 46.6-0-0
Diammonium phosphate	= 21-53-0
K_2SO_4	= 0-0-54

282. The principal component of the majority of phosphate rocks is Fluorapatite $[Ca_3(PO_4)_2 \cdot 3CaF_2]$. Other substances that may be present include small amounts of $CaCO_3$, $CaCl_2$, $CaSO_4$, $MgCO_3$, silica, Fe_2O_3, Al_2O_3, combined water and organic matter.

283. B.P.L. stands for Bone Phosphate of Lime and is the trade name for the $Ca_3(PO_4)_2$ content of phosphate rock. A rock with high B.P.L. content fetches higher price.

284. Phosphorus in soils and in fertilisers is generally present as orthophosphates – water soluble ammonium phosphate and monocalcium phosphate $[Ca(H_2PO_4)_2]$, less soluble basic calcium phosphate $[Ca_3(PO_4)_2]$ – though other forms of phosphate and organophosphorus compounds are also encountered. Available or assimilable phosphorus includes only that part of total phosphorus which the plants can absorb from the soil. There is no general agreement on the best chemical method of measuring phosphorus availability and the various procedures proposed for available soil-phosphorus are:

(a) Extraction with water followed by extraction with neutral ammonium citrate solution

(b) Extraction with dilute acid

(c) Extraction with fluoride

(d) Extraction with bicarbonate

285. A known weight of the powdered sample is first extracted with several successive small portions of distilled water. The residue is then extracted with a neutral solution of ammonium citrate under prescribed conditions. The residue is then decomposed with aqua regia and the phosphorus content determined and reported as citrate-insoluble phosphorus. The difference between the total phosphorus and citrate-insoluble phosphorus is a measure of the available phosphorus.

286. Nitrogen – NH_4^+, NO_3^-

Phosphorus – $H_2PO_4^-$

Potassium – K^+

287.
$$4OH^- + Cl_2 \longrightarrow 2OCl^- + 2H_2O + 2e$$
$$Cl_2 + 2e \longrightarrow 2Cl^-$$
$$\overline{4OH^- + 2Cl_2 \longrightarrow 2OCl^- + 2Cl^- + 2H_2O}$$

288. When moist bleaching powder is allowed to stand, some chlorite (ClO_2^-) is formed which reacts very slowly with iodide in presence of acetic acid:

$$ClO_2^- + 4H^+ + 4e \longrightarrow Cl^- + 2H_2O$$

$$2I^- \longrightarrow I_2 + 2e] \times 2$$

$$ClO_2^- + 4H^+ + 4I^- \longrightarrow Cl^- + 2I_2 + 2H_2O$$

This slow liberation of iodine causes the colour to reappear at the end-point. In presence of sulphuric acid, however, the above reaction is complete in 2-3 minutes and a stable end-point can be obtained. Since chlorite is also a bleaching agent, it need not be differentiated from hypochlorite.

289. When strongly acidified with HCl, calcium chlorate, which is usually present in small amounts and is not a bleaching agent, also reacts slowly with iodide liberating additional iodine and

$$ClO_3^- + 6H^+ + 6e \longrightarrow Cl^- + 3H_2O$$

$$2I^- \longrightarrow I_2 + 2e] \times 3$$

$$ClO_3^- + 6H^+ + 6I^- \longrightarrow Cl^- + 3I_2 + 3H_2O$$

thus higher results are obtained.

290. The arsenite method is considered to be more accurate than the iodometric method as there is no risk of the interference from chlorate. To 50 ml of the sample solution in a conical flask, add a known excess of 0.1 N As_2O_3 solution. Mix well by shaking and titrate the excess As_2O_3 with 0.1 N iodine solution, after adding 2 ml of freshly prepared starch solution. Calculate available chlorine from the amount of As_2O_3 consumed.

291. Bleach liquor is the trade name for a clear 'Hypochlorite solution' generally used for bleaching purposes. It is made by shaking bleaching powder with distilled water, allowing the 'mud' to settle down and decanting off the clear supernatant liquid. It generally contains from 20 to 35% available chlorine. A relatively weaker (10–15% available chlorine) but more stable bleach liquor is prepared by passing Cl_2 gas into a cold solution of NaOH.

292. The cloth is dipped into the bleach liquor and then drawn through vats containing dilute HCl or H_2SO_4. To remove excess of chlorine from cloth, it is drawn through an antichlor bath containing $Na_2S_2O_3$ or $NaHSO_3$ solution. Finally, it is freely washed with water, squeezed, dried and calendered.

293. 1. It should contain at least 35% available chlorine.

2. A 6% suspension of it in distilled water, thoroughly stirred for 15–20 minutes, should settle clear within one hour so that bleach liquor may be prepared without much loss of Cl_2.

294. The strength of a bleaching powder sample in French or Gay-Lussac Degrees indicates the number of litres of Cl_2 gas, measured at 0°C and 760 mm pressure, that can be obtained from one kg of the sample (100 French Degrees = 31.78% Cl_2).

295. (1) Hydrogen peroxide

(2) Ozone

(3) Chlorine gas

(4) Benzoyl peroxide $(C_6H_5{\cdot}COO)_2$

(5) Nitrosyl chloride (NOCl).

296. The lower end of the capillary tube is flattened, ground and polished carefully to get a large dropping surface with a sharp boundary.
297. The capillary slows down the flow of the liquid through the tip. The final adjustment of the rate of fall of the drops however is made by regulating the screw-pinch-cock on the rubber tubing.
298. The passage of the liquid meniscus may not exactly coincide with the marks X and Y when the drop falls. The consequent error can be eliminated with the help of graduation. The number of divisions through which the liquid meniscus passes for the formation of one drop is determined. Let it be 5. Also, let the first drop formation commence at the 4th division below mark X and the last drop fall when the liquid meniscus is 2 divisions below the mark Y. Then the correction factor for the number of drops is $(+4 - 2)\cdot\frac{1}{5} = +\frac{2}{5}$ drops, that is 2/5 (= 0.4) should be added to the number of drops counted to get the true number of drops formed from a volume of liquid between marks X and Y.
299. The glass tube serves as an air-vent so that the pressure in the receiving bottle remains the same as the atmospheric pressure.
300. Rise in temperature of a liquid increases the kinetic energy of its molecules. This decreases intermolecular forces of attraction, resulting in a decrease in surface tension. At the critical temperature of a liquid, its surface tension vanishes altogether since, at this temperature, the surface of separation between the liquid and its vapour disappears completely.
301. It is the force of tension that always exists along the surface of separation between any two phases of different nature, e.g. the interface between immiscible liquids, between solids and gases, and between solids and non-dissolving liquids. Surface tension is only a specific case of interfacial tension when a liquid is in contact with air saturated with the liquid vapour.
302. The unbalanced forces of attraction acting along the surface of each liquid (individual surface tensions) are partly compensated by the forces of attraction that the molecules of one liquid have for the molecules of the other.
303. Parachor is equal to the molecular volume of a liquid at a temperature at which its surface tension is unity. It is both an additive as well as a constitutive property.
304. All such substances which when added to a solvent (particularly water) decrease its surface tension are called surface active agents or surfactants. Examples – soap and alcohol decrease the surface tension of water.
305. As a liquid tends to have minimum surface area, any attempt to increase the surface area will require some work to be done. The amount of work required to increase the surface area of a liquid by 1 cm^2 is known as the

surface energy of the liquid. It is expressed in ergs/cm^2. Surface energy is numerically and dimensionally equal to surface tension (dynes/cm).

306. The impurities are usually precipatated as sludge but may affect the coating characteristics due to change in the composition of the bath and due to codeposition.

307. Should the anode be of a non-corroding or noble metal, it will not dissolve and so the composition of the electrolyte will change.
Example: If a platinum anode is used in copper plating, the reaction at the anode is the oxidation of water:

$$2H_2O \rightleftharpoons 4H^+ + O_2 + 4e$$

The cathode reaction (8.23) being the deposition of copper ($Cu^{2+} + 2e \rightleftharpoons Cu$), the concentration of Cu^{2+} ions in solution will decrease while that of H^+ ions will increase. In order to maintain uniform conditions, appropriate amounts of copper sulphate and alkali will have to be added from time to time to make up the shortage of Cu^{2+} and to neutralise the extra H^+ produced.

308. Put some water on the surface. it will not bead (separate into drops) if the surface is clean.

309 (i) Agitation helps the acid or alkali to penetrate all crevices and cracks on the surface.

(ii) It improves cleaning or pickling effect by mechanical rubbing action.

(iii) It prevents local dilution of the solution near the surface being cleaned.

310. For achieving a real good finish, it is essential that plating should be done on a polished base. Polishing is accomplished by holding the article against a rapidly revolving buff made of many thicknesses of linen, cotton cloth, wool or felt. The buff receives frequent applications of a polishing composition of wax or grease and fine grit — tripoli, emery or rouge — depending on the nature of surface and the degree of polishing desired.

311. It is an electrolytic operation used for polishing the base metal surface or for obtaining a high luster on finished articles for decorative purpose. The metal surface is made the anode in a suitable polishing solution (usually acetic acid, phosphoric/chromic acid, phosphoric/sulphuric acid or sulphuric/citric acid mixtures). Dissolution of protuberant points or areas leads to marked smoothening and brightening comparable to that resulting from buffing, in many cases at much lower cost. Electrolytic polishing is particularly suitable for removing small scratches and imperfections.

312. Periodic short-time reversal of the direction of current during electroplating is termed as Periodic Reverse Plating. The resulting alternate plating and deplating (polishing) process greatly improves uniformity and smoothness of the plated film.

313. Brighteners are complex proprietary materials added to the plating bath for depositing bright coatings on matt surfaces. They are usually organic compounds such as aldehydes, ketones, formates, citrates, tartrates, lactose,

dextrose, saccharine, thiourea and its derivatives, organic sulphonic acids, etc., or colloidal substances such as glue, gelatin, albumin, etc. Reactions involved are very complex and the brightener's function is not fully understood. An accepted theory is that these substances reduce grain size due to simultaneous deposition.

An overdosage of brighteners may result in a brittle, stressed, less corrosion-resistant deposit that may easily peel off the base metal.

In copper plating, the common brightener is 1 g of phenol (as sulphonic acid) per litre of the plating solution.

314. Chemical attack on a metallic surface by the corrosive environment (electrolyte), or electrochemical displacement during electro-deposition of more noble metals on less noble-metal-articles (such as deposition of gold and silver on copper or copper and nickel on iron) result in non-adherent coatings and contamination of the bath. 'Quicking' and 'Striking' are the two processes generally employed to overcome these difficulties.

(i) In the 'Quicking' process, the article (made of copper, brass or German silver) to be electroplated is first dipped in a quicking solution (usually containing mercuric cyanide in sodium or potassium cyanide solution). Due to electro-chemical displacement, a very thin film of mercury is deposited on the surface of the article. This film satisfactorily resists the corrosive action of the plating bath during electrodeposition.

(ii) In the 'Striking' process, the article to be plated is made cathodic immediately on immersion into the bath which is initially operated for a short duration at high current density. This step may be accomplished in a separate 'strike' bath, containing a low concentration of ions to be deposited, and then the article is shifted to the main plating bath. For deposition of copper on iron articles, the 'striking power' of the cyanide bath is greatly enhanced by the addition of Rochelle salt (sodium potassium tartrate, $NaKC_4H_4O_6 \cdot 4H_2O$).

315. Alloy deposition is codeposition or simultaneous deposition of two or more metals. The necessary conditions are that

(i) the deposition potentials of the metals should be close,

(ii) the polarisation curves (cathode potential versus current density) of the metals concerned should be similar, and

(iii) the metal ions should be replenished in a proportion to their rates of deposition.

Deposition of brass

Deposition potentials of copper (+ 0.34 V) and zinc (– 0.77 V) are brought closer together by adjusting their ion concentrations through complexing with KCN. Due to the difference in the dissociation constants of $\left[Cu(CN)_4\right]^{3-}$ and $\left[Zn(CN)_4\right]^{2-}$, the concentrations of Cu^+ and Zn^{2+} in solution are around 10^{-27} mole/litre and 10^{-18} mole/litre, respectively. At these concentrations, the deposit potentials for both the metals are about –1.30 V.

316. 'Throwing power' of an electroplating system is a qualitative measure of its ability to deposit metal on the entire surface of the article. The percentage ratio of the thinnest and thickest deposit is usually taken as a rough measure of Throwing Power.

It depends on the current density distribution which is a function of the geometry of the cell (i.e., the size and shape of the electrodes and the intervening distance) and polarisation (influenced by current density, concentration of dischargeable ions and their transport velocity).

The throwing power is said to be micro (Micro-throwing Power) if the metal deposit occurs preferentially in pores and scratches and macro (Macrothrowing Power) when a deposit of relatively uniform thickness is produced on an irregularly-shaped cathode.

Conditions which improve one kind of throwing power lead to worsening of the other. Low polarisation due to high concentration of dischargeable ions in acid copper baths improves microthrowing power (levelling and defect hiding characteristics) rendering these baths suitable for depositing copper in holes and pores as for printed circuit boards. Low concentration of dischargeable ions in cyanide copper baths, on the other hand, increases polarisation resulting in improved macrothrowing power. These baths are therefore employed to produce uniform copper deposits on irregularly shaped cathode surfaces, e.g., production of an undercoat of copper on a car bumper.

317. The object is properly cleaned and suspended by iron wires from the anode bus bar into an electrolytic bath composed of NaCN (90 g/l) and NaOH (10 g/l). An iron sheet is used as cathode and current from a 6 V battery is passed through the circuit. Copper starts dissolving from the surface of the object which is carefully watched. When the stripping is complete, the object is quickly removed from the bath, washed in running tap water and dried.

318. Electrosalvaging is a process of repairing parts worn out due to continual use, or mis-machined and undersized parts. The part is built up through electroplating Fe, Cu, Ni or Cr, usually to a thickness greater than necessary, and then machined to size.

319. Three types of electroless coatings are distinguished:

(i) *Immersion of Displacement Coatings*

Simple immersion of an object of more active metal than ions in solution can produce deposits that are acceptable for certain purposes. *Example:* Copper is almost instantaneously deposited on small iron or steel articles or steel wires when immersed in a copper sulphate – sulphuric acid bath (5 wt% each).

(ii) *Non-catalytic Chemically-reduced Coatings*

Solutions of a salt of coating metal, and a reducing agent are sprayed on the object from a dual spray gun. Silver mirrors are produced by using the following solutions:

(a) Silver nitrate 10 g/l
Aqueous ammonia (28 wt %) 4.5 g/l

(b) Hydrazine sulphate 20 g/l
Sodium hydroxide 5 g/l.

(iii) *Catalytic Chemically Reduced or Catalytic Electroless Coatings*

The coating obtained is uniform over the entire surface even if the shape of the object is highly complex (deeply recessed areas, the inside of tank cars, pipes, etc.) and even non-conductors such as glass and plastics can be plated. The process is however limited by the requirement that the metal being deposited should itself act as catalyst. Nickel plating has received maximum commercial applications (e.g., black nickel plating on aluminium sheets for solar panels) though several other metals (such as Co, Cu, Au, Ge, etc.) can be plated in this manner.

An electroless acid nickel bath (pH 4-6, 90–100°C) consists of a nickel salt (usually chloride), a reducing agent (usually sodium hypophosphite) and an organic acid buffer to prevent deposition of nickel phosphite.

An alkaline bath (pH 10–11, 65–75°C) consisting of nickel sulphate, sodium hypophosphite and pyrophosphate produces corrosion resistant coatings, semi-bright in appearance, that are somewhat superior to electrodeposited coatings.

320. Electroplating is an electrolysis process brought about by passage of external current through an electrolyte. Only reduction reaction occurs at the object to be plated (Cathode) while the oxidation reaction occurs at the other electrode (Anode). Electroless plating is a redox process in which no current is involved. Both reduction and oxidation reactions occur at the surface of the object to be plated.

321. Plastics are electroplated for making reflectors, electrical condensers, gramophone records and for use in electroforming. Phenol formaldehyde and urea formaldehyde are two of the common plastics most suited for electroplating. Electroplated plastics are very attractive and light in weight. Electroplating of plastics involves the following steps:

(i) Treatment with an organic solvent followed by conditioning or etching by dipping, for a few seconds, in a bath of chromic-sulphuric acid (for achieving good adhesion between plastic and metal).

(ii) Activation by immersion in a colloidal solution of tin or palladium (nuclei of tin or palladium get deposited on the surface).

(iii) Electroless plating of Ni, Co or Cu by immersion for 20–30 minutes, in the respective electroless bath (the material becomes electrically conducting).

(iv) Electroplating with the desired metal.

322. (i) *Copper*: Cyanide copper baths have high macro-throwing power. They are also used as strike baths (p. 296). The Rochelle salt bath produces a semi-lusterous, finer and denser plating.

Formulation

Copper cyanide (CuCN)	25 g/l	Rochelle salt	45 g/l
Sodium cyanide	30 g/l	Temperature	50–70° C
Sodium carbonate	25 g/l	Voltage	6 V
Sodium hydroxide	adjust to pH 10–11	Anode	Copper
		Cathode Current density	0.02–0.04 A/cm^2.

(ii) *Silver*: It is electroplated mostly for decorative purposes on jewelry, table ware, reflectors and art works, and for engineering purposes an electronic equipment, for electrical contacts and for aircraft bearings.

Formulation

KCN	30–45 g/l	Temperature	25–30° C
AgCN	25–30 g/l	Cathode C.D.	0.005–0015 A/cm^2
K_2CO_3	30–60 g/l	Anode	Silver

(iii) *Rhodium*: It provides bluey white, absolutely untarnishable coatings of high decorative value. Very thin coatings (0.025–0.1 μm) are deposited on nickel or silver plated objects — jewelry, tableware, reflectors and electrical contacts.

Formulation

Rhodium sulphate	1 g/l	Temperature	15° C
Ammonium sulphate	30 g/l	Cathode C.D.	0.06–0.1 A/cm^2
H_2SO_4	60 g/l	Voltage	2 V
Cathode	Ni or Ag plated object	Anode	Thin Pt strip or carbon rod
Time	20–30 seconds		

(iv) *Cadmium*: It is a soft metal used for rust-resistant coatings on hand tools and marine applications. The following formulation has good macrothrowing power and produces coatings which appear like that of silver:

$Cd(CN)_2$	40 g/l	Voltage	1–3 V
NaCN	50 g/l	Cathode C.D.	0.1–0.04 A/cm^2
NaOH	25 g/l	Temperature	35°C
Anode	Cadmium in combination with iron.		

(v) *Nickel*: Nickel provides rust resistant and very hard coatings for corrosion prevention, electro-salvaging, electro-typing and electro-forming. Coatings from Watt type formulation

$NiSO_4 \cdot 6H_2O$	300 g/l	Temperature	45–60°C
$NiCl_2 \cdot 6\,H_2O$	45 g/l	Voltage	1.5–4.5 V
Boric acid	40 g/l	Cathode C. D.	0.03–0.1 A/cm^2

have low internal stresses and high ductility. They are dull grey in colour but can be brightened by buffing or by adding naphthalene disulphonic acids and diphenyl sulphonates as brighteners to the plating bath.

(vi) *Chromium*: Chrome plating is done, usually on an undercoat of nickel, for both decorative and industrial (for longer life and resistance to corrosion) purposes. The best formulation is

Chromic acid ($Cr\,O_3$)	250–300 g/l	Temperature	30–40° C
H_2SO_4	2.5–3 g/l	Cathode C.D.	0.008–0.01 A/cm^2
Anode	Lead	Efficiency	15%

with periodic addition of chromic acid.

323. (i) *Hot dipping*: The method is limited to the production of coatings of low-melting metals such as Zn, Sn, Pb, Al. The metallic work or object is dipped for a short time in molten bath of the coating metal. The composition of the coating is not uniform and its thickness is much greater than electro-plated coatings. Coated parts are heat treated to form an alloy between the coating and substrate. Galvanising (with Zn) of iron and steel articles and tinning of cooking utensils is most common.

(ii) *Diffusion or cementation*: The method is well adapted to small pieces (nuts, bolts, screws, etc.) which are packed in large numbers in coating-metal powder, the system made gas-tight and rotated about one axis so as to tumble the powder over the surface of the article. Heating the system, below the melting points of the two metals involved, during tumbling leads to alloy formation by diffusion. The coating is of uniform thickness but uneven composition. It is hard and more brittle than pure metal coating. Sherardising (Zn), calorising or alonising (Al) and chromising (Cr) are most common.

(iii) *Cladding* : In order to combine the strength of alloys and corrosion resistance of pure metals, alloy sheets are covered, on one or both sides, with very thin sheets of pure metal, in the form of a sandwich, and hot-rolled together to achieve firm and permanent bonding. Ni, Cu, Al and Ti are often used to clad mild steel, Cu, Al and their alloys.

(iv) *Flame spraying or metallizing* : The coating metal (usually Zn, Sn or Pb) in the form of wire is fed through a melting flame (e.g., oxy-acetylene) and blown, in the form of finely divided liquid particles, on to the base metal surface roughened by sand blasting. Bridges, shiphulls, large surfaces of irregular shape are coated by this method.

(v) *Vapour deposition*: Coating metal is vapourised by heating electrically in high vacuum chamber and vapours led to the surface to be coated. The method, being very expensive, is limited to the coating of high strength parts for missiles and rockets.
An alternative method is the decomposition of vapours of volatile metal compounds on the base metal surface, e.g., tungsten is coated by the decomposition of its hexachloride and nickel by decomposition of nickel carbonyl.

324. (i) Coating thickness can be very easily controlled to fraction of a μm.
(ii) It is the most convenient method for applying coatings of high melting metals such as Cu, Ni, Cr, Ag, Au, Pt, etc.
(iii) The intermediate layer between the coating and the base metal (formed in the case of hot dip or diffusion process) is absent.
(iv) The electroplated coatings have fine structure and improved hardness, water resistance, electrical and thermal conductivity, solderability, reflectivity, etc.
(v) The process adds beauty to the product. An electroplated article is inexpensive when compared with the cost of the same article made of the metal used for plating.

325. In most cases of electroplating, the current does not produce the theoretical amount of metal deposit as calculated from equation (8.26). Some current is lost in liberation of hydrogen along with metal. The ratio of the actual metal deposit to the theoretical (expressed in %) is called the plating efficiency.

326. Weight of copper deposited, w = 1.5875 g

Current strength, I = 1 A

Electrochemical equivalent of copper, Z $= \frac{\text{Eq. Wt. of copper}}{\text{1 Faraday}}$

$$= \frac{31.74}{96500} \text{ g/coulomb}$$

From equation (8.26),

time, $$t = \frac{w}{Z \times I} = \frac{1.5875}{31.75} \times 96500$$

$$= 4825 \text{ sec.}$$

$$= 1 \text{ h } 20 \text{ min } 25 \text{ sec.}$$

327. Cathode current density = 0.1 A/cm^2

Let the surface area of cathode = $x \text{ cm}^2$

Current strength, I = $0.1 \times x$ A

Time, t = 1 hour = 60×60 sec

Electrochemical equivalent, Z $= \frac{63.5}{96500}$ g/coulomb

Theoretical weight of copper deposit

$$= ZIt = \frac{63.5 \times 0.1 \times x \times 60 \times 60}{96500} \text{ g}$$

Plating efficiency $= 90\%$

$\therefore$ Actual copper deposited, w $= \dfrac{63.5 \times 0.1 x \times 60 \times 60 \times 90}{96500 \times 100}$ g

Density of copper, ρ $= 8.9 \text{ g/cm}^3$

Volume of deposit $= \dfrac{\text{weight}}{\text{Density}} = w/\rho$

Thickness of deposit $= \dfrac{\text{Volume}}{\text{Surface area}} = \dfrac{w}{\rho}/x$

$$= \frac{63.5 \times 0.1 \times x \times 60 \times 60 \times 90}{96500 \times 100 \times 8.9 \times x} \text{ cm}$$

$$= \frac{63.5 \times 0.1 \times 60 \times 60 \times 90}{96500 \times 100 \times 8.9} \text{ cm}$$

$$= 0.02396 \text{ cm}$$

328. When the same quantity of electricity is passed through a number of electrolytes, the weights of different substances deposited or liberated at the electrodes are directly proportional to the equivalent weights of the substances.

Wt. of silver deposited $= 2.697$ g

Eq. wt. of silver $= 107.88$

Eq. wt. of copper $= 31.80$

$$\frac{\text{Wt. of copper deposited}}{\text{Wt. of silver deposited}} = \frac{\text{Eq. wt. of copper}}{\text{Eq. wt. of silver}}$$

$\therefore$ Wt. of copper deposited $= \dfrac{31.80}{107.88} \times 2.697$

$= 0.795$ g.

329. A silver coulometer consists of a weighted platinum or silver vessel containing an aqueous solution of pure silver nitrate. While the vessel itself acts as cathode, the anode is pure silver rod. For measuring the quantity of electricity passing through a circuit in a given time, the coulometer is connected in series with the circuit. After electrolysis, the vessel is emptied, washed, dried and weighed.

The quantity of electricity passed through the circuit

$$= \frac{1 \text{ Faraday}}{\text{Eq. Wt. of silver}} \times \text{Increase in weight of vessel}$$

$$= \text{Increase in weight of the vessel} \times \frac{96500}{107.88} \text{ C}$$

APPENDIX

Table I Density, surface tension and viscosity of water at some temperatures

Temperature (°C)	Density (g/ml)	Surface tension (dryness/cm)	Viscosity (centipoise)
5	0.99999	74.92	1.519
10	0.9997	74.22	1.308
15	0.9991	73.49	1.140
16	0.9990	73.34	1.110
17	0.9988	73.19	1.082
18	0.9986	73.05	1.056
19	0.9984	72.90	1.024
20	0.9982	72.75	1.002
20.20	–	–	1.000
21	0.9980	72.59	0.981
22	0.9978	72.44	0.958
23	0.9976	72.28	0.936
24	0.9973	72.13	0.914
25	0.9971	71.97	0.894
26	0.9968	71.82	0.874
27	0.9965	71.66	0.855
28	0.9962	71.50	0.836
29	0.9960	71.35	0.818
30	0.9957	71.18	0.801
31	0.9954	71.02	0.784
32	0.9950	70.86	0.768
33	0.9947	70.71	0.752
34	0.9944	70.54	0.737
35	0.9941	70.38	0.723
40	0.9922	69.56	0.656

Table II Solubility of oxygen (mg/l) in fresh and saline water in equilibrium with air at atmospheric pressure

Temp. (°C) / Chloride content (mg/l)	0	200	500	1000	5000	10000	25000
0	14.62	14.55	14.43	14.24	13.73	12.89	10.66
5	12.77	12.74	12.70	12.62	12.02	11.32	9.44
10	11.29	11.26	11.23	11.16	10.66	10.06	8.45
15	10.08	10.06	10.03	9.97	9.54	9.03	7.64
16	9.87	9.85	9.83	9.76	9.34	8.84	7.50
17	9.67	9.65	9.61	9.56	9.15	8.67	7.36
18	9.47	9.45	9.42	9.37	8.97	8.50	7.22
19	9.28	9.26	9.23	9.18	8.79	8.33	7.09
20	9.09	9.08	9.04	8.99	8.62	8.17	6.96
21	8.92	8.90	8.87	8.83	8.46	8.02	6.84
22	8.74	8.73	8.70	8.66	8.30	7.87	6.72
23	8.58	8.57	8.53	8.49	8.14	7.73	6.61
24	8.42	8.40	8.38	8.34	8.00	7.59	6.50
25	8.26	8.25	8.22	8.18	7.85	7.46	6.39
26	8.11	8.10	8.07	8.03	7.71	7.33	6.29
27	7.97	7.95	7.93	7.89	7.58	7.20	6.18
28	7.83	7.81	7.79	7.75	7.44	7.08	6.09
29	7.69	7.68	7.65	7.62	7.32	6.96	5.99
30	7.56	7.54	7.52	7.49	7.19	6.85	5.90
31	7.43	7.42	7.40	7.36	7.07	6.73	5.81
32	7.31	7.29	7.27	7.24	6.96	6.66	5.72

Temp. (°C) / Chloride content (mg/l)	0	200	500	1000	5000	10000	25000
33	7.18	7.17	7.15	7.11	6.84	6.52	5.63
34	7.07	7.05	7.03	7.00	6.73	6.42	5.55
35	6.95	6.94	6.92	6.88	6.62	6.31	5.46
40	6.41	6.40	6.39	6.36	6.12	5.84	5.08
50	5.48	5.47	5.45	5.43	5.24	5.02	4.39

Table III Specific gravity (ρ_4^{20}) of some common liquids

Liquid	Sp. Gr. (ρ_4^{20})	Liquid	Sp. Gr. (ρ_4^{20})
Methyl alcohol	0.810	Acetic acid	1.049
Ethyl alcohol	0.791	Benzene	0.8707
Glycol	1.113	Toluene	0.866
Glycerol	1.261	Aniline	1.022
Methyl acetate	0.933	Carbon disulphide	1.263
Ethyl acetate	0.901	Carbon tetrachloride	1.594
Diethyl ether	0.708	Chloroform	1.483
Acetone	0.791		

Table IV Surface Tension (dynes/cm) of some common liquids in contact with air

Liquid / Temperature (°C)	10	20	30	40
Methyl alcohol	–	22.6	–	–
Ethyl alcohol	23.14	22.27	21.43	20.60
n-Propyl alcohol	–	23.78	–	–
Glycol	–	47.7	–	–
Glycerol	–	63.4	–	–
Methyl acetate	–	24.6	–	–
Ethyl acetate	–	23.9	–	–
Diethyl ether	–	17.0	15.93	–
Acetone	28.8	23.70	–	21.16
Acetic acid	28.6	27.60	–	–
Benzene	30.22	28.88	27.56	26.26
Toluene	29.7	28.44	27.32	26.13
Aniline	–	42.9	–	–
Carbon disulphide	44.10	32.3	–	26.5
Carbon tetrachloride	28.4	26.77	25.53	24.41
Chloroform	28.5	27.10	–	–

Table VI Specific gravity, acid value, saponification value, iodine value, melting points and drying characteristics of some oils, fats and waxes

	Sp.gr 60/60 F°	Acid value	saponification value	Iodine value	Melting point (°C)
Fats					
1. Beef tallow	0.943–0.952	0.25	195–200	35–45	42–48
2. Butter	0.935–0.940	0.5–35	220–230	28–45	28–35
3. Coconut	0.925–0.927	2.5–10	245–270	8–10	23–27
4. Palm	0.920–0.925	9–10	195–205	50–60	27–43
Waxes					
1. Bees wax	0.958–0.970	5–10	75–125	6–10	60–70
2. Carnauba wax	0.99–1.0	2.5–3	80–90	13–15	80–90
3. Chinese wax	0.96–0.98	–	80.90	1–2	80–85
4. Paraffin wax	0.867–0.910	–	–	3–4	50–70
Oils					***Drying characteristics***
1. Anwala	0.918–0.921	–	165–185	95–100	Non-drying
2. Castor	0.960–0.970	0.1–0.3	175–185	80–90	Non-drying
3. Chinese wood (tung)	0.940–0.945	2	190–200	150–170	Drying
4. Cod liver	0.925–930	5–6	170–190	135–170	–
5. Corn (Maize)	0.915–925	–	190–195	105–125	Semi-drying
6. Cotton seed	0.922–0.925	0.6–0.9	190–197	100–115	Semi-drying
7. Groundnut (peanut)	0.910–0.915	–	185–195	85–95	–
8. Linseed	0.930–0.940	1–3.5	190–195	165–195	Drying
9. Mustard	0.915–0.920	–	170–180	95–120	Semi-drying
10. Olive	0.915–920	0.3–1.0	185–200	75–95	Non-drying

(*Contd.*)

	Sp.gr 60/60 F°	Acid value	saponification value	Iodine value	Drying Characteristics
11. Perilla	0.930–0.937	–	190–195	185–200	Drying
12. Rapeseed	0.913–0.918	0.4–1.0	170–180	95–105	Semi-drying
13. Seasame	0.920–0.925	9–10	185–195	105–115	Semi-drying
14. Soyabean	0.924–0.927	0.3–1.8	190–195	120–125	Semi-drying
15. Sunflower	0.924–0.926	10–11	185–195	120–135	Semi-drying
16. Whale oil	0.916–0.925	1.2–2.0	190–200	110–145	–

Table V Viscosity (centipoise) of some common liquids

Liquid / Temperature (°C)	10	20	30	40
Methyl alcohol	0.69	0.593	0.510	0.449
Ethyl alcohol	1.466	1.200	1.005	0.834
Glycol	–	19.9	–	9.3
Glycerol	–	10.69 poise	6.29 poise	–
Methyl acetate	–	0.384	0.356	0.320
Ethyl acetate	0.512	0.455	0.410	0.367
Diethyl ether	0.268	0.240	0.220	0.199
Acetone	–	0.331	0.293	0.270
Acetic acid	–	1.222	1.040	–
Benzene	0.757	0.647	0.561	0.495
Toluene	0.710	0.590	0.525	0.471
Aniline	6.5	4.46	3.16	–
Carbon disulphide	–	0.366	0.348	0.330
Carbon tetrachloride	–	0.968	0.847	0.736
Chloroform	–	0.563	0.510	0.464

Table VII Flash Points and Boiling points/Boiling ranges of some liquids

Liquid	Flash Point closed* cup(°F)	Boiling point/ Boiling range (°F)
A. *Organic compounds*		
Acetic Acid (glacial)	105 (110)	245
Acetone	0 (15)	134
Diethyl ether	–20	95
Ethyl alcohol	55	173
Methyl alcohol	55 (60)	147
Ethylene glycol	230 (240)	387
Chloroform	Non-flammable	142
Benzene	12	176
Toluene	40(45)	232
n-Hexane	–7	156

(Contd.)

Table Contd.

Liquid	Flash Point closed* cup(°F)	Boiling point/ Boiling range (°F)
n-Heptane	25	208
n-Decane	115	344
B. *Petroleum and its fractions*		
Petroleum crude	20–90	–
Petroleum ether	–50	100–160
Gasoline	–50	100–400
Naphtha (painters)	20–45	210–325
Naphtha (solvent)	100–110	300–400
Kerosene	110–120	350–550
Diesel	130–150	400–600
Gas oil	+ 150	400–750
C. *Lubricating oils*		
Spindle	170(200)	–
Light Machine	320(370)	–
Turbine	(400)	–
Air Craft Engine	(420–470)	–
Motor	(450)	–
Cylinder	(535)	–
Transformer Oil	340	550–750
D. *Vegetable oils*		
Castor	450(540)	600
Coconut	420(510)	
Corn	490	
Cotton seed	590	
Linseed	435(535)	
Olive	440	
Palm	420	
Ground nut	540	
Perryilla	520	
Rapeseed	550	
Tung	550	
Turpentine	95	300
E. *Waxes*		
Peraffin	390(430)	> 700
Carnauba	540(600)	

* Figures in parenthesis, wherever given, represent the open cup Flash Point

Table VIII Pour points (°F) of some oils

Oil	Pour point (°F)	Oil	Pour point (°F)
Castor	–15	Kerosene	–50 to –60
Olive	20	Diesel	–22 to –30
Rapeseed	10	Lubricating oil for aircraft engine	0 to 10
Tallow	40		
Whale	30	Transformer	20

Table IX Intrinsic viscosity–molecular weight constants (K and *a* in equation 7.7) for some polymer/solvent systems

Polymer	Solvent	Temperature (°C)	$K \times 10^4$	a
Polystyrene	Benzene	25	1.02	0.74
		30	1.10	0.735
	Toluene	25	1.1	0.72
		30	1.1	0.725
Polyisobutylene	Benzene	30	6.1	0.56
	Toluene	30	2.0	0.67
Polyvinyl acetate	Acetone	25	1.88	0.69
		30	1.02	0.72
Natural rubber	Benzene	30	1.85	0.74
	Toluene	25	5.0	0.67
Polymethyl methacrylate	Acetone	25	0.75	0.7
		30	0.7	0.7
	Toluene	25	0.75	0.71
		30	0.7	0.72

Table X Answer to Exercise No. 211

S. No.	Polymer	Abbreviation	Trade name(s)	Starting material(s)	Structural/ Repeating unit	Class*	Important application(s)
1.	Cellulose acetate ($R=COCH_3$)	CA	Plastacele Kodapak Tenite	Cellulose/ Acetic-anhydride	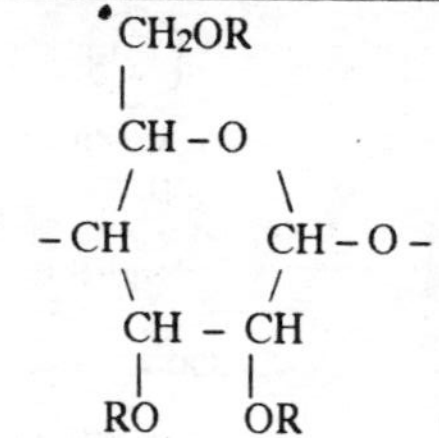	TP/F	Spectacle frames, semi-permeable membranes, toys, textiles, photographic films.
2.	Cellulose nitrate ($R = NO_2$)	CN	Celluloid Nitron	Celluose/ HNO_3/H_2SO_4		TP	Photographic films, sheeting, toys, lacquers, explosive (gun cotton), solid propellant for rockets.
3.	Ethyl celluose ($R = C_2H_5$)	EC	Ethocel Campco	Cellulose/ $NaOH/C_2H_5Cl$		TP/A	Coatings, adhesive, wire insulation, toughening agent for plastics.
4.	Epoxy resins	EP	Araldite Epon	Bisphenol-A/ Epichlorohydrin	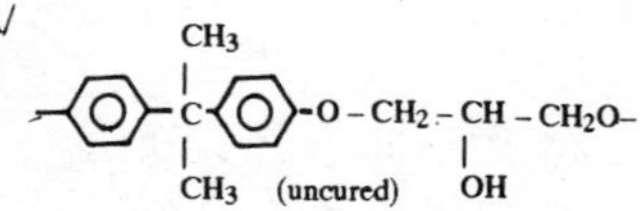	TS	Adhesives, surface coatings, laminates, foams.
5.	Phenol formaldehyde	PF	Bakelite Formica	Phenol/ Formaldehyde	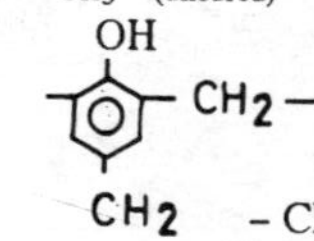	TS/A	Radio cabinets, impregnating resins, electrical components, adhesive.
6.	Poly (acrylonitrile)	PAN	Orlon Acrilan Zefran	Vinyl cyanide	$-CH_2-CH(CN)-$	F	Textiles, strengthening agent for cements.
7.	Polyamides	PA	Nylons	Diamine/diacid or aminoacid			
	Nvlon 6	PA6	Capron	ε-Aminocaproic acid	$-NH(CH_2)_5CO-$	TP/F	Food packaging, ropes, tyre cords, unlubricated bearings, gears, textiles.
	Nylon 6, 10	PA6, 10	Thermo-comp	Hexamethylene diamine/ sebacic acid	$-NH(CH_2)_6NHCO(CH_2)_8CO-$		
	Nylon 66	PA66	Zytel	Hexamethylene diamine/adipic acid	$-NH(CH_2)_6NHCO(CH_2)_4CO-$		

8.	Polycarbonate	PC	Merlon Lexan	Bisphenol-A/ diphenyl carbonate	$-O-C_6H_4-C(CH_3)_2-C_6H_4-O-\underset{\overset{\parallel}{O}}{C}-$	TP	Safety helmets, ball bearings, lenses, insulators, photographic film.
9.	Polychloroprene	PCP	Neoprene Perbunan	Chloroprene	$-CH_2-C(Cl)=CH-CH_2-$	E/A	Automotive parts, tank linings, adhesive, coatings, binder for rocket fuels, conveyer belts.
10.	Polyethylene	PE	Pentothene Alkathene Poly-Eth	Ethylene	$-CH_2-CH_2-$	TP/A	Pipes, packaging, radiation shields, linings, coatings, housewares, fibres.
11.	Poly (ethylene-terephthalate)	PETP	Terylene Mylar Dacron Econol	Ethylene-glycol/tere-phthalic acid	$-CH_2-CH_2-C-\underset{\overset{\parallel}{O}}{C}-C_6H_4-\underset{\overset{\parallel}{O}}{C}-O-$	F	Textiles, tyre cords, electrical insulators, magnetic recording, conveyor belts, tapes (films).
12.	Polyiso-butylene	PIB	Vistanex	Isobutylene	$-CH_2-C(CH_3)_2-$	TP/E	Electrical insulators, viscosity index improver, pipes and tubes.
13.	Polyisoprene (*cis*)	PIP	Natural rubber	Latex of hevea tree or Isoprene	$-H_2C(H_3C)C=C(H)CH_2-$	E	Adhesive, gaskets, tyres, hose, cable covering, footwear.
14.	Polymethyl-methacrylate	PMMA	Plexiglass Lucite Perspex	Methylmeth-acrylate	$-CH_2-C(CH_3)(COOCH_3)-$	TP	Lenses, wind screens, attractive sign-boards, TV-screen guards.
15.	Polypropylene	PP	Pro-Fax PP Poly-Pro	Propylene	$-CH_2-CH(CH_3)-$	TP/F	Sterilizable medical equipment, pipes and tubes, fibres and filaments, electronic components.
16.	Polystyrene	PS	Lustrex Styron Dylene	Styrene	$-CH_2-CH(C_6H_5)-$	TP	Pipes, foams, lenses, containers, insulation for refrigerator and air conditioners.
17.	Poly (tetra-fluoroethylene)	PTFE	Teflon Fluon Halon	Tetrafluoro ethylene	$-CF_2-CF_2-$	TP	Electrical insulation at high temperature, gaskets, tubes, seals, non-stick cookware finishing, unlubricated gears.

(*Contd.*)

(Table Contd.)

18.	Poly (vinyl-acetate)	PVAc	Vinylite Gelva	Vinyl acetate	$-CH_2-CH(OOCCH_3)-$	TP/A	Adhesive, lacquers, component of inks, gramaphone records.
19.	Poly (vinyl-alcohol)	PVAl	Gelvatol	Polyvinyl acetate	$-CH_2-CH(OH)-$	TP/F	Textile fibres, protective colloid, adhesive, photosensitive films.
20.	Poly (vinyl-chloride)	PVC	Kohinor Geon	Vinyl chloride	$-CH_2-CH(Cl)-$	TP	Conveyer belts, electrical insulations, pipes for chemical plants, gas & oils, shoes, raincoats.
21.	Poly(vinylidene-chloride)	PVDC	Saran	Vinylidene chloride	$-CH_2-CCl_2-$	F/TP	Pipes & fittings for hot corrosive materials, packaging film, screens, upholstery, carpets, filter cloth.
22.	Silicones	SI	Pyrotex Dow-Corning	CH_3SiCl_3/ $(CH_3)_2SiCl_2$/ $(CH_3)_3SiCl$	$-Si(CH_3)_2-O-Si-O-$	L/E/TS	Lubricants, water repellents, antifoams, rubbers, high temperature electrical and electronic insulators, encapsulating resins.
23.	Urea-formal-dehyde resins	UF		Urea/ formaldehyde	$-NH-CO-N(CH_2-)-CH_2-NH-CO-$	TS/A	Textile finishing, paper coating, adhesive for furniture, laminates, electrical accessories, foams.
24.	Unsaturated polyesters	UP	Dupon Aropol	Unsaturated diacids/ diols.	$-O-OC-CH=CH-COOCH_2-CH_2-$	TS	Fibre reinforced plastics, boats, hulls, aircraft components, laminates.

*A – Adhesive. E – Elastomer. F – Fibre. L – Lubricant. TP – Thermoplastic. TS – Thermoset.

BIBLIOGRAPHY

Chemistry of Engineering Materials, Fourth Edition, Robert B. Leighou; International Chemical Series: McGraw-Hill Book Co. Inc., New York,1953.

Applied Chemistry for Engineers, Third Edition, Eric S. Gyngell; Edward Arnold (Publishers) Ltd., London, 1960.

Chemistry in Engineering and Technology, J.C. Kuriacose and J. Rajaram; Tata McGraw-Hill Publishing Company Ltd., New Delhi, 1984.

Chemistry in Industry, S.K. Chatterjee; Book Syndicate Private Ltd., Calcutta, 1987.

Chemistry — An Experimental Science, Chemical Education Material Study: W.H. Freeman and Company, Cooperating Publishers, San Francisco, 1963.

Chemistry for Environmental Engineering, Third Edition, Clair N. Sawyer and Perry L. McCarty; McGraw-Hill Book Company, New York, 1978.

Environmental Pollution, Second Edition, Laurent Hodges; Holt, Rine hart & Winston, U.S.A., 1977.

Experiments in Inorganic Chemistry, Third Edition, Department of Chemistry, South Dakota College, Brookings; Burgess Publishing House, Minnesota, 1965.

A Text-Book of Quantitative Inorganic Analysis, Third Edition, Arthur I. Vogel; The English Language Book Society and Longmans Green & Co., Ltd., London, 1962.

A Systematic Handbook of Volumetric Analysis, Francis Sutton; Butterworths Scientific Publications, London, 1955.

Technical Methods of Analysis, R.C. Griffin; McGraw-Hill Book Company, Inc., New York, 1955.

Standard Methods of Chemical Analysis, Sixth Edition, Frank J. Welcher (Ed); D.Van Nostrand Company, Inc., Princeton, New Jersey, 1963.

Standard Methods for the Examination of Water and Wastewater, 16th Edition, APHA-AWWA-WPCF, Washington, 1985.

Water Quality and Treatment, Third Edition, The American Water Works Association. Inc., McGraw Hill Book Company, Inc. New York, 1971.

Mannual of Public Health Engineers, Vol. I, V.S. Sharma; International Book Traders, Delhi, 1985.

Standard Handbook of Lubrication Engineering, American Society of Lubrication Engineers, McGraw-Hill Book Company, Inc. New York, 1968.

The Lubrication Engineers Mannual, First Edition, Charles A. Bailey & Joseph S. Aarons (Eds); United States Steel, 1981.

The Performance of Lubricating Oils, Second Edition. H.H. Zuidema; Reinhold Publishing Corporation, New York, 1959.

Lubricating and Allied Oils, Fourth Edition, E.A. Evans; Champman & Hall Ltd., London, 1963.

Theory & Practice of Lubrication Systems, Alexandru Nica & P.A.J. Scott; Scientific Publications (Great Britain) Limited, London, 1969.

Effective Lubrication, Allen F. Brewer; Robert Krieger Publishing Co., Huntington, New York, 1974.

Handbook of Greases, Lubricants and Refining of Petro Chemicals, First Edition, R.K. Malik and K.C. Dhingra; Small Industries Research Institute, Delhi, 1975-76.

Fuels – Solid, Liquid & Gaseous, Fifth Edition, J.S.S. Brame and J.G. King; Edward Arnold (Publishers) Ltd., London, 1961.

Physical Chemistry (A Modern Laboratory Course), Hugh W. Salzberg, Jack I. Morrow, Stephen R. Cohen, Michael E. Green; Academic Press, Inc., New York, 1969.

Polymer Science, V.R. Gowariker, N.V. Vishwanathan, Jayadev Sreedhar; Wiley Eastern Limited, New Delhi, 1988.

Polymer Technology, D.C. Miles and J.H. Briston; Temple Press Books, London, 1965.

Preparative Methods of Polymer Chemistry, Wayre R. Sorenson and Tod W. Campbell; Interscience Publishers, Inc., New York, 1962.

Simple Methods for Identification of Plastics, Dietri Ch Brown; McMillan Publishing Co. Inc., New York, 1982.

Success in Chemistry, John Bandtock and Paul Hanson; John Murrary, London, 1986.

Handbook of Electroplating, Anodizing and Metal treatments, O.N. Tandon, V.K. Aggarwal and K.C. Dhingra; Small Industry Research Institute, New Delhi.

Subject Index